Practical Surveying and Computations

Practical Surveying and Computations

Revised second edition

A. L. Allan

Laxton's
An imprint of Butterworth-Heinemann
Linacre House, Jordan Hill, Oxford OX2 8DP
A division of Reed Educational and Professional Publishing Ltd

℞ A member of the Reed Elsevier plc group

OXFORD BOSTON JOHANNESBURG
MELBOURNE NEW DELHI SINGAPORE

First published as *Practical Field Suvrveying and Computations* 1963
Reprinted 1971, 1973
Second edition 1993
Revised second edition 1997

11563621

Learning Resources
Centre

British Library Cataloguing in Publication Data
A catalogue record for this book is available from the British Library

ISBN 0 7506 3655 6

Library of Congress Cataloguing in Publication Data
A catalogue record for this book is available from the Library of Congress

Printed and bound in Great Britain by Hartnolls Ltd, Bodmin, Cornwall

To Daphne

Contents

Appendices

References

Bibliography

Index

Preface to the second edition

More than twenty years have elapsed since the first edition of this book was published. John Hollwey, who had prepared the initial stages of this edition, has died, and the third author, John Maynes, has retired from an active interest in surveying. In the interests of continuity and economy, I have completed and virtually rewritten the whole book.

Every effort has been made to retain the character of the successful first edition: a selection of key material; emphasis on fundamentals; outlines of proofs; a feeling for quality; and some indication of the relevance of various techniques.

Although the book is primarily directed towards land and geodetic surveyors, civil engineers, field scientists, and archaeologists, it should also prove useful to mechanical engineers concerned with large scale metrology, and the rising breed of computer analysts involved in programming survey problems for software houses.

The massive changes in electronic technology which have taken place since 1968 need hardly be ennumerated, and although the rate of change has slowed down, developments continue which will render current technology obsolete. These developments, however, are unlikely to use hitherto unexploited principles of geometry. By concentrating on these geometrical principles, this text will have an enduring function. For this reason it is only lightly based on current technology and professional practice.

In the past much time in the education of the geodetic surveyor was given over to a study of instrument design. Today only fundamental principles need be treated, since the detailed manner in which an instrument functions lies at the forefront of electronic technology. As always, the surveyor must be able to detect when the instrument is not performing to standard. Performance tests can be more readily designed and executed if basic operational principles are understood.

Now, as ever, proper procedures must be adopted for the gathering of data, and these must be linked efficiently to data management and processing. A quality assurance system in which efficiency is seen to be achieved, is now expected by clients. It is therefore most important not to view the data gathering stage in isolation from the equally important stages of organisation, processing and presentation. Data by itself is of little value: conversely an elaborate data management system is useless without quality-assured data. Although this book concentrates on field methods of data acquisition, the growing links with photographic and television technology, and the need for cartographic and data presentation in digital form are not ignored.

The impact of rapid processing has greatly increased the complexity and flexibility of the setting-out and building processes and of the manner in which they can be carried out. Quality control can also be assured with such processing.

Since the first edition was published, the numerical handling of data has undergone most change. The procedure in which all measured data are combined to yield the best estimates of parameters, traditionally known as "least squares adjustment", has become the norm instead of the exception. Because much of the processing is done without human intervention, statistical methods have replaced eyeball methods of blunder detection. A vast literature dealing with least squares estimation has emerged over the last twenty years. I have selected and condensed from this to give, within manageable space, a level of treatment which should provide the essentials for further reading. In this respect the reader will find the greatest changes from the first edition. Generally problems are treated from a parametric stand-point using observation equations, with conditions amd constraints if necessary. The use of condition equations is eschewed for all but levelling problems and data snooping.

Although the subject of "survey computation" has undergone many remarkable transformations, the surveyor still needs a basic understanding to avoid being misled by computer systems. The judgement needed to evaluate complex results can only be based upon a good understanding of the elementary principles of measurement and geometry. The role of simple semi-graphic methods is a powerful one which provides this basic alternative approach to verify the order of magnitude of results.

The revision reflects some radical changes which have been made in the syllabuses of courses in mathematics over the last twenty years, particularly in the use of matrix methods and vector ideas. The aim has been to restrict the use of "new" mathematical languages to a minimum, even although such topics as vector calculus and complex variables do have their place to treat dynamic force field problems and map projection theory. I prefer to use the minimum of mathematical knowledge required, rather than the most efficient or most elegant. Students with extended mathematical knowledge will be able to treat some topics more elegantly if they so wish.

The treatment of coordinate systems follows the dual approach of "plan plus height" for simple plane surveying, and truly three-dimensional for geodesy and industrial surveying.

How to deal with computations and explanatory exercises posed the most difficult problem. On the one hand, simple numbers have the merit of arithmetic convenience and illustrate principles well. On the other hand, most professional problems do not deal with convenient numbers, and the student must learn to deal with them as they arise in reality. Modern automated instrumental systems yield a plethora of numbers which can only by processed realistically by computer software. The emergence of the spread-sheet computation system has brought together computational power with a requirement for the user to understand the theoretical stucture of a problem. For many tasks it is an excellent mixture and especially so in education. From time to time we develop a computer algorithm for a problem, leaving the detailed manner of execution to whatever system is at the reader's disposal. This is not to say that the time for programming in a high level language is over: professional software will require such software for some time to come. However, it is a matter of some conjecture to predict the future direction of computational processes. Whatever its direction, we cannot escape the need to understand basic geometrical principles.

Inevitably some old material, which may be useful to some readers, has been omitted to save space. At least it can be found in the first edition. The thirteen chapters are presented in an order advocated for beginners, though it is impossible to learn complex subjects in a purely linear manner. Cross references between chapters are inevitable. The reader is advised to become familiar with the notation used for the "adjustment" processes by reference to Appendix 5. I have avoided the term "adjustment" wherever possible, since the only adjustment made is to the physical components of an instrument. Observations are used to find the "best estimates" of parameters which satisfy mathematical models. The criterion which defines these estimates is the least squares principle.

Although every effort has been made to remove mistakes from the text and examples, it is likely that some remain. I must apologise for these and hope that readers will notify me so that a reprint may be corrected.

University College London, 1993   Arthur L Allan

Acknowledgements

I am indebted to many people for assistance with this book. Although a complete list of their names would be impractical, I must express my very sincere thanks to the people mentioned below,without whom the book would never have been completed.

The late John Hollwey deserves all the credit for initiating the re-write in the first place, and for making the administrative arrangements with the publisher and advisers who supported the venture at the outset. Fortunately I was able to discuss with John the general restructuring of the book and the level of treatment that would be given. I only hope he would have been satisfied with the outcome. I am also grateful to Mrs Hollwey for her interest and support.

Professor Ian Harley, of University College London, has given me endless assistance with the intricacies of the Macintosh computer word-processing and its attendant software. He has also given much technical advice and encouragement about detailed aspects of the text, especially on the mutual pointing of theodolites and least squares theory.

To Mrs Jacqueline Chapman, I am greatly indebted for the accurate re-typing of large portions of the original text of the first edition, which have formed the basis of much of the book, and also to Kevin Grist who drew copies of many of the original diagrams in Macdraw software. Although I have had to make many revisions of both text and diagrams, these draft versions made the enormous task practicable.

I am also grateful to my current colleagues at University College London for assistance with ideas and some parts of the text.They are J V Arthur, K B Atkinson, D P Chapman, J C Iliffe, G Bentley, J Pearson and Richard J Hibbert, who deserves special mention for checking some of the computations.

Many persons have helped indirectly to sharpen my ideas and select material both for inclusion and exclusion. From among such persons pride of place must be given to the late D P Mason and my former colleague T O Crompton and to a lesser extent to C Tomlinson, S Longdin, J R Smith and R C H Smith.

To others with whom I have worked I owe my gratitude: the late Keith Rylance; B Pearn of Rank Taylor Hobson; T Davis, P Range, S Kyle, M Grist and N Vancans of the former Wild and Kern Companies; G Pask and B Davis of Tellumat; M Paine of Westland Helicopters; D

Jones of Thames Water Utilities, and M Mayoud, C Lasseur and D Veal of CERN.

It would be difficult to overstress the role played by my former students in fashioning my approach to this text. Although some quite specific names come to mind, it would be invidious to mention them and not others. Therefore I express my grateful appreciation to them all for their collective role in assisting me with their questions, comments and criticisms spread over the years.

I also owe a great debt to the authors of the texts which I have cited as references, and to all other authors not mentioned by name, who have assisted my understanding in some way.

Although I have personally typed the text of this book, I have been immensely assisted from the publisher's end, both with proof reading and general layout, by Bridget Buckley. She deserves very special thanks for her efficiency and tact. None the less, I apologise for any residual errors in the text which may have passed undetected.

Finally I wish to apologise to my wife, family and friends for my neglect of them whilst writing this book. I am most grateful for their understanding and continued support.

Whilst I acknowledge assistance from many named and unnamed persons, I accept full responsibility for the text, and trust that any criticisms will be directed in my direction only, and that I can be informed of them.

ALA 1993

I am most grateful to all readers and reviewers who have pointed out a number of errors in the original text, thus enabling me to correct this reprint of the book. Should there still be some undetected errors, I apologise to readers, and trust that they will once again inform me of them.

ALA 1997

1 Introduction to surveying

Chapter summary

This book deals with that branch of surveying which is concerned with the geometry of three dimensions, in the Euclidean sense.Three-dimensional objects, varying in size from small industrial components to the Earth itself, often need to be measured. For subsequent manipulation, the results of these measurements have to be represented in the form of a *mathematical model.* The results are prepared and presented, either graphically or numerically, for presentation to a client for subsequent use. What is more, the quality of the measurements, usually expressed in statistical form, is often of as much concern as the results themselves.

The book restricts its scope to those topics which form part of ground measurement techniques, and to their relationships with other relevant forms of measurement such as photogrammetry or satellite geodesy.

This chapter deals in turn with the elementary aspects of project planning and finance; raises the issues of quality control, legal liability, project management, the needs of clients, the need for archival records; and ends with a broad overview of measurement methods.

1.1 Clients and their needs

Surveying and setting out are activities which exist only because of the needs of an ever increasing group of clients. It is imperative that the special needs of each client be viewed with sympathy and that their opinions are given credence. The recent history of surveying is populated with instances of lost business opportunities because of the seemingly unsympathetic attitudes of the surveyor to the wishes of potential clients.

Also, the surveyor has taken too narrow a view of his competence and has failed to recognize problems of geometrical measurement within his expertise. Fortunately this situation is changing as the following list of applications shows.

Cadastral and land tenure problems; data input to geographical information systems; the surveys of interiors and buildings; industrial applications such as ship and aircraft construction; de-comissioning of nuclear reactors; measurements in sport; mining and quarry excavation; medical and surgical use; dimensional evidence in police cases; map and plan making; scientific studies of ice and tectonic movements ; oil platform location and construction; control for remote sensing and photographic imagery; monitoring of dam and structural deformations; the location of large engineering components and structures; and so on.

However there needs to be a dramatic change of emphasis in the role of the surveyor. It is not sufficient to measure, to process, and to present data. An understanding of the subsequent use of the data in the computer chain of processing is also needed if the full potential of the value which is added to the surveyor's work is to be realised. This is particularly so in land, building and engineering information systems.

Specifications
It is impossible to carry out the wishes of the client if these are not clearly understood by both parties and written into a sensible specification. Dangers lie in overspecifying a task just as much as underspecifying it. Issues that should be made clear include:

(1) The time scale for the task and any penalty arrangements if it is not kept. The likely influence of the weather in outdoor work should not be ignored.

(2) The financial arrangements and any sequential payments.

(3) The technical specification. This should concentrate on results and the quality assurance of these results, and not on specific methods of attaining these. Too tight a technical constraint may prohibit the use of new technology or may stem from advice from a consultant whose knowledge is out of date.

(4) Reasonable provision for sample inspection of the contractor's performance.

(5) Prior arrangements for settling disputes, to avoid the need for litigation.

1.2 Planning and documentation

Once the purposes of the work have been agreed, the technical work may begin. This involves thorough planning and costing. Usually this is left to an experienced surveyor who understands the complete time sequence of events and the client's needs.

The first stage is to examine existing information about previous survey work such as old maps and points, or knowledge of prior problems such any mine workings or previous litigation.

The second stage is to visit the site to see the environment, identify restrictions on access, likely problems of intervisibility, location of permanent marks, and above all to propose a detailed technical plan of execution. The degree of detail described in a report depends on the calibre of the surveyors who will actually carry out the task. All documentation should be clearly written and in accordance with the needs of quality assurance, with the expectation that it may have to be produced in court as evidence of proper practice.

In addition to technical work, especially in tasks of a major nature, administrative information should also be obtained, such as the availablity of accommodation, and sources of fuel, materials, and possible labour.

Field organization
The organisation of field work forms a very important part of the surveyor's work. It entails the preparation of an efficient technical programme, which is both cost effective and acceptable to the workforce. Because surveying practice is still weather dependent, work cannot always be scheduled to normal working hours. It is also affected by the site environment. Surveying near an operating railway or busy motorway is particularly hazardous; work in tropical forests is a danger to health; sub-zero temperatures call for special clothing; and efficient transport requires the proper care and maintenance of vehicles or boats. The provision of food, accommodation and fuel has to be attended to, as has the recruitment and payment of temporary staff and care for their health. Technical operations are often the easiest part of any surveying task.

Technical procedures
The surveyor must be fully conversant with his instrument if the best results are to be obtained. The centering of instruments and targets over reference marks needs continual vigilance, and all tripods and plummets need to be secure and adjusted.

For the best results on control work, horizontal and vertical angles should be taken separately; the former on various arcs or zeros. Independent pointings are necessary to remove systematic bias. A regular field booking procedure should be adopted to avoid omission of vital dimensions, such as heights of instruments and targets.

In the case of EDM, *electro-magnetic distance measurement*, proper pointing procedures and meteorological measurements are essential, as is the attention to batteries and their charging.

A technical programme must have sufficient flexibility to respond to weather and other unpredictable factors. Radio communication adds greatly to efficiency and general security.

Archives
The merit of keeping old records is not always appreciated. Quite apart from the obvious advantages of leaving permanent marks and clear descriptions of their whereabouts for further use, long term needs such as evidence of ground subsidence, historical information on development, and above all, evidence as to land ownership, can be most important. With the increasing use of computer data and reduced production of good quality paper maps, problems are looming for the future unless a proper archival policy is adopted.

The problem is not confined to mapping. Many large engineering structures, such as boilers and reactors, have to be assembled with the assistance of a complex sequence of surveying measurements. If records of these are lost it can be extremely difficult, if not impossible, to dismantle the structure at a later date. This is particularly true in the hostile environments of nuclear reactors.

1.3 Objectives of surveying

The immediate objectives of surveying are twofold.

(1) To create a *mathematical model* of a three dimensional solid from various forms of measurement, to assess its quality, and to present the results to a client for his purposes. This is usually called *surveying*.

(2) To control the *construction* of a three dimensional solid, described by its design model, and to ensure it meets a specified accuracy standard. This is usually called *setting out* or *building*.

Technology gives much scope for choice of methods to meet various criteria: such as the size of project, the speed with which it has to be carried out, the amount of added processing to be made on the results, the technical and manpower resources available, and above all, the finances involved.

Surveying
To help focus our attention on key aspects of surveying and setting-out, consider the idealized tower shown in Figure 1.1. This figure shows how to construct two views of the tower for stereoscopic viewing. If the two images B and C are copied on separate pieces of paper and viewed by stereoscope, a three dimensional illusion is created. Two separate drawings are required to enable their separation to be adjusted for viewing by each observer to suit the eye-base. Such an illusion is often required by a client to inspect a proposed construction, and its impact on the surrounding

environment. It is not difficult to create such images by computer for presentation on a computer screen or on hard copy.

The general shape of the tower can be seen from the two views labelled B and C. To obtain its size, length and angle measurements need to be made from points around it. These measurements are made to key points such as the corners. Next they are recorded as coordinate values (x, y, z) with respect to some arbitrary datum and coordinate axes. From these coordinates we construct a plan view

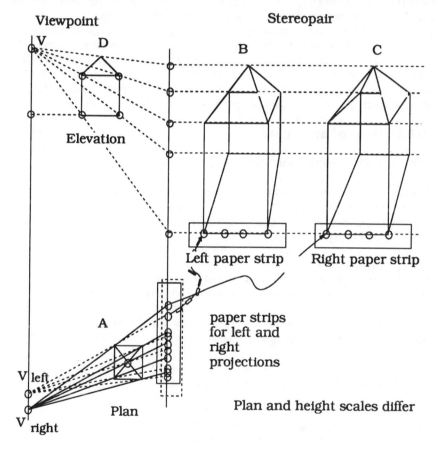

Figure 1.1

such as A, an elevation such as D, or two oblique views such as B and C. All these graphical representations need to be at a smaller scale than the original. In this case the vertical and horizontal scales are different for visual effect. Often they are constructed on a computer screen for temporary inspection, or are drawn on a plotter and printed for permanent use.

Later, the user may wish to modify the original building, or record its ownership in some register. To do these efficiently some form of permanent marking of control points and of permanent data storage is needed.

Setting out

By contrast to surveying an existing tower, it is possible that another client may wish to *construct* an identical tower to this. In this case a suitable site has to be selected; a geotechnic survey has to be carried out to see what foundations are required: the selection of possible routes for services, such as water and power, access to roads has to be made; legal opinion as to ownership and any objections from obscuring sunlight from other properties will be sought; and aesthetic and financial considerations will need to be settled.

Eventually the setting out will commence, with the selection of a suitable datum and coordinate reference system related to surrounding structures or to a national coordinate reference system. Survey measurements of lengths and angles will be required at various stages in the construction: to ensure that the tower is in the correct place; to ensure that it is vertical; and to ensure that it is of the correct dimensions. The construction process should not be delayed because of inaccurate or slow surveying work. The surveyor is often subjected to demands for speed which may impair the accuracy, and the success of the whole project. Tunnelling work is an obvious example.

It is the task of the geodetic surveyor to understand and control all geometrical operations in such projects. The term 'geodetic surveyor' is used to distinguish these activities from other forms of 'surveying', such as those dealing with valuation, or cost control. This title is preferred to the older 'land surveyor' which is much too restricting a term for modern acceptance.

1.4 Methodology of surveying and setting out

We now present a brief outline of the methodology adopted by geodetic surveyors in the course of their work. This includes the selection of a datum: the control of length standards; angle, direction and time measurements; the selection of a coordinate reference system; and the adoption of suitable mathematical models for computations.

Datum selection

No matter what the survey may be, a suitable datum of reference is required on which to make calculations and present results. At its simplest, the datum will be an arbitrary starting point, and an arbitrary reference direction.

A more sophisticated system will use the direction of north for the reference direction. In this case the directions are called 'bearings'. Because there are three types of north, magnetic, true or projection, a distinction has to be made between *magnetic bearings, true bearings* and *grid bearings.*

A magnetic bearing is obtained by a magnetic compass, the true bearing from stellar observations or from a gyroscope, and the grid bearing by calculation from either of these two, according to the map projection used.

It is also usual to align one of the axes of the adopted coordinate system with the direction of gravity, the local *vertical*. In some special cases, such as on a floating platform, this is not practicable, and an arbitrary direction has to be adopted instead. Any time a gravity sensor, such as a bubble, is used to set up an instrument, the direction of gravity is controlling the geometrical calculations. Over large distances, the direction of gravity varies because of Earth curvature and therefore has to be allowed for.

In many surveys, the absolute height of the land above sea level is of paramount importance. This is particularly serious when dealing with the complex infrastructure of water supply and sewage disposal in maritime countries. In other countries far from the sea, it may be that irrigation and lake levels are more significant. Thus it is clear that careful attention must be given to the selection of a suitable height datum. Usually a long-period average height of the sea at some selected point or points is adopted as the datum for heights in a national system, although efforts are now being made to adopt a satellite-based height datum for global studies, because of the inconsistencies which arise in levels of the sea over long periods of time.

Because the results of surveys depend on the adopted datum, it is vital that they are only used and interpreted within the proper context. Great care must be taken when the results of two or more surveys are combined together. This combination usually means that some transformation of coordinates is required. With the increasing use of satellite position-fixing systems, this datum care is ever more important.

Scale and length standards

Another vital factor in surveying is standard of length. A fundamental part of any measurement is its size, or scale. Clearly this will require that any measurement of length is traceable to an acceptable length standard, usually the international metre. This metre has to be defined in an unambiguous way and by a process which is reasonably practicable to repeat anywhere in the world. Today, the metre is defined, in terms of time and the speed of light, to an accuracy of about one part in ten million as follows:

A metre is the distance travelled by a light beam in vacuo during
one 299 792 458 th of a second of time.

This definition requires advice from a suitable physical laboratory
on how to establish the unit of time.In the UK this work is carried out
by the National Physical Laboratory, which also maintains a service
to enable surveyors to trace the scale of their instruments back to a
standard. Most countries provide a similar service, which many now
extend to the calibration of satellite based systems.

In many countries, units of length other than the metre are in
current use, or were used in the past for survey work. The US foot is
one such unit. There were also many different versions of the metre
in common use. It is vital that the surveyor uses the correct
conversion factor when relating old to new work.

Of equal importance is the need for the surveyor to check his tape or
measuring instrument for length, at periodic intervals, for use in
ordinary work.

Angles and directions

Many surveying instruments are designed to measure angles and
directions. Any angle is the difference between two directions, one of
which is taken as the reference direction. Angles are recorded in a
plane, selected by the observer as with a sextant or theodolite, or
measured from a map or photograph. They can be obtained to an
accuracy of a second of arc sexagesimal, or even better, with a modern
theodolite. The units of measurement are:

sexagesimal degrees, i.e. 360° to a circle

centesimal gons, i.e. 400 gons to the circle

radians, i e. 2π to the circle.

Because the arguments of most computer angle functions and the
angle units of the differential calculus are in radians, conversions
from sexagesimal or centesimal units to radians and vice versa are
necessary. For mental work the following relationships are
convenient:

One radian = 57.3° = 3438' = 206 265" sexagesimal.

One radian = 63.7^g = 6370^c = $637\,000^{cc}$ centesimal.

Time

Intervals of time can be measured to accuracies of better than one
part in ten million by electronic methods and atomic clocks.
Traditional time standards, such as the rotating Earth, are not

regular to such accuracy. All time-keepers are calibrated by the International Time Service available from National Physical Laboratories or other sources, such as the BIH Paris, France.

There are approximately 366.2422 sidereal days in the solar year which contains 365.2422 solar days.The one day difference is due to the fact the Earth rotates once round the Sun in a year.Thus each sidereal day is about 4 minutes shorter than the solar day.

The *period* T of an oscillation is related to the *frequency* f in units of cycles per second, or hertz (Hz), by

$$T = 1 / f$$

The speed of light c is related to the frequency and wavelength λ by the relationship

$$c = f \lambda$$

where $c \approx 300 \times 10^6$ metres per second approx.

The distance s travelled by light on a forward and back trip in time t (double transit time) from one end of a line to another is given by

$$s = 0.5 \, c \, t$$

Because many distance measurement systems are really timing systems, the following approximate relationships are useful in mental work:

(1) A frequency of 1500 megahertz (MHz) implies a wavelength of λ = 0.2 m.

(2) The length equivalent of a double transit time of 10^{-9} second (a nanosecond) is 300 mm of the distance measured.

1.5 Coordinate systems

The coordinate systems used to model geometrical shapes in three dimensions are illustrated in Figures 1.2, 1.3 and 1.4.These are

(1) Cartesian coordinates	(x, y, z)
(2) Polar coordinates	(φ, ω, r)
(3) Cylindrical coordinates	(φ, u, z)
(4) The two-plus-one hybrid	(x, y, h)

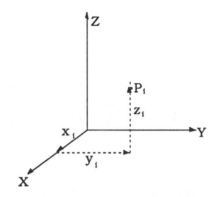

Figure 1.2 Cartesian coordinates

Cartesian coordinates

Cartesian coordinates employing a set of orthogonal axes, i.e. at right angles to each, other are most common. Not only does such a system generate obviously convenient mathematical forms, but it is easy to create in practice. Since early times it has been simple to set out or draw a right angle, to set up a grid of squares, and to take measurements in such a system.

It is equally practicable to construct rectangular plotting devices, such as coordinatographs and digitizing tables, using such methods. Because a grid is so fundamental to practical work it is advisable to check it for size and orthogonality. This can be done by a reversal procedure, such as drawing two copies of a grid, and seeing if they match when rotated through a right angle.

Three-axis coordinate measuring machines (CMMs) are a mechanical way of establishing three-dimensional grids in the industrial sector. Other systems create the necessary right angles optically via pentagonal prisms or mechanically.

The map grid, not only valuable as a reference system and a necessity for plotting, also provides a key to any distortion of the paper on which the map is drawn. It is often vital to allow for this distortion when working with old maps, as in legal disputes or in the building up of a Geographical Information System (GIS) for the computer handling of map and other data.

Current computational strategy is tending towards a rigorous cartesian three-dimensional approach, even if this results in more complex mathematical models.

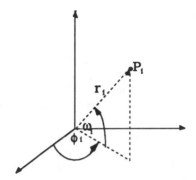

Figure 1.3 Polar coordinates

Polar coordinates

The polar system is basic to the modern theodolite with its integral distance measurement capablity, and to the traditional instrument using optical distance measurement. In two dimensions the drafting protractor still has some advantages over cartesian methods. Mechanical polar coordinate measurement systems still have some advantages in that they are more compact, as seen for example in an industrial robot.

Polar coordinates are basic to many mathematical models such as vector geometry and spherical trigonometry, and the *total-station* (theodolite plus distance measurer) is their mechanical analogue.

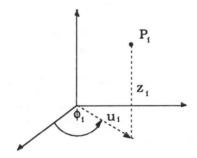

Figure 1.4 Cylindrical coordinates

Cylindrical coordinates

Although not as widely used as the other two systems, the cylindrical system has advantages when measuring cylindrical and rectangular objects, such as ships, large pipes, tunnels and oil storage tanks.

The two-plus-one system
Since the Earth's surface is comparatively flat, it can be modelled conveniently in two separate coordinate systems, plan and height. The coordinate system for plan is reduced to two dimensions (x, y) from which the z, or height h, is kept separate. This is often described as a 'two plus one' coordinate system. This system is necessary when a three-dimensional surface is represented on a map or plan. In this case the heights are represented in a conventional way as isolines of equal height (contours), or by spot heights written on the map. For example, the plan of the tower of Figure 1.1 would have the ground and roof heights written on it as in Figure 1.5, together with some suitably labelled contours.

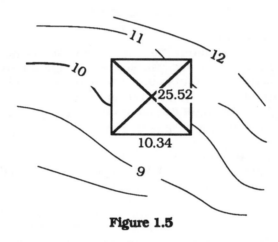

Figure 1.5

Traditionally, geographical or geodetic coordinates were also treated by the two plus one system of (latitude, longitude, height) in which the first two coordinates give positions on a reference surface (an ellipsoid) and the third gives the departure from this surface in a direction normal to it. Current practice is to calculate measurements in a three-dimensional cartesian system, before converting them to the suitable two-plus-one projection system and contour lines of the conventional topographic map.

1.6 Geometrical principles

In three dimensional Euclidean space, a point can be described uniquely by three relevant independent pieces of information.This information consists ultimately of some form of measurement such as length, or length difference, or length sum; or direction, or direction difference, or combinations of them. The position in space is ultimately determined by the intersection of at least two vectors (lines), and the accuracy of this fixation depends on two factors:

(1) their angle of intersection,

(2) their likely relative lateral shift.

The former depends on the *design* of the measurement system, and the latter on the design and *quality* of the measurement system. Quality may be improved either by taking more measurements, or by changing to a better instrument, or by changing to a better designed measurement system using the same instruments.

A preliminary graphical treatment of a problem can often yield a good idea of the potential accuracy of a system. For example, lines meeting at right angles are preferable to those intersecting at an acute angle.

Mathematical modelling

It is not practicable to determine the position of every point on an object, nor will it even be necessary. For example, a straight line is described fully by two points, one at each end; a circle by three suitable points and so on. It is therefore more economical to model a surface or line with the minimum of data.This modelling may be done by the field surveyor by selecting discrete points at changes of direction or slope on the ground, or by the cartographer when digitizing a map. The process of tracing lines in a photogrammetric plotter is, by contrast, a non-selective one. Even in this process, some data condensation has to take place when a record of the trace is kept in computer form.

Clearly, a limit to accuracy is implied by the modelling process. In the traditional line-map a plotting accuracy of 0.2 mm was the accepted error bound. Hence the real accuracy depends on the map scale. For example at a scale of 1 to 500 the accuracy limit is 0.1 metre. With the increased use of computer drafting, such conventional concepts of accuracy have tended to be ignored with serious results.

Again, it is reasonable to collect more than the minimum amount of data, to provide a means of checking against mistakes, or to control the quality of the work. A redundancy of data imposes the need to reconcile inconsistencies by a simple method or by the method of *least squares*, which in turn allows a statistical estimate of precision to be evaluated and affords an opportunity to detect blunders.

Algebraic and algorithmic methods

Many of the fundamental geometrical concepts can be described in simple graphical terms, and many results can be depicted in graphical form. On the other hand, algebraic methods are required if complex problems are to be treated.

Although conventional, matrix and vector algebras are each of value in expressing concepts, and in the increasing use of spread-sheet computer systems, many numerical problems have to be recast in suitable algorithmic form to avoid errors arising from the number crunching itself. These numerical methods need to be robust. Hence it is vital to check any computer software against standard sets of data to ensure that there are no 'bugs' in the system and that proper theory has been used. In the first analysis, it is the professional surveyor who is liable for his results, not the software house which wrote the computer package.

In the last analysis, there is no substitute for the use of common sense methods to check all work, be it field work, computations or plotting and setting out. Often a simple graphical treatment will be the basis of this common sense.

2 Measurement and mathematical models

Chapter Summary

In this chapter the geometry of various measurement systems is treated in as general a manner as possible, with applications reserved for later chapters. It deals with the geometrical entities, such as length and direction, which in turn are used to generate the geometrical primitives such as planes, spheres and conic sections.

In addition, some discussion is made of the principal mathematical models used to represent these measurement systems in analogue and numerical processing.

The chapter ends with a detailed consideration of the problems of coordinate transformations so fundamental to all branches of surveying.

2.1 Geometrical entities and primitives

The measurement processes used by the surveyor include the following entities as the bases of mathematical models usually selected to determine and describe the relative positions of points.

(1) Length.
(2) Length difference.
(3) Length sum.
(4) Alignment.
(5) Direction.
(6) Angle.

These measured quantities may be treated in two distinct ways :

(1) Figurally.
(2) Referenced to a coordinate system.

Figural treatment always remains close to the essential geometry of a problem in a way that coordinate systems do not. Tests for figural fidelity, for example that the three angles of a plane triangle sum to 180°, should form the basis of robust computer software. Figural treatment can also indicate the effects of errors, with a clarity not exhibited by coordinate systems. However, mathematical modelling by figural methods tends to be complex, and lacks both simplicity and generality of approach. Both methods have useful places in the computational toolbox.

Length

Length is fundamental to geometrical measurement; although shape alone can sometimes be of interest. The various ways in which length is measured are described in Chapter 3.

Length is usually taken to be the straight line distance between two points. Lengths are combined figurally into triangles. To form a triangle, three lengths are required, and furthermore, the sum of any two sides must be greater than the third. This is called a necessary figural condition which should be tested for in robust computer software.

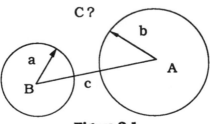

Figure 2.1

A triangle ABC can be viewed in a different way. See Figure 2.1. In two dimensions, if one of the sides c is considered the base, the other two sides a and b can be thought of as the radii of circles drawn from the base to meet each other. If they do not meet, there is no solution. If they do meet, they do so in two points forming not one triangle but two, as shown in the Figure 2.2. In the vicinity of the point to be fixed, the circumferences of the circles can be replaced by their tangents, i.e. by straight *position lines*.

Thus a computer solution will give two solutions in a plane. The correct solution has to be resolved by external evidence. The enlarged portion of Figure 2.2 in the vicinity of C shows quite clearly the direction of the tangents to the circles. The intersection is weak because these tangents meet at a narrow angle t. If each radius is changed by a small amount we can see the change in the solution at once, C moving to C'. Most accuracy problems in surveying boil down to a simple question of the intersection of two lines.This matter is considered more formally later.

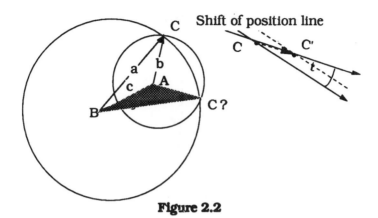

Figure 2.2

If one end of a line is considered fixed in space, the other is free to describe a sphere. Thus the measurement of distances, to fix a point in three dimensions, can be described mathematically by spheres centred on fixed points. If two spheres intersect as in Figure 2.3, they

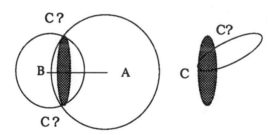

Figure 2.3

do so in a circle, shown shaded. A third intersecting sphere is required to give another circle to meet the first. These two circles will intersect in two points. Often the correct one can be resolved on the basis of other information, such as the fact that only one point lies on the surface of the Earth.The process of fixing a point by three distances, or three spheres, is called *trispheration.*

Clearly the matter of ambiguity can be resolved if a fourth appropriate line is measured giving yet another sphere, and yet another circle to intersect the first two shown in Figure 2.3 at the correct point C. Thus the geometry of the fixation ends finally in the intersection of three lines in space.

Length difference
Some measurement systems, like Doppler, derive the difference between two distances. Such a system generates an hyperbola in two

dimensions, or an hyperboloid in three. The geometry of a two dimensional problem is illustrated in Figure 2.4.

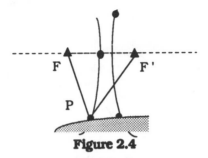

Figure 2.4

The length difference $\Delta = PF' - PF$ is measured by the system in which the points F and F ' are known. This Δ generates the two sheets of the hyperbola, the curved lines, showing the locus of all solutions, two of which are on the Earth. It is possible to draw this figure knowing the relative positions of F and F '. their relationship to the surface of the Earth, and Δ. A check would be given by another hyperbola meeting at a good angle. In three dimensions this second hyperbola would also need to be drawn in a plane inclined to that of the paper, as shown in Figure 2.5. As before, the problem is ultimately resolved into the intersection of two lines in the vicinity of the point P.

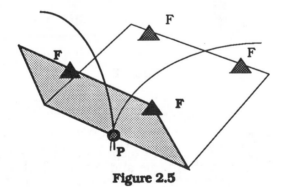

Figure 2.5

Length sum

The measured sum of two distances generates an ellipse in two dimensions and an ellipsoid in three. The ellipse and ellipsoid feature widely in surveying, geodesy, engineering and error theory. Figure 2.6 shows the basis of these two figures. The point P is related to two fixed points F and F ' by the fact that the sum $S = PF + PF'$ is known from measurement. This principle has applications in EDM, and in setting out works. It is unusual to fix position entirely by length sum.

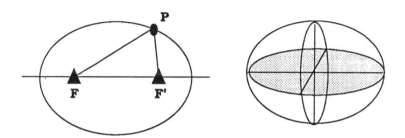

Figure 2.6

The basis of an ellipsoid of revolution is the plane ellipse, rotated about one of its axes; if rotated about the minor axis, an *oblate* ellipsoid is produced, and if about the major axis, a *prolate* ellipsoid results. An ellipsoid which is elliptical relative to all three axes is called *tri–axial*.

Alignment
Alignment is a way of describing that at least three points lie on a straight line. It is of great value to surveying because it can be achieved to great accuracy with very little effort. For example the human eye can detect alignment to about 20" arc. Methods of achieving alignment are either visual, optical by light beam, or mechanical by wire or straight edge. Visual and optical methods can suffer from refraction, and mechanical methods by shock waves or imperfections in the edge.

Geometrically, collinearity is represented by a freely moving straight line connecting at least three points in space.

Direction and angle
Direction is similar to alignment except that the line is related to a measuring system through a scale and rotational axis. The general name for any *angle* measuring device is *goniometer*. The common goniometers of surveying are the sextant and the theodolite. Although the idea of direction is useful in theory, a direction cannot be used by itself. It has to be combined with another into an *angle*.

When one of these directions has some physical significance, an angle becomes either a *bearing* from north in the horizontal plane, or a *zenith distance* from the vertical, in a vertical plane. The use of the word ' distance ' is an unfortunate convention in this case.

Instead of a physical references, such as north, an arbitrary line may be used. Some convention of sign has to be adopted to avoid confusion. Bearings are always clockwise from north (except in South Africa where south is used instead). Zenith distances are reckoned positive clockwise from the zenith. *Vertical angles* are the complement of zenith distances, usually reckoned positive upwards and negative downwards.

Angles can be combined to give the relative shapes of bodies. Length is needed if size is required.

Observed angles generate cones and cylinders. Consider the case of one direction considered fixed in space as a reference line; the other direction generates a cone in three dimensions about this reference line. For example a vertical angle V to a fixed high point A generates a cone about the vertical through A, as shown in the left–hand diagram of Figure 2.7

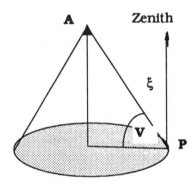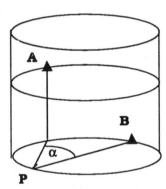

Figure 2.7

An angle α observed in a horizontal plane from a point P to two other fixed high points A and B, generates a cylinder through all three points, as shown in the right of Figure 2.7. Usually these two angles combine to give a solution for P, provided the intersection angle between of the tangents, to the cone and cylinder at P, is good. If B is also at a different height from P, a redundant check measure is available from the second cone whose vertical axis is through B.

This ability to establish a position from observations to unoccupied stations,*resection*, is very convenient, especially in astronomy, in hydrography and in engineering work.

Combined methods
It is often the case that several of the above geometrical methods are combined together to give a strong fixation. Care must be taken to ensure that the fixing lines cut at good angles of intersection, that is at angles better that 30˚, if possible. For this reason, the various methods of measurement are required. For example, it would be

impossible to fix by the intersection of two directions inside a large pipe; a direction and distance are required. This common combination yields the most satisfactory relationship of two position lines perpendicular to one another.

Although much less important than in former times, the ability to visualize the intersection of position lines aids the surveyor to design a measurement system suited to the task, and during its execution, helps to ensure that the maximum accuracy is achieved.

The figural approach also has a vital role to play as an independent thought process, useful for checking the results of computer software.

2.2 Mathematical models and algorithms

By contrast to the figural approach, is the coordinate treatment of geometry. This alternative approach has greater generality and is usually simpler when forming computer algorithms. It does however hide some geometrical relationships better seen in the figural approach.

Of fundamental importance to surveying is the topic of coordinate transformation: therefore considerable attention is paid to it here. The beginner is advised to concentrate on this important matter which has applications in almost every branch of surveying.

Plane coordinates

Fundamental to surveying are two or three–dimensional plane cartesian coordinates. Problems arise with the conventions in use. In a two dimensional system (x, y) or (n, e), it has become customary to direct the x axis to the north and the y axis to the east, because a bearing t is reckoned clockwise from north in surveying. This convention accords with mathematics.

It is, however, in direct contrast with the conventions of map reading and cartography, where it is customary to quote easting (e) before northing (n).

However, the convention to quote geographical coordinates, latitude (ϕ) before longitude (λ), once more accords with x and y respectively of our convention. To avoid potential confusion, all listed coordinates should clearly indicate which definition is being used, as well as the units of measurement.

Thus a point may have coordinates

$$x = 2.5, \; y = 5.5$$
$$e = 5.5, \; n = 2.5$$

which correspond to values (r, t) in a polar system, of

$$r = 6.041$$
$$t = 65.556°$$

The bearings t are reckoned in sexagesimal degrees.

The conversion of coordinates from cartesian to polar values is very important in surveying. Some care is needed to cater for all quadrants. For example converting the point (–2.5, –5.5) we obtain from a calculator with a rectangular–to–polar function key, the following values

$$r = 6.041$$
$$t = -114.443°$$

Because all survey bearings are reckoned clockwise from north, this t, obtained from a calculator, has to be added to 360° giving the survey bearing

$$t = 245.556°$$

The mathematical functions being used are

$$r^2 = x^2 + y^2$$
$$\arctan t = y/x$$

The arctan function has taken note of the negative signs of both x and y, thus ensuring that an erroneous first quadrant angle is not returned, because

$$-5.5 / -2.5 = 2.2$$

which is the tangent of 65.556°. Usually this function is listed as *ATAN2* in contrast with *ATAN* which does not deal with quadrants properly. If no *ATAN2* function is available, the following algorithm might be used

$$r = SQRT(x*x + y*y)$$

if $r + x = 0$ then t = 180° else

$$t = 2*ATAN(y/ (r + x))$$

adding 360° if necessary.

Because not all algorithms, to compute the reverse trigonometrical functions, deal properly with the various quadrants, a check should be made when using a calculator or computer algorithm for the first time. Also, a check should be made to see if a correct result is given for the limiting cases of points on the coordinate axes. In other words, we could check that the correct related data are

Point	1	2	3	4
x	6	0	−6	0
y	0	6	0	−6
t	0	90	180	−90

The last result has to be added to 360° to give the whole circle bearing of 270°.

Again, most computer algorithms require angles to be converted to radians before calling a trigonometric function, and in the reverse mode, yield results also in radians. Thus a suitable algorithm to calculate the sine of 12° 34′ 56″ is

$$\text{SIN(PI ()} * (12 + (34 + 56/60)/60)/180) = 0.217\ 840\ 422$$

and to convert 12.582 222 22° to traditional form might be

$$x = 12.582\ 222\ 22$$

$$\text{INT}\ (x) = 12°$$

$$y = (x - \text{INT}\ (x)) * 60 = 34.933\ 333$$

$$\text{INT}\ (y) = 34′$$

$$\text{INT}\ ((y - \text{INT}\ (y)) * 60 + 0.5) = 56″$$

If no function for PI is available a convenient algorithm is

$$\text{PI} = 4 * \text{ATAN}(1)$$

Accuracy of plane coordinates
The whole question of the accuracies of computed results is treated throughout this book, and particularly in Chapter 5. The practical creation of coordinated positions either by drawing a map, or setting out a structure, is treated in Chapter 10.

The treatment of coordinate conversions in geodesy requires special care because of the very large numbers and high precision required. Again, any computer system should be tested to ensure that the required precison is being achieved.

2.3 Principal mathematical models

Because most computation requires the adoption of a system of rectangular cartesian coordinates and the processing of data by the method of least squares, most mathematical analysis uses a coordinate approach for its final processing. It operates by a differential technique based on provisional coordinates which are estimated by inspection, or are calculated by an initial figural approach, such as the use of the sine rule in a plane triangle to derive side lengths from angles. The method of calculation of provisional coordinates is special to each problem or application.

Weights

Accompanying each mathematical model is the need to assign weights to the observations. Single observations, such as distances and independent theodolite pointings, are generally assumed to be uncorrelated. However, combinations of these cannot be considered uncorrelated. Allowance has to be made by use of the proper *dispersion matrix* (inverse of the weight matrix).

For example the co–variance of two adjacent angles in a round is minus the variance of a direction. See Chapter 5 for more details.

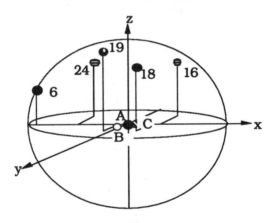

Figure 2.8

Standard data set

For ease of reference we have gathered the principal models together to be illustrated by a set of data points, shown in Figure 2.8, and whose coordinates are given in Table 2.1.

The high points, 18 to 6, are the simultaneous positions of GPS satellites at a moment of observation. The origin is at the centre of the Earth, the x axis is through the Greenwich meridian and the z axis coincides with the Earth's spin axis. A spherical Earth of radius 6.4 units is assumed, and the GPS satellites are taken to be in circular orbits of radius 26.67 units. For details see Chapter 6.

These same points can be scaled to represent most typical surveying applications. Because all the mathematical models are essentially the same, they have been brought together to stress this generality.

Table 2.1

Cartesian coordinates of data points

Point	x (N)	y (E)	z
18	3431.47	3730.53	1343.97
19	2946.97	3011.82	1827.87
24	1824.27	739.19	1396.02
16	3521.06	829.67	1700.06
6	885.13	2000.00	986.18
A	3000.00	2000.00	0.00
B	2994.40	2003.55	−0.03
C	2999.99	1996.41	−0.01

The bearings U, the vertical angles V, and the slope distances S, between the points of Table 2.1 are listed in Tables 2.2, 2.3, and 2.4. The reader may be surprised at the number of figures listed. They are needed in satellite work where the distances involved are very large. In the scaled down applications of these figures, we will use only a truncated set appropriate to the problem.

Table 2.2

	Slope distances from		
To point	A	B	C
18	2233.1941	2231.5524	2235.9851
19	2089.9044	2088.0796	2091.6534
24	2218.3000	2217.3782	2216.2625
16	2128.7038	2132.0579	2126.7426
6	2333.5008	2328.4421	2333.4988
A	0	6.6305	3.5900
B	6.6305	0	9.0680
C	3.5901	9.0680	0

Table 2.3

| To point | Bearings in degrees from North from | | |
	A	B	C
18	75.999 99	75.797 60	76.0275 25
19	93.000 16	92.693 26	92.989 00
24	−133.000 14	−132.783 40	−133.081 39
16	−66.000 18	−65.836 66	−65.934 30
6	180	−179.903 57	179.902 74
A		−32.371 78	89.840 40
B	147.628 22		128.057 86
C	−90.159 60	−51.942 14	

Table 2.4

| To point | Verts in degrees from | | |
	A	B	C
18	37.000 97	37.032 74	36.946 45
19	61.001 60	61.092 07	60.914 16
24	39.000 95	39.020 24	39.042 95
16	53.001 41	52.881 95	53.070 69
6	25.000 81	25.058 87	25.000 29
A		0.2592 39	0.1595 97
B	0		−0.1263 69
C	0.3191 96	0.1263 70	

Mathematical models for length

In three dimensions the most common mathematical model is

$$S^2 = (x_2 - x_1)^2 + (y_2 - y_1)^2 + (z_2 - z_1)^2$$

which on partial differentiation reduces to

$$dS = L(dx_2 - dx_1) + M(dy_2 - dy_1) + N(dz_2 - dz_1)$$

where L, M, and N are the direction cosines of the ray, ie are given by

$$L = \frac{(x_2 - x_1)}{S} = \cos V \cos U$$

$$M = \frac{(y_2 - y_1)}{S} = \cos V \sin U$$

$$N = \frac{(z_2 - z_1)}{S} = \sin V$$

Strictly these direction cosines should be written as $\overset{*}{L}$ etc because they are evaluated from the provisional coordinates. As this will always be the case we will use the simpler notation in this chapter.

The quantitiy dS may be expressed as

$$dS = 1 + v$$

if no instrument index is being modelled; or as

$$dS = 1 + E + v$$

if an index E is modelled. In both cases the vector 1 is the O–C value of the distance S, and v is the residual.

Example
Consider the ray A 18 of Table 2.1. If these are provisional coordinates and the observed slant length was 2233.39, working to four decimal places only we have

$$L = 0.1932 \quad M = 0.7749 \quad N = 0.6018 \quad 1 = 0.20$$

and the check that

$$L^2 + M^2 + N^2 = 1$$

A fully worked example in two dimensions is given in Chapter 8.

Length difference or length sum equations
Length differences arise in the Doppler method of measurement and by choice from GPS and high precision EDM work. In the latter two, the measured lengths, burdened by an unknown index or clock error, are differenced to remove this unwanted effect.

The two equations for each length of the form already derived in the last section are

$$_1dS_2 = {}_1L_2 (d x_2 - d x_1) + {}_1M_2 (d y_2 - d y_1) + {}_1N_2 (d z_2 - d z_1)$$

$$_1dS_3 = {}_1L_3 (d x_3 - d x_1) + {}_1M_3 (d y_3 - d y_1) + {}_1N_3 (d z_3 - d z_1)$$

These are subtracted from each other to give an equation in the nine variables. A worked example of this technique is given for a two dimensional problem in Chapter 8.

A length sum equation, for the little-used technique of line crossing, is obtained by adding these equations.

Length ratio equations
To eliminate scale effects arising from uncertain refraction, length ratios are sometimes used in monitoring work; thus we have

$$\frac{{}_1S_2}{{}_1S_3} = R$$

giving the differential equation

$$\frac{{}_1\,dS_2}{{}_1S_2} - \frac{{}_1\,dS_3}{{}_1S_3} = \frac{dR}{R}$$

which is expressed in coordinate terms by substituting for the length equations.

Alignment equations
One mathematical model for three points (1), (2) and (3) to be collinear is given by the following three equations

$$x_3 = x_1 + L\,t_1$$
$$y_3 = y_1 + M\,t_1$$
$$z_3 = z_1 + N\,t_1$$

where L, M an N are the direction cosines of the line evaluated from points (1) and (2) and t_1 is a scale factor the ratio of lengths given by

$$t_1 = \frac{(1)-(3)}{(1)-(2)}.$$

Direction equations
Although a direction has no meaning by itself, the concept is a useful basis for a mathematical model. The direction U is related to the coordinates in plan (x, y) by the equation

$$\tan U = (y_2 - y_1) / (x_2 - x_1)$$

which gives the differential equation

$$dU = P\,(d\,x_2 - d\,x_1) + Q\,(d\,y_2 - d\,y_1)$$

where the coefficients P and Q are given by the following expressions if dU is in sexagesimal seconds of arc:

$$P = -\,206265\,(y_2 - y_1)\,\sin U\,/\,S\,\cos V = -\,206265\,(y_2 - y_1)\,\sin U\,/\,D$$

$$Q = +\,206265\,(x_2 - x_1)\,\cos U\,/\,S\,\cos V = +\,206265\,(x_2 - x_1)\,\cos U\,/\,D$$

where D is the horizontal distance, and

$$dU = 1 + E + v$$

The additional parameter E is to allow for the error of orientation of the reference direction which usually depends on the provisional coordinates. If the orientation is obtained either from a magnetic compass or from an astronomical azimuth, the *direction* becomes an *azimuth or true bearing* subject only to observational error. Thus the parameter E is omitted from the azimuth equation. Otherwise, the equation is identical to that for a direction.

Example
Working with the data for lines A 18 and A 19 gives respectively

$$_AP_{18} = 112.25 \qquad _AQ_{18} = -27.99$$
$$_AP_{19} = 203.34 \qquad _AQ_{19} = +10.65$$

A fully worked example is given in Chapter 8, where it is shown how the unwanted orientation parameter E may be eliminated from the direction equations by Schreiber's method.

Angle equations
Since an angle is the difference between two directions, a differential angle equation is obtained for the two direction equations from point (1) to (2) and from (1) to (3) of the form

$$_1dU_2 = {_1}l_2 + E + {_1}v_2 = {_1}P_2\,(dx_2 - dx_1) + {_1}Q_2\,(dy_2 - dy_1)$$

$$_1dU_3 = {_1}l_3 + E + {_1}v_3 = {_1}P_3\,(dx_3 - dx_1) + {_1}Q_3\,(dy_3 - dy_1)$$

These are subtracted, thus eliminating the common orientation parameter E at the station and an equation for the angle at point (1).

Because directions are likely to be correlated, the appropriate dispersion matrix for angles must be calculated from the directions. Since Schreiber's method avoids this complication, the once-popular angles method is little used.

Vertical angle equations
The mathematical model for vertical angles is

$$\tan V = \frac{(z_2 - z_1)}{D}$$

where the horizontal distance D is given by

$$D^2 = (x_2 - x_1)^2 + (y_2 - y_1)^2$$

Differentiation gives the equation

$$[\frac{x_2 - x_1}{D}(dx_2 - dx_1) + \frac{y_2 - y_1}{D}(dy_2 - dy_1)] \tan V + D \sec^2 V \, dV$$
$$= dz_2 - dz_1$$

which may be recast as

$$P\,dx + Q\,dy + R\,dz = -dV$$

where the coefficients are

$$P = \sin 2V \cos U / 2D \qquad Q = \sin 2V \sin U / 2D \qquad R = -\cos^2 V / D$$

Vertical angles are only of value for position fixing when they exceed 30°, such as in engineering and industrial work, or in the special case of field astronomy. An example is given in Chapter 8.

2.4 Transformation of plane coordinates

A major problem of surveying and geodesy, if not *the* major problem, is the establishment of relationships between different sets of coordinates. There are two cases to be considered :

(a) When the datum shift, the rotation of axes, and the scale changes are known.

(b) When these parameters are unknown and have to be calculated from two sets of points whose coordinates on both systems are known.

The most common case of the direct transformation in two dimensions will be considered first. The coordinates used by surveyors are generally orthogonal and conformal. Because bearings t are reckoned clockwise from north, the convention is to align the x axis north and the y axis east. In this way the standard formulae of mathematics are preserved

$$x = r \cos t$$
$$y = r \sin t$$

In matrix notation we have

$$\mathbf{x} = \begin{bmatrix} x \\ y \end{bmatrix} = \begin{bmatrix} r \cos t \\ r \sin t \end{bmatrix}$$

To clarify notation, and to ensure that the beginner fully understands this most important of procedures, the simple data sets shown in Tables 2.5 and 2.6 are used for a series of examples. These

Table 2.5

Coordinate conversion data set – One

Point	A	B	C	O	G
x (n)	2.5	4.5	3.5	2.0	3.5
y (e)	5.5	4.5	8.5	2.5	6.2
r	6.04	6.36	9.19	3.20	7.09
t'	65.56	45.00	67.62	51.34	60.42

Table 2.6

Coordinate conversion data set –Two

Point	A'	B'	C'	o	G
X(N)	1.89	2.91	3.20	-2.73	2.66
Y(E)	2.74	1.66	4.39	-1.66	2.93
r'	3.33	3.35	5.43	3.20	3.96
t'˙	55.5	29.71	53.97	238.66	47.75

---

data sets are the coordinates scaled from Figure 2.9. Since we cannot predict the final scale of the figure when printed in this book, a scale of units has been drawn on the diagram to allow for scale change. Should the reader wish to use a ruler to measure the coordinates, a scale factor will need to be used to convert his or her figures to those given in Tables 2.5 and 2.6. No great precision should be expected of these figures.

The distance and bearing from the origin, (r, t), is given for each point in rows three and four of the tables.

The Figure 2.9 shows the triangles A B C and A' B' C', and their common centre of gravity G, together with two sets of axes referred to two origins, o and O. For separate diagrams of each system see Figures 2.10 and 2.11.

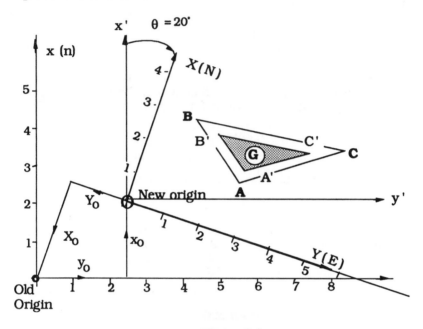

Figure 2.9

Figure 2.9, shows the triangles ABC and A'B'C', and their common centre of gravity G, plotted with respect to two sets of coordinate axes (x o y) and (X O Y). Their cartesian coordinates are listed in Tables 2.5 and 2.6. together with their respective polar coordinates.

The beginner is encouraged to verify the calculations and plotting before proceeding further.

Conversion to new system
For various reasons, there is often a requirement to change the origin of coordinates from o to a new point such as O, and change the orientation of the axes by *rotation* through an angle θ. In the example θ = 20°. The datum shift from o to O is brought about by a *translation* of the axes through

$$\begin{bmatrix} x_0 \\ y_0 \end{bmatrix} = \begin{bmatrix} 2.0 \\ 2.5 \end{bmatrix}$$

Thus we obtain a new set of coordinates relative to the new origin, but still with the same orientation. This operation can be expressed as

$$\begin{bmatrix} x'_i \\ y'_i \end{bmatrix} = \begin{bmatrix} x_i \\ y_i \end{bmatrix} - \begin{bmatrix} x_o \\ y_o \end{bmatrix} \tag{1}$$

or in matrix algebra as

$$\mathbf{x}' = \mathbf{x} - \mathbf{x_o}$$

Notice that the vector **x,** when expressed in bold type, represents both the x and the y coordinates.

The subscript i refers to the points A, B, C, and G. The reader should verify that the values now are as in Table 2.7. The coordinates of the old origin o with respect to the new origin O should be noted. This point is treated just as any other in the new system.

Table 2.7

Point	A	B	C	o	G
x'	0.5	2.5	1.5	-2.0	1.5
y'	3.0	2.0	6.0	-2.5	3.7

We now rotate the axes through an angle $\theta = 20°$ to give coordinates in the final (X,Y) system. This rotation is achieved as follows

The relationship between the two systems is written

$$\mathbf{X} = \mathbf{R}\,\mathbf{x}'$$

or in full

$$\begin{bmatrix} X \\ Y \end{bmatrix} = \begin{bmatrix} \cos\theta & \sin\theta \\ -\sin\theta & \cos\theta \end{bmatrix} \begin{bmatrix} x' \\ y' \end{bmatrix} \tag{2}$$

Inserting the values of $\theta = 20°$ and coordinates of point A we have the transformation using equations (2)

$$\begin{bmatrix} X_A \\ Y_A \end{bmatrix} = \begin{bmatrix} 0.9397 & 0.3420 \\ -0.3420 & 0.9397 \end{bmatrix} \begin{bmatrix} 0.5 \\ 3.0 \end{bmatrix}$$

$$= \begin{bmatrix} 1.5 \\ 2.65 \end{bmatrix}$$

Table 2.8 gives the results for all the transformed points of Table 2.7.

These can be verified graphically from the Figure 2.9. It is important to notice two things about the equations (2)

(a) They refer to a rotation of the *axes* .
(b) The angle θ is *positive* for a clockwise rotation about a z axis into the plane of the paper.

By contrast, sometimes a rotation refers to the *body* within fixed axes. In such a case the signs of $\sin \theta$ are interchanged in the above rotation matrix. See Chapter 3 for an application.

Table 2.8

Point	A	B	C	o	G
X	1.50	3.03	3.46	−2.73	2.67
Y	2.65	1.02	5.13	−1.67	2.96

These two operations of translation and rotation could be written

$$\mathbf{X} = \mathbf{R}\,(\mathbf{x} - \mathbf{x_0}) = \mathbf{R}\,\mathbf{x} \ - \ \mathbf{R}\,\mathbf{x_0}$$

But

$$\mathbf{R}\,\mathbf{x_0} = \mathbf{X_0}$$

so we may also write

$$\mathbf{X} = \mathbf{R}\mathbf{x} - \mathbf{X_0}$$

or in full

$$\begin{bmatrix} X \\ Y \end{bmatrix} = \begin{bmatrix} \cos \theta & \sin \theta \\ -\sin \theta & \cos \theta \end{bmatrix} \begin{bmatrix} x \\ y \end{bmatrix} - \begin{bmatrix} X_0 \\ Y_0 \end{bmatrix} \qquad (3)$$

The reader should verify that the alternative transformation from (x, y) to (X, Y) by formula (3) gives the same results as from the application of equations (1) and (2) in succession. The transformation for point A is as follows:

$$\begin{bmatrix} X_A \\ Y_A \end{bmatrix} = \begin{bmatrix} 0.9397 & 0.3420 \\ -0.3420 & 0.9397 \end{bmatrix} \begin{bmatrix} 2.5 \\ 5.5 \end{bmatrix} - \begin{bmatrix} 2.73 \\ 1.67 \end{bmatrix} \qquad (4)$$

$$= \begin{bmatrix} 1.50 \\ 2.65 \end{bmatrix}$$

Transformation by polar coordinates

The third way in which the coordinates of points in the old (x, y) system could be transformed into the new (X, Y) system is through the polar coordinates, by directly altering the survey bearings and recalculating the new coordinates. For example, beginning with the old coordinates (x, y) we could obtain the distances and bearings (r, t) as in Table 2.5., then change all bearings using

$$t'' = t - \theta$$

Thus we can alter all the bearings to the new system. Distances could also be re-scaled if necessary. The coordinates are then moved to the datum at O via (X_0, Y_0). It is always possible to treat these problems by polar coordinates instead of cartesian coordinates.

Change of scale

When a change of scale is required, a scale factor k is introduced in such a way that all lengths, including coordinates, are altered in proportion. For example, if we wish to plot the map at a scale of 1: 500 all sizes are scaled down by this amount. Formally we state

$$r' = kr; \quad x' = kx; \quad y' = ky$$

where k = 1/500. This is usually called the *nominal scale* of the map.

A change of units of coordinates from metres to feet is another example of nominal scale, with a scale factor of 1/ 0.3048.

There is also another way in which the scale may alter, this time by a small amount. This happens when a reduced distance is plotted on some map projection, or the real distance is computed on a map projection. See Chapter 4 for more details.

However a scale change can arise from the use of incorrect lengths in the survey itself. The use of an unstandardized tape, or the wrong refractive index in electromagnetic distance measurement (EDM) will cause such a scale change. In such a case the scale factor k is approximately equal to one, and we can put

$$k = 1 + e$$

where e is called the *scale error*. This is an unfortunate term because it can be confused with random errors of observation. It is in fact a deliberate *systematic error*. A scale factor of this type is often found when an old survey is repeated with modern instruments capable of better accuracy and the two have to be combined into a homogeneous or consistent system.

2.5 Reverse transformations

Suppose the triangle ABC of Figure 2.9 is scaled down by a factor of k = 2/3, the result is triangle A' B' C'. Suppose the old triangle ABC was plotted on the (x, y) system on an assume north (n) as shown in Figure 2.10, and the new triangle A'B'C' is derived on the (X, Y) system, shown in Figure 2.11 with orientation on true north (N). Each system of coordinates is recorded in Tables 2.5 and 2.6.

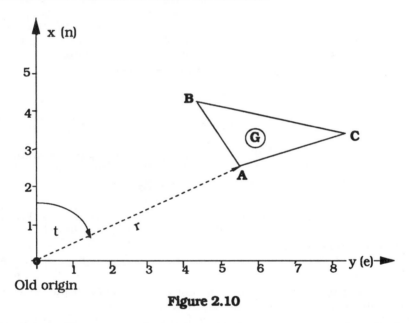

Figure 2.10

To bring them into a homogeneous system, we need to find the transformation parameters which link them. This is the reverse of the problems dealt with by equation (3).

Empirical transformation

The empirical method assumes that we do not wish to know the nature of the transformation, but merely to carry it out. Thus we assume some mathematical model to relate the two sets. The simplest is a linear transformation of the form

$$\mathbf{X} = \mathbf{R}\,\mathbf{x}$$

or in full

$$\begin{bmatrix} X \\ Y \end{bmatrix} = \begin{bmatrix} r_{11} & r_{12} \\ r_{21} & r_{22} \end{bmatrix} \begin{bmatrix} x \\ y \end{bmatrix}$$

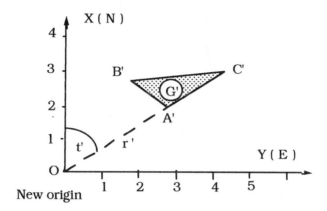

Figure 2.11

In this equation, at each point both sets of coordinates **X** and **x** are known; the unknowns are the coefficients of the **R** matrix. When multiplied out and rearranged we have for one point;

$$
\begin{bmatrix} X \\ \overline{Y} \end{bmatrix} = \begin{bmatrix} x & y & 0 & 0 \\ 0 & 0 & x & y \end{bmatrix} \begin{bmatrix} r_{11} \\ r_{12} \\ r_{21} \\ r_{22} \end{bmatrix}
$$

This is insufficient information to solve for the four variables. The coordinates from a second point are required, to give

$$
\begin{matrix} \text{Values for} \\ \text{point 1} \\ \\ \text{Values for} \\ \text{point 2} \end{matrix} \begin{bmatrix} X \\ Y \\ \overline{X} \\ Y \end{bmatrix} = \begin{bmatrix} x & y & 0 & 0 \\ 0 & 0 & x & y \\ x & y & 0 & 0 \\ 0 & 0 & x & y \end{bmatrix} \begin{bmatrix} r_{11} \\ r_{12} \\ r_{21} \\ r_{22} \end{bmatrix}
$$

If in the example point 1 is A and point 2 is B we obtain the equations

$$
\begin{matrix} \text{Values for} \\ \text{point A} \\ \\ \text{Values for} \\ \text{point B} \end{matrix} \begin{bmatrix} 1.89 \\ 2.74 \\ \overline{2.91} \end{bmatrix} = \begin{bmatrix} 2.5 & 5.5 & 0 & 0 \\ 0 & 0 & 2.5 & 5.5 \\ 4.5 & 4.5 & 0 & 0 \end{bmatrix} \begin{bmatrix} r_{11} \\ r_{12} \\ r_{21} \end{bmatrix}
$$

The solution of these equations is

$$
\begin{bmatrix} r_{11} \\ r_{12} \\ r_{21} \\ r_{22} \end{bmatrix} = \begin{bmatrix} 0.5556 \\ 0.0911 \\ -0.2370 \\ 0.6059 \end{bmatrix}
$$

giving the transformation equations

$$
\begin{bmatrix} X \\ Y \end{bmatrix} = \begin{bmatrix} 0.5556 & 0.0911 \\ -0.2370 & 0.6059 \end{bmatrix} \begin{bmatrix} x \\ y \end{bmatrix}
$$

Applying these equations to A, B, C and G gives the results

	A'	B'	C'	G'
X	1.89	2.91	2.72	2.51
Y	2.74	1.66	4.32	2.93

Not surprisingly, there is perfect agreement with the correct values in Table 2.6 at points A' and B', because they were used to derive the transformation equations; but there are discrepancies at points C' and G', away from the reference line AB.

This simple transformation method may be sufficient for the purpose in hand. We shall assume that this is not the case and that another and better approach is required; the method of centroids.

Method of centroids
Although there are other methods of empirical transformations which will give better results than the above, we move straight away to the best general method, involving the centroids of figures.

The concept is to move all scaling and rotations to the middle of the areas concerned. The method has two stages; a translation, and a rotation and scaling together.

First the coordinate system is moved to the centroid in each case, giving new coordinates expressed by

$$
\begin{bmatrix} x^G \\ y^G \end{bmatrix} = \begin{bmatrix} x_i \\ y_i \end{bmatrix} - \begin{bmatrix} x_G \\ y_G \end{bmatrix}
$$

$$
\begin{bmatrix} X^G \\ Y^G \end{bmatrix} = \begin{bmatrix} X_i \\ Y_i \end{bmatrix} - \begin{bmatrix} X_G \\ Y_G \end{bmatrix} \tag{6}
$$

Thus we have for the example the figures of Tables 2.9 and 2.10.

Table 2.9

Point	A	B	C	G
x	2.50	4.50	3.50	3.50
y	5.50	4.50	8.50	6.20
x^G	−1.00	+1.00	0.00	0.00
y^G	−0.67	−1.67	2.33	0.00

Table 2.10

Point	A'	B'	C'	G'
X	1.89	2.91	3.2	2.67
Y	2.74	1.66	4.39	2.93
X^G	−0.78	0.24	0.53	0.00
Y^G	−0.19	−1.27	1.46	0.00

These coordinates, reduced to the centroid indicated by the ^{G}, are next transformed by the empirical formula of the form

$$\mathbf{X}^G = \mathbf{R}\mathbf{x}^G$$

Which is recast in terms of the **R** matrix elements as before into

$$\begin{array}{l}\text{Values for point 1}\\ \\ \text{Values for point 2}\\ \\ \end{array}\left[\begin{array}{c} X^G \\ Y^G \\ \hline X^G \\ Y^G \end{array}\right] = \left[\begin{array}{cccc} x^G & y^G & 0 & 0 \\ 0 & 0 & x^G & y^G \\ \hline x^G & y^G & 0 & 0 \\ 0 & 0 & x^G & y^G \end{array}\right]\left[\begin{array}{c} r_{11}\\ r_{12}\\ r_{21}\\ r_{22}\end{array}\right]$$

The numerical values for points A and B are

$$\begin{array}{l}\text{Values for point A}\\ \\ \text{Values for point B}\\ \\ \end{array}\left[\begin{array}{c} -0.78 \\ -0.19 \\ \hline 0.24 \\ -1.27 \end{array}\right] = \left[\begin{array}{cccc} -1 & -0.67 & 0 & 0 \\ 0 & 0 & -1 & -0.67 \\ \hline +1 & -1.67 & 0 & 0 \\ 0 & 0 & +1.0 & -1.67 \end{array}\right]\left[\begin{array}{c} r_{11}\\ r_{12}\\ r_{21}\\ r_{22}\end{array}\right]$$

The solution for the coefficients of the **R** matrix are

$$\begin{bmatrix} r_{11} \\ r_{12} \\ r_{21} \\ r_{22} \end{bmatrix} = \begin{bmatrix} 0.6243 \\ 0.2286 \\ -0.2271 \\ 0.6257 \end{bmatrix}$$

The final transformation formula then becomes

$$\begin{aligned} \mathbf{X} &= \mathbf{X_G} + \mathbf{x}^G \\ &= \mathbf{X_G} + \mathbf{R}\,\mathbf{x}^G \\ &= \mathbf{X_G} + \mathbf{R}(\mathbf{x} - \mathbf{x_G}) \end{aligned} \tag{8}$$

$$\begin{bmatrix} X \\ Y \end{bmatrix} = \begin{bmatrix} X_G \\ Y_G \end{bmatrix} + \begin{bmatrix} 0.6243 & 0.2286 \\ -0.2271 & 0.6257 \end{bmatrix} \begin{bmatrix} x - x_G \\ y - y_G \end{bmatrix}$$

$$\begin{bmatrix} X \\ Y \end{bmatrix} = \begin{bmatrix} 2.67 \\ 2.93 \end{bmatrix} + \begin{bmatrix} 0.6243 & 0.2286 \\ -0.2271 & 0.6257 \end{bmatrix} \begin{bmatrix} x - 3.5 \\ y - 6.2 \end{bmatrix}$$

Comment
The application of this formula to points A, B and C gives perfect agreement with the original values of Tables 2.5 and 2.6 as expected. The agreement of point C comes about because the two sets of values are entirely consistent. In practice, the two surveys of the points of the triangle would contain errors which would give rise to slight inconsistencies. In this case a least squares approach is used.

2.6 General case of centroids with least squares

We shall now deal with the problem in which the data are inconsistent, and both sets of coordinates are associated with known dispersion matrices. The reader who is unfamiliar with the least squares process should read Chapter 5, and/or Appendix 5, which explain these matters and the notation being used. Although the example is for a two-dimensional coordinate problem, the method is identical in three dimensions.

The situation is that the triangles ABC and A'B'C' are not exactly similar in shape as well as being of different size. In addition, the coordinates in both systems have associated dispersion matrices $\mathbf{D}_x$ and $\mathbf{D}_X$. The method follows similar lines to the centroid method, except that each centroid is obtained as a weighted operation. To see how this is done we must first set up the observation equations for each set of coordinates separately in the form

$$\mathbf{A}\,\mathbf{x} = \mathbf{l} + \mathbf{v}$$

For the (x, y) system these are in full

$$
\begin{bmatrix} 1 & 0 \\ 0 & 1 \\ 1 & 0 \\ 0 & 1 \\ 1 & 0 \\ 0 & 1 \end{bmatrix}
\begin{bmatrix} x_G \\ y_G \end{bmatrix} =
\begin{bmatrix} 2.5 \\ 5.5 \\ 4.5 \\ 4.5 \\ 3.5 \\ 8.5 \end{bmatrix} + \mathbf{v}
$$

The dispersion matrix $\mathbf{D_x}$ could be full with many non zero covariances. However, to simplify the calculations for the reader we assume the upper triangle of this symmetric matrix as

$$
\begin{bmatrix}
0.1 & 0.05 & 0 & 0 & 0 & 0 \\
 & 0.1 & 0 & 0 & 0 & 0 \\
 & & 0.1 & 0.05 & 0 & 0 \\
 & & & 0.1 & 0 & 0 \\
\text{symmetric} & & & & 0.1 & 0.05 \\
 & & & & & 0.1
\end{bmatrix}
$$

The coordinates of the centroid are then given by

$$
\mathbf{x_G} = (\mathbf{A^T W A})^{-1} \mathbf{A^T W x} \tag{9}
$$

where

$$
\mathbf{W} = (\mathbf{D_x})^{-1}
$$

Thus the coordinates of the centroid are

$$
\begin{bmatrix} x_G \\ y_G \end{bmatrix} =
\begin{bmatrix} 3.5 \\ 6.17 \end{bmatrix}
$$

In an exactly similar way, using the coordinates in the (X,Y) system, where it should be noted a change has been made to the ordinate of point C to ensure that the two triangles are no longer similar, we have the equations

$$
\begin{bmatrix} 1 & 0 \\ 0 & 1 \\ 1 & 0 \\ 0 & 1 \\ 1 & 0 \\ 0 & 1 \end{bmatrix}
\begin{bmatrix} X_G \\ Y_G \end{bmatrix} =
\begin{bmatrix} 1.89 \\ 2.74 \\ 2.91 \\ 1.66 \\ 3.20 \\ 4.00 \end{bmatrix} + \mathbf{v}
$$

and a dispersion matrix which, we assume for ease of working, is the same as for the (x, y) system, that is

$$\mathbf{D}_X = \mathbf{D}_X$$

We obtain the centroid coordinates of

$$
\begin{bmatrix} X_G \\ Y_G \end{bmatrix} = \begin{bmatrix} 2.67 \\ 2.80 \end{bmatrix}
$$

New coordinate systems $\mathbf{x}^G$ and $\mathbf{X}^G$ with respect to these centroids give respective values for the points in columns two and three of Table 2.11. The items of the other columns will be explained later.

As previously, we wish to find the matrix **R** which effects the following transformation

$$\mathbf{X}^G = \mathbf{R}\ \mathbf{x}^G$$

Because both sets of coordinates are observed and have dispersion matrices, and the elements of the matrix **R** are not linear, we proceed by the combined least squares method. This assumes an initial value

Table 2.11

Point		$\mathbf{x}^G$	Observed $\mathbf{X}^G$	Provisional $\mathbf{X}^G$	l vector
A	x	−1.00	−0.777	−0.777	0.00
	y	−0.67	−0.060	−0.190	0.13
B	x	1.00	0.243	0.243	0.00
	y	−1.67	−1.140	−1.270	0.13
C	x	0.00	0.530	0.530	0.00
	y	2.33	1.200	1.460	−0.26

for the **R** matrix, and for one of the sets of coordinates. We assume that

$$\mathbf{R} = \overset{*}{\mathbf{R}}$$

For this assumed value we use the **R** matrix found previously by the direct method, that is

$$
\overset{*}{\mathbf{R}} = \begin{bmatrix} r_{11} & r_{12} \\ r_{21} & r_{22} \end{bmatrix}
$$

$$= \begin{bmatrix} 0.6243 & 0.2286 \\ -0.2271 & 0.6257 \end{bmatrix}$$

From the values of x in Table 2.11 we compute the provisional values of $\mathbf{x}^G$ in column four of the same table. On subtraction of these from the observed values in column three we obtain the l vector in column five.

Next, we derive the combined least squares model of the form

$$\mathbf{A\,x} + \mathbf{C\,v} = \mathbf{C\,l}$$

In this case the **x** vector consists of the four small corrections to the elements of the provisional **R** matrix used to calculate the l vector.

The mathematical model is as before

$$\begin{bmatrix} X \\ Y \end{bmatrix} = \begin{bmatrix} r_{11} & r_{12} \\ r_{21} & r_{22} \end{bmatrix} \begin{bmatrix} x \\ y \end{bmatrix}$$

For simplicity we have dropped the superscripts G. Differentiating with respect to all elements, we obtain

$$\begin{bmatrix} dX \\ dY \end{bmatrix} = \begin{bmatrix} dr_{11} & dr_{12} \\ dr_{21} & dr_{22} \end{bmatrix} \begin{bmatrix} x \\ y \end{bmatrix} + \begin{bmatrix} r_{11} & r_{12} \\ r_{21} & r_{22} \end{bmatrix} \begin{bmatrix} dx \\ dy \end{bmatrix}$$

or recasting we have

$$[\mathbf{0}] = [\qquad \mathbf{A} \qquad][\quad \mathbf{x} \quad] + [\qquad \mathbf{C} \qquad][\mathbf{dx}]$$

$$\begin{bmatrix} 0 \\ 0 \\ 0 \\ 0 \end{bmatrix} = \begin{bmatrix} x & y & 0 & 0 \\ 0 & 0 & x & y \end{bmatrix} \begin{bmatrix} dr_{11} \\ dr_{12} \\ dr_{21} \\ dr_{22} \end{bmatrix} + \begin{bmatrix} r_{11} & r_{12} & -1 & 0 \\ r_{21} & r_{22} & 0 & -1 \end{bmatrix} \begin{bmatrix} dx \\ dy \\ dX \\ dY \end{bmatrix}$$

This is the data available from each point. The full matrix for all three points of the example, with numerical values is therefore

$$\mathbf{A\,x} = \begin{bmatrix} -1 & -0.667 & 0 & 0 \\ 0 & 0 & -1 & -0.667 \\ 1 & -1.667 & 0 & 0 \\ 0 & 0 & 1 & -1.667 \\ 0 & 2.333 & 0 & 0 \\ 0 & 0 & 0 & 2.333 \end{bmatrix} \begin{bmatrix} dr_{11} \\ dr_{12} \\ dr_{21} \\ dr_{22} \end{bmatrix}$$

The vector **dx** is split into the two parts 1 + **v** where the 1 vector consists of the observed-minus-provisional values of the observed parameters. Because we selected the observed values of the **x** system to be their provisional values, and from them computed the provsional values of the **x** system, the top half of the 1 vector will be zero. The transpose of this vector is in full

$$1^T = (0,0,0,0,0,0:0,0.13,0,0.13,0,-0.26)$$

The **C** matrix has a particular structure for three points as follows

$$C = \begin{bmatrix} R & 0 & 0 & -I & 0 & 0 \\ 0 & R & 0 & 0 & -I & 0 \\ 0 & 0 & R & 0 & 0 & -I \end{bmatrix}$$

The submatrices **R** and **I** are of (2 x 2) dimensions in a two dimensional coordinate system (x, y), as here. In our example **R** is the provisional transformation matrix

$$R = \begin{bmatrix} r_{11} & r_{12} \\ r_{21} & r_{22} \end{bmatrix}$$

$$= \begin{bmatrix} 0.6243 & 0.2286 \\ -0.2271 & 0.6257 \end{bmatrix}$$

The full (12 x 12) dispersion matrix of the two observed sets of two dimensional coordinates for the three points has the structure

$$W^{-1} = \begin{bmatrix} D & 0 & 0 & 0 & 0 & 0 \\ & D & 0 & 0 & 0 & 0 \\ & & D & 0 & 0 & 0 \\ & \text{Symmetric} & & D & 0 & 0 \\ & & & & D & 0 \\ & & & & & D \end{bmatrix}$$

Where each submatrix **D** is

$$D = \begin{bmatrix} 0.10 & 0.05 \\ 0.05 & 0.10 \end{bmatrix}$$

This gives $C\,W^{-1}\,C^T$ of the structure

$$\begin{bmatrix} P & 0 & 0 \\ 0 & P & 0 \\ 0 & 0 & P \end{bmatrix}$$

where each submatrix **P** is

$$\mathbf{P} = \begin{bmatrix} 17.32 & -8.91 \\ -8.91 & 17.32 \end{bmatrix}$$

and finally

$$\mathbf{A}^T (\mathbf{C} \ \mathbf{W}^{-1} \mathbf{C}^T)^{-1} \mathbf{A} \ \mathbf{x} = \mathbf{A}^T (\mathbf{C} \ \mathbf{W}^{-1} \mathbf{C}^T)^{-1} \ \mathbf{C} \mathbf{l}$$

$$\begin{bmatrix} 0.1474 & -0.0737 & 0.0621 & -0.0310 \\ & 0.6387 & -0.0311 & 0.2692 \\ & & 0.1208 & -0.0604 \\ & \text{Symmetric} & & 0.5235 \end{bmatrix} \mathbf{x} = \begin{bmatrix} 0 \\ -0.0283 \\ 0 \\ -0.0550 \end{bmatrix}$$

Which gives the solution **x** and the final values of the **R** matrix shown in Table 2.12

Table 2.12

Solution x	R prov	R final	R matrix	
-1.429E-05	0.6243	0.624 286	0.624 286	0.228 571
-2.857E-05	0.2286	0.228 571	-0.282 857	0.514 286
-0.0557571	-0.2271	-0.282 857		
-0.1114143	0.6257	0.514 286		

Added constraints

In the above transformation it has not been assumed that the matrix **R** consists of a scaled orthogonal matrix. If we wish this to be so, for example in photogrammetry or cartography, we can do so by imposing constraints on the solution, such as

$$\hat{r}_{11} = \hat{r}_{22}$$
$$\hat{r}_{12} = -\hat{r}_{21}$$

therefore

$$\dot{r}_{11} + dr_{11} = \dot{r}_{22} + dr_{22}$$
$$\dot{r}_{12} + dr_{12} = -\dot{r}_{21} - dr_{21}$$

giving the constraint equations

$$dr_{11} - dr_{22} = \dot{r}_{22} - \dot{r}_{11} = 0.0014$$
$$dr_{12} + dr_{22} = \dot{r}_{12} - \dot{r}_{21} = -0.0015$$

When these equations are added to the normal equations we obtain the hypermatrix, where p is the Lagrangian vector

$$\begin{bmatrix} 0.1474 & -0.0737 & 0.0621 & -0.0310 & 1 & 0 \\ & 0.6387 & -0.0311 & 0.2692 & 0 & 1 \\ & & 0.1208 & -0.0604 & 0 & 1 \\ & & & 0.5235 & -1 & 0 \\ & \text{Symmetric} & & & 0 & 0 \\ & & & & & 0 \end{bmatrix} \mathbf{x} = \begin{bmatrix} 0 \\ -0.0283 \\ 0 \\ -0.0550 \\ 0.0014 \\ -0.0015 \end{bmatrix}$$

The solution and the final values of the coefficients of **R** are

Prov R	Soln	Final R	Matrix R	
0.6243	−0.084 650	0.539 650	0.539 650	0.214 440
0.2286	−0.014 160	0.214 440	−0.214 440	0.539 650
−0.2271	0.012 660	−0.214 440		
0.6257	−0.086 050	0.539 650		

The matrix coefficients are then split into the scale and rotation elements k and θ from

$$k^2 = 0.539\ 650^2 + 0.214\ 440^2$$
$$k = 0.580\ 696$$
$$\theta = \arctan(0.214\ 440 / 0.539\ 650) = 21.67°$$

Summary
The various methods of coordinate transformation have their uses according to circumstances which depend on the accuracy required and the distortions that exist between the two sets of coordinates.

The full least squares method is worth programming because of its general application. It is also possible to obtain error estimates of the parameters, and to carry out tests for the location of possible blunders in the data.

These matters are discussed in Chapter 5 and Appendix 5. To reduce the working, the data of the example was kept to a minimum and therefore will not yield statistical data. If at least one other data point is introduced which is incompatible with the three points used, the subsequent redundancy will be capable of statistical analysis.

Transformations in three dimensions
The transformation of coordinates in three dimensions is generally a non-linear problem. Because of this, it requires the application of the general differential least squares method for solution. The method is similar to that just explained for the two-dimensional problem, with an **R** matrix of dimensions (3 x 3).

3 Instrumentation

Chapter Summary

In this chapter, we outline the various forms of instrumentation commonly used in ground surveying. The objective is to provide a basic understanding of the physical and geometrical principles involved in instrument design, and of those factors which limit system performance. Because the current manifestation of these fundamental principles in instrumental form depends on the current state of rapidly changing technology, the treatment is kept to a schematic minimum.

Instrumentation

Most instruments employ optical and mechanical devices to measure length, angle, time, or attitude. They each have a point of reference which has to be related to some physical mark on the ground or structure. Instruments are imperfect measurers which suffer from systematic and random errors. Most instrument designs and procedural use are based on efforts to reduce the systematic errors to a minimum, usually by some form of reversal or balancing. Most instruments still depend on observation skills to achieve their best performance.

We concentrate on the theodolite and its components, since it is the principal surveying instrument, which, with the addition of a distance measuring device, becomes a *total station*. The surveyor's level has some similar characteristics to the theodolite. Ancillary instruments, and simple methods of determining lengths, are also considered.

3.1 The theodolite

In Figure 3.1 the essential design features of a theodolite are depicted. The instrument has three axes mounted orthogonally in paired sequence.

(a) The *tertiary axis*, or line of sight PT, is orthogonal to the secondary axis PS, about which it can rotate.

(b) The *secondary axis* PS is orthogonal to the primary axis PZ, about which it can rotate. Rigidly attached to the tertiary axis is a circle for measuring angles V. These angles are usually measured relative to the *vertical*, giving *vertical angles*. The vertical is established by an independent bubble or compensator to which is attached a measuring index. Some instruments rely solely on the primary axis to define the vertical.

(c) The *primary axis* PZ, about which the other two axes can rotate, is usually set vertically, thus allowing *horizontal angles* to be measured on a circle. Because there is no natural reference line equivalent to the vertical, a horizontal *angle* is purely the relative difference between two *directions.*

The terms *primary, secondary* and *tertiary* are to be preferred to all others since they convey the correct information even when the theodolite is stowed in its box. They also describe the rotational hierarchy of each axis. The part of the instrument carrying the secondary and tertiary axes, mounted about the primary axis, is called the *alidade.*

Occasionally the horizontal circle is fitted with a magnetic compass to define the *north* direction, to about 10'. This contrasts with the *vertical* which can be accurate to 0.5".

When the vertical circle is to the observer's left, as in Figure 3.1, the theodolite is said to be on *face left.* After the telescope is transitted through the zenith and the alidade rotated through 180° to sight back on an object, the theodolite is then on *face right.* This *change of face* or reversal almost completely eliminates most errors of construction.

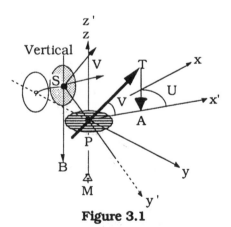

Figure 3.1

Because some cheaper instruments have no independent bubble or compensator, the angle V is measured with respect to the primary axis PZ.

Note that vertical lines through different points on the earth's surface are not parallel to each other; for two points a distance of

30 m apart in latitude, the verticals differ by an angle of 1 ", or 1 in 200 000. This can be significant even in engineering work.

Sometimes at industrial sites or on floating platforms or ships, the vertical direction is not useful, and the primary axis is set relative to an arbitrary direction which moves with the structure.

Modern theodolites are capable of accuracies of 0.5 " or 1: 400 000, a remarkable achievement for circles of 150 mm diameter. This accuracy is achieved by optical Moire fringe or by electrostatic-field techniques. Ordinary optical circle instuments are accurate to 10 ".

Establishment of the vertical direction

The vertical direction is established by a plumb line or pendulum, or by a bubble, or by the normal to a free liquid surface such as mercury. Visual methods of detection are capable of 0.5 " accuracy whilst electronic systems can reach 0.01 ".The bubble attached to the alidade (see Figure 3.3) is typical of all systems. Figures 3.2 (a) and (b) show the bubble LR inside its glass vial when the primary axis Z is vertical. Notice that the bubble is not usually central to the tube. When the alidade is rotated about the Z axis the bubble does not move from its levelling position, provided the primary axis is vertical. Notice also that the ends of the bubble are always labelled L and R *as viewed from the observer's position.*

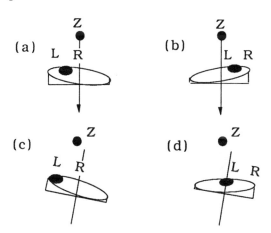

Figure 3.2

To set up a theodolite, the first part of the process is to determine this bubble levelling position from the situation shown in Figure 3.2 (c) and (d), where the primary axis is tilted from the vertical. Diagram (c) shows the bubble LR with the alidade at any direction; diagram (d) shows it again with the alidade rotated through 180°. The mean of these two bubble positions is the required levelling position. When this is known, the Z axis is tilted by means of the footscrews until the bubble takes up this position; that is until it assumes the circumstances of diagrams (a) or (b).

If the bubble levelling position is too far off centre for convenience, the vial-holding screw can be adjusted to bring the position more central. There is no point wasting time trying to achieve exact centrality.

Levelling procedure

The procedure for levelling is illustrated in Figures 3.3 which show the three footscrews A, B and C, the alidade bubble and the Z axis pointing out of the page. For simplicity the bubble is assumed in perfect adjustment. In position (a) the bubble has been aligned parallel to any two footscrews A and B, and in (b) in a direction at 90° to this. In (c) the Z axis has been tilted over to the right so that the bubble moves to the left. And in (d) it has been further tilted up the page with the bubble moving down. Bubble vials are graduated in angular terms to enable these tilts to be recorded if necessary.

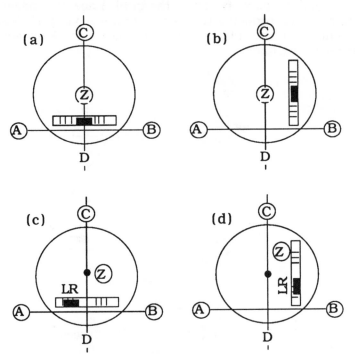

Figure 3.3

The process of levelling is the reverse of that just described. The alidade is placed first in position (c) ie with the bubble parallel to foot screws AB. It is then rotated through 180° to discover the levelling position, which is then used in position (d) to achieve position (b) by turning footscrew C only. Finally position (a) is achieved by moving footscrews A an B in opposition. The practical procedure is quite simple when demonstrated.

If very precise levelling is needed, the vertical bubble or compensator can be used instead of the alidade bubble. Additionally, if the glass vial containing the bubble is graduated in angular terms, the tilt I can be measured. Modern optical sensors detect this bubble movement, and record it digitally for application by hard-wired software within the instrument, or off-line as required. Such devices require the theodolite to be set vertically by conventional bubbles to within a few minutes of arc, to allow the compensators to operate.

Footscrews
For a number of good reasons instruments are normally fitted with three footscrews. Two or four footscrews have the advantage of preserving the instrument height during the levelling. The two screw instruments usually rest on a central ball about which the instrument pivots, thus preserving the height of instrument.

Determination of primary axis tilt I
It must be stressed that there is no substitute for proper levelling of the instrument, which should be mounted on a stable stand. However there are cases when the tilt of the axis has to be determined during observations, which are then corrected for its effect. Of particular significance is the correction to horizontal directions for the effect of axis tilt. The most accurate way to determine tilt is to use the vertical

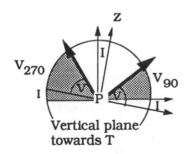

Figure 3.4

circle bubble or compensator, instead of the alidade bubble. Consider Figure 3.4 which shows the primary axis PZ tilted to the right by an angle I.

With the tertiary axis kept at any fixed angle V, the alidade is turned through 90˚, the independent vertical bubble levelled, or compensator allowed to operate, and the vertical angle V_{90} recorded. The alidade is then turned through a further 180˚, the bubble relevelled and V_{270} recorded. It can be seen by inspection that the required tilt angle I is given in magnitude and sign by

$$2\,I = V_{270} - V_{90}$$

The angle V can be any value, but is best not close to zero.

Effect of tilt I on a horizontal direction U

When the line of sight points to an object at a vertical angle V the effect dU of a cross tilt I on the horizontal reading U is given by

$$dU = I \tan V$$

For example if V= 45˚ and I = 20", then dU = 20" : a serious amount in precise work.

We will derive this important result and other similar formulae by treating the theodolite as a solid body rotated within a coordinate system.

From Figure 3.1 we can write down the coordinates of point T with respect to axes Px ', Py ' and Pz ', assuming PT = 1, as

$$x' = \cos V$$
$$y' = 0$$
$$z' = \sin V$$

A tilt, I, clockwise about the axis Px ' produces new coordinates given by

$$\begin{bmatrix} x'' \\ y'' \\ z'' \end{bmatrix} = \begin{bmatrix} 1 & 0 & 0 \\ 0 & \cos I & -\sin I \\ 0 & \sin I & \cos I \end{bmatrix} \begin{bmatrix} x' \\ y' \\ z' \end{bmatrix}$$

The horizontal direction dU which results from this tilt is given by

$$\tan dU = y'' / x'' = (y' \cos I - z' \sin I) / x'$$

Since I is usually quite small, and normally so is dU, we can put cos I = 1 and sin I = I, giving approx

$$dU = -I \sin V / \cos V = -I \tan V$$

In a practical case the reading from an already tilted theodolite is too small by dU. Hence the correction to give the reading as if from an untilted instrument is

$$+ I \tan V$$

Notice that the sign of I is given already from

$$2 I = V_{270} - V_{90}$$

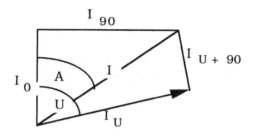

Figure 3.5

There is however no need to measure the tilt I for every pointing, because it can be calculated if the magnitude I and direction A of the maximum tilt are known. Consider Figure 3.5, which shows the maximum tilt I at direction A. Its resolved small components along the axes are I_0, I_{90} given by

$$I_0 = I \cos A \qquad\qquad I_{90} = I \sin A$$

The tilt I_U at any direction U is given by

$$I_U = I \cos (U - A) = I_0 \cos U + I_{90} \sin U$$

and in the direction at right angles to U, ie at direction $U + 90$, is

$$I_{U+90} = - I_0 \sin U + I_{90} \cos U$$

Thus the correction dU to a direction U with vertical pointing V is

$$dU = (- I_0 \sin U + I_{90} \cos U) \tan V$$

It must be stressed that this is a correction to a *horizontal* reading on a tilted theodolite, not to a vertical angle. The equation is used whenever high accuracy work with steep sights is involved. It is also the basis of Black's method of determining a geodetic azimuth from astronomical observations; the tilt components being the deviation of the vertical from the ellipsoidal normal. See section 6.6.

The effect of this *primary axis tilt* is not cancelled by changing face. However, an inclination of the *secondary axis* to the primary axis *is* cancelled by changing face. In precise astronomy, a special bubble, the *striding level*, is hung directly from the secondary axis to allow for both sources of tilt.

Effect of collimation c on a horizontal angle reading U
In a similar way to a tilted axis, the component c, of the collimation in the horizontal plane, also affects the direction U. The change dU is given by

$$dU = c \sec V$$

The collimation error c means that the tertiary axis lies along Px' instead of Px. (See Figure 3.6). When the telescope shows V = 0 or the line of sight lies in the plane Pxy, the point T has coordinates

$$x = \cos c$$
$$y = \sin c$$
$$z = 0$$

or since c is small

$$x = 1$$
$$y = c$$
$$z = 0$$

Pointing the telescope at an angle V is achieved by a rotation of the point T about the y axis by angle V giving new coordinates given by

$$
\begin{bmatrix} x" \\ y" \\ z" \end{bmatrix} =
\begin{bmatrix} \cos V & 0 & \sin V \\ 0 & 1 & 0 \\ -\sin V & 0 & \cos V \end{bmatrix}
\begin{bmatrix} x \\ y \\ z \end{bmatrix}
$$

Thus the change in direction is given by

$$\tan dU = y"/x" = y/\cos V = c \sec V$$

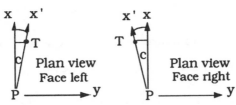

Figure 3.6

When the telescope is changed to the other face, c changes sign while V does not. The mean of the two directions is therefore free of error. This is illustrated in Figure 3.6, which shows the case when V = 0.

However when a moving target, such as a star, is observed, V cannot be exactly balanced, thus leaving, as a residual effect, the difference of the two corrections.

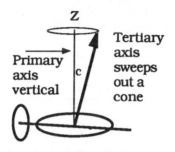

Figure 3.7

Since sec V tends to infinity as V tends to 90°, the effect is very great on steep sights. It also means that when the primary axis is vertical the line of sight cannot point to the zenith. If the alidade is rotated, the tertiary axis traces out a cone of collimation centered on the vertical. Hence the observer can locate the zenith at the centre of the circle traced out when the alidade is rotated in azimuth. See Figure 3.7.

Eccentricity of circles
Although the accuracy of instrument construction is phenomenal, it is impossible to centre a circle, relative to its corresponding axis of rotation, without some error. The left diagram of Figure 3.8 shows

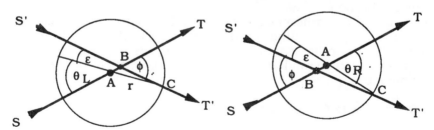

Figure 3.8

the main eccentricity problem. The line of sight is centred on its axis B whilst the centre of circle graduations is at A such that AB = d. The angle between the telescope pointings to T and T' is ϕ, but the angle recorded is θ_L, indicating the face left position. The error is given by

$$\varepsilon = \phi - \theta_L$$

When the telescope is transitted and the alidade rotated through 180° on change of face, the right hand diagram results. The centre of graduations at A is now 180° from its first position. Thus the error now is given by

$$\varepsilon = -\phi + \theta_R$$

Thus we obtain

$$\phi = \frac{(\theta_R + \theta_L)}{2}$$

Therefore the mean of a face left and a face right reading eliminates the error ε. This procedure also applies to the vertical readings where

$$V = \frac{(V_R + V_L)}{2}$$

In addition, most modern instruments combine readings from both sides of the circle to reduce the eccentricity error. One way to do this is to create Moire fringes by combining images from diametrically opposite parts of a finely ruled circular grating.

For the record we can express the eccentricity error as

$$\sin \varepsilon = (d/r) \sin \phi$$

This formula occurs in a different context when a theodolite is set up eccentrically to a survey mark, or an eccentric target is bisected. See Chapter 8 for more detail.

In practice the difference between a face left and a face right horizontal circle reading is the net effect of the collimation, eccentricity and other minor errors. Taking the mean reduces their systematic effects to random noise, not including, of course, the effect of primary axis tilt.

3.2 Instruments for levelling

Instruments for levelling are of two types:

(a)*Tilting levels*, in which the line of sight may be tilted separately from the primary axis as in a normal theodolite.

(b)*Dumpy and laser sweep levels*, in which the line of sight is fixed rigidly at right angles to the primary axis.

Figure 3.9 shows the principal elements of the tilting and automatic levels. To make a reading, the line of sight has to be horizontal. This is achieved in the tilting level by moving the telescope, in response to the bubble, about its secondary axis at the pivot, which incidentally does not usually lie on the primary axis.

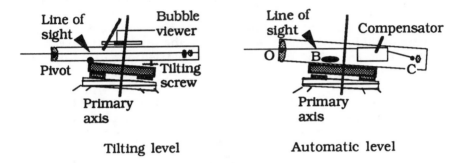

Figure 3.9

In the *automatic level*, a horizontal line through the objective centre at O is directed by a pendulous compensator to pass through the centre of the crosshairs at C. The compensator will only operate in this fashion if the instrument is approximately levelled by a crude circular bubble B.

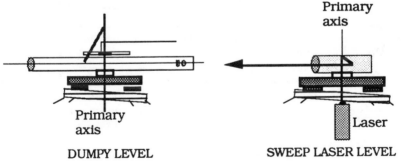

Figure 3.10

In the dumpy and sweep laser levels, the primary axis has to be set vertically, so that the tertiary axis sweeps out a horizontal plane when the alidade is rotated about the primary axis. In the sweep laser instrument this rotation is motorized. Although convenient on a construction site, the ability to sweep out such a plane is unnecessary for line levelling.

All systems are subject to index error, which has to be found by the *two peg test*, and either allowed for or adjusted.

Two peg test for levels

All levels, including sweep lasers, are subject to index error. Index error is the amount by which the system produces a collimation effect. The first step is to determine the magnitude of this error to see if it is significant. This procedure should be carried out before all field work begins, and from time to time during work, especially if the instrument has been bumped in transit. We describe the test for a conventional tilting bubble-level. The test is normally carried out using only one levelling staff, because it too can have an index error.

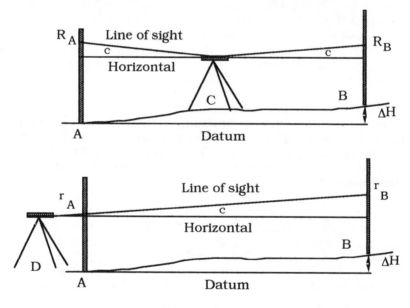

Figure 3.11

The level is set up at C, as in the top diagram of Figure 3.11, midway between two points A and B, located on solid ground about 50 m apart, and the staff readings taken in succession. Because the index error c cancels out, the true difference in height is given by

$$\Delta H = R_B - R_A = 1.930 \text{ say}$$

The instrument is moved to D of the lower diagram, as close as possible to the near staff at A as the minimum focal distance will allow, usually about 1.5 m. Readings are taken to the staff in the two successive positions. In this case the height difference, affected by the collimation error, is given by

$$\Delta H' = r_B - r_A = 1.937 \text{ say}$$

The correction over the distance DB is − 0.007. This correction can be applied to all readings in proportion to distances. In other than precise levelling, it is better to adjust the level to reduce the effect to negligible proportions. The adjustment differs from instrument to instrument. In the case of a conventional tilting level, it is made by altering the bubble adjustment screw as follows.

With instrument still at D, the line of sight is tilted to give the correct height difference from a staff reading at B, 1.950 say. The bubble will then be off centre. This bubble position could be noted and always used when working with the level, or it can be corrected by altering the bubble setting screws; a procedure which needs patience. After the adjustment, the height difference is checked by readings again taken at D.

Clearly the best practice is always to use the level with equal back and fore sights. See Chapter 9 for more details. Dumpy, automatic and sweep laser levels required the adjustment to be made to the crosshairs or laser alignment: an even more delicate task.

Clinometers
Clinometers are instruments which measure slope alone. They may be hand-held for approximate work, such as the Abney level, good to 10', or very precise industrial clinometers accurate to 0.01": some using electronic sensors with digital output. They too need to be tested for index by the principle of reversal.

3.3 Optical systems

The principal optical system of many instruments is the telescope. Its function is to gather light, from a target or scale, bring it to form a real image in the plane of a reference mark, such as crosshairs etched on glass, which together are viewed by the human eye, or electronic sensor, by means of an eyepiece. Telescopes have three functions in most instruments including the theodolite.

(a) As a sighting device to view a distant target.
(b) In a folded form as a plummet to centre the instrument.
(c) As part of the angle reading system.

Unlike in photogrammetric instruments and cameras, theodolite telescopes function in the paraxial region, ie never far from the optical axis. The optical axis, defined by the line joining the effective centre of the lens and a physical mark such as a cross hair, has to be maintained at a fixed relationship to the mechanical axes, to which circles are attached. It is part of the adjustment process to correct any misalignment of the optical axes, and a feature of the observing procedure to reduce any residual measurement error by reversal of the telescope and reading systems, principally by changing face.

An efficient instrument matches the performance of its optical and mechanical components with the precision of its reading system, to achieve consistent accuracy for its purpose.

We now consider the various factors involved in the optical performance of an instrument.

Resolving power and magnification

The detection system of rods and cones in the human eye, is able to resolve an angle of about 30" arc sexagesimal, or 1: 7000. This means that to resolve an angle of 1" a telescope with 30X magnification is required to aid the eye, and that the diameter of the circular aperture must be at least 50 mm.

Magnification is contributed by the eyepiece, while the resolving power depends on the diameter d of the aperture. The formula to give the minimum resolved angle α, Rayleigh's criterion for a circular aperture, is

$$\alpha = 1.22 \ \lambda/d$$

where λ is the wavelength of the electromagnetic waves being used, and the resolved angle α is in radians. Since visible light has a wavelength of about 0.5 μm or 0.0005 mm, it will be seen that a normal telescope can resolve about 1".

To exploit this accuracy the telescope needs to be held still on some form of rigid stand. Conversely the magnification limit for hand-held binoculars is about 8X.

The telescope and its use

Figure 3.12 shows the essentials of a classical telescope in use by the surveyor. Tracing two incident rays is sufficient to explain the nature of the transformations involved. One of these rays passes through the optical centres, while the other, initially parallel to the axis, passes through the foci of lenses, such as F.

Light from the sun or from an artificial source, such as a lamp, is needed to illuminate the target A. Some of the reflected light travels towards the telescope, passing through its lenses until it emerges through a small aperture at E, the *exit pupil* of the telescope. The observer places the eye so that its *entrance pupil* coincides with E, and thus light passes into the eye and is focussed on the retina at D with a resolution of about 30". To achieve a sharp image, the eyepiece lens at B has to focus a virtual or apparent image of the target and crosshairs on the retina. This image of the target has in turn been focussed on the glass plate in front of B by the internal lens L.

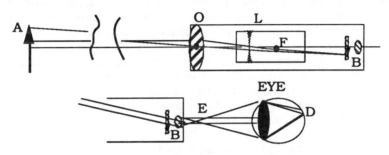

Figure 3.12

It is important to see a sharp image of the crosshairs at B when first looking into the telescope. If this is not so, eye strain and headaches will result. The eyepiece focus is special to each person.

For each target distance AO the internal lens L has to be moved forward or back to achieve a sharp image on the glass plate at B. If this is not so the observer will see a relative movement of the target against the cross hairs: an effect called *parallax*, which if not removed causes bad observation.

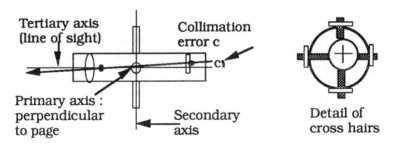

Figure 3.13

Most modern telescopes use an extra lens to present an erect image to the observer; which improves convenience at the expense of optical performance.

It is most important to preserve a fixed relationship between the line of sight, and the secondary axis about which the telescope rotates. In a perfect instrument, the angle between these should be right. The small departure from 90˚ is called the collimation angle c.

It has components in and perpendicular to the plane formed by the tertiary and secondary axes.

For example, the Figure 3.13 could represent either the horizontal view shown, or a vertical view if we interchange the primary and secondary axes labels. The component of collimation in the vertical plane is combined with its bubble or sensor error into the vertical index E. The detail diagram of the crosshairs shows the opposing screws which can position the crosshairs central to the telescope tube.

In a good telescope the collimation angle will not change by more than 20" at different focussing distances; that is from the minimum, usually about 1.5 m, to infinity. To overcome its effect, the telescope is reversed in attitude. Special expensive tooling telescopes preserve collimation, from zero to infinity distances, to within 2 ".

Should the collimation prove to be too large, the cross hairs can be moved laterally by opposing screws to reduce its effect. A task requiring patience and a shade of luck.

Stadia lines

The angle of field of most theodolites and levels is about 1.5°, sufficient to accommodate the Sun's 32 minutes of arc and a stadia angle of about 34 minutes of arc. This stadia angle is created by a pair of lines set at distance from the central cross such that they subtend a fixed angle at the optical centre usually, but not always, of about 34' arc. See Figure 3.14. These stadia lines enable distances to a measuring staff to be calculated by a form of optical triangulation, called *optical tachymetry*. as explained below.

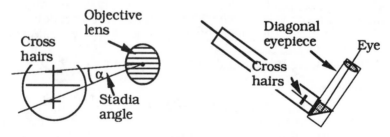

Figure 3.14

Diagonal eyepiece

Steep sights, possibly vertical, are viewed with the aid of a diagonal eyepiece, shown in Figure 3.14, which bends the light through a right angle.

Sun observations

When observing the Sun, its image and that of the crosshairs is brought to a focus on a piece of white card held where the eye is normally placed. On no account should the Sun be viewed directly without a dark filter in position.

Optical plummet

The Figure 3.15 shows the optical arrangement of the plummet which consists of a folded telecope. Some have full focussing of both the objective and the eyepiece. Most just have the latter.

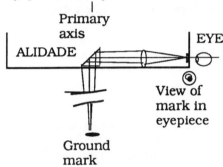

Figure 3.15

Most plummets are mounted in the alidade as shown in Figure 3.15. When the alidade is rotated, the ground mark should not appear to move. If it does so, the telescope has a collimation error which can be reduced by adjusting the circular reference mark. Ultimately the small residual error is eliminated by centering the theodolite within the collimation cone. Recollect that the centering is only as good as the ability to level the instrument.

Plummets not mounted in an alidade cannot as easily be checked for collimation. A simple technique for this test is as follows.

Draw round the triangular outline of the footscrew base (the tribrach) on the tripod top. Centre and level the instrument with the plummet over a dot marked on a piece of white card stuck to a floor. Then rotate the whole tribrach through 120° to fit the triangular drawing on the tripod top, level up and view again through the plummet to see if the dot is still central. If it is not, mark another dot on the card in the apparently central position. Repeat for the third position. The centre of the triangle formed by the three dots on the card is the correct position vertically below instrument. Now use the adusting screws of the plummet eyepiece to view this centre through the plummet.

The result is checked from all three positions 120° apart. The same technique can be used when centering the tripod in the field.

For very precise centering a mercury pool is placed on the ground mark to be viewed in the plummet. Since the surface of mercury levels to about 1 " such centering is highly accurate.

3.4 Ancillary optical components

The following optical components augment the performance of a theodolite or level, and have uses in their own right.

(a) Parallel plate
(b) Pentagonal prism
(c) Optical wedge
(d) Corner cube

Parallel plate

The glass plate with parallel plane sides shown in Figure 3.16 is used to measure small offsets d (up to 5 mm) from a line of sight. The offset d is calibrated in terms of the thickness of the plate t, its refractive index n, and the rotation angle β by the expression

$$d = t\beta \frac{n-1}{n}$$

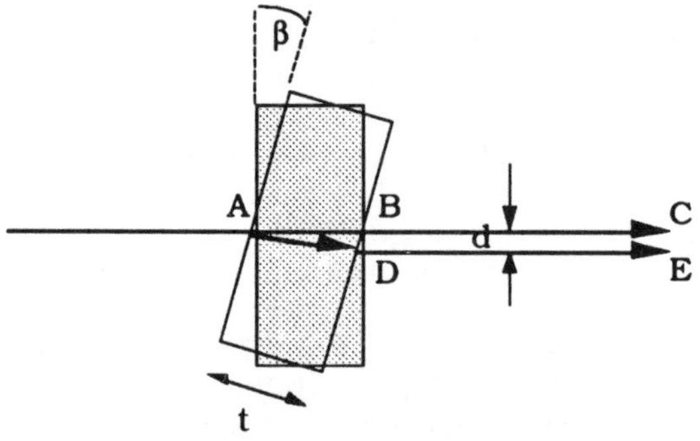

Figure 3.16

The parallel plate is used within instruments to interpolate scale readings, and externally as a telescope attachment, usually to a level telescope, to interpolate staff readings. Notice that, unlike the wedge, the line of sight is *shifted* sideways. In the tooling telescope, the plate can be tilted both horizontally and vertically to measure offsets.

Pentagonal prism

The pentagonal prism of Figure 3.17 deviates a line of sight through a right angle irrespective of the the attitude of the prism, provided the light rays and principal prism section all lie in the same plane, as shown.

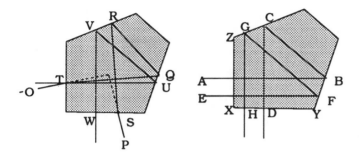

Figure 3.17

Ray OT, meeting a face at an angle of more than 45°, is refracted by the prism to Q, ultimately emerging as SP perpendicular to OT. The optical paths taken by orthogonal rays from A and E are each of the same length L , given, where d = XZ = XY, by

$$L = (2 + \sqrt{2})\, d$$

The optical path length, as measured by EDM, is given by

$$n\,L = (2 + \sqrt{2})\, n\, d$$

where n is the refractive index of the glass for the carrier wavelength of the EDM. It is in the region of 1.5.

The prism has many uses: as an optical square; as an attachment to a telescope for special steep sights, as well as in instruments such as the laser level and tunnel profiling equipment, to sweep out planes. Like all instruments its accuracy can be checked by the reversal principle, as indicated in Figure 3.18.

From A the point C is located looking through the prism at B, and D located similarly. If the prism angle is 90° points CBD are collinear. If not the angular error c is given by

$$\tan c = \frac{EB}{EC}$$

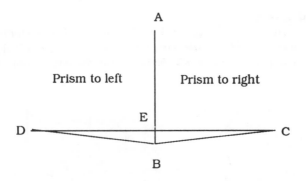

Figure 3.18

The optical square
The optical square is a small hand-held instrument with two pentagonal prisms mounted at right angles to each other. The observer's field of vision is large enough to see through both prisms and straight ahead at the same time. In Figure 3.18 if the observer's eye is at point E he can see all three points A, D and C together, and thus place himself on the line DC at right angles to EA. In this way he is resecting position using only right angles. The device is of great value in detail surveying and setting out small works. See Chapter 10.

Optical wedge
The optical wedge is commonly used with another to form a matched pair, capable of rotation relative to each other as shown in Figure 3.19. There are two reasons for the use of a pair of wedges.

    (a)The deviation angle may be varied.
    (b)The deviation angle is confined to only one plane.

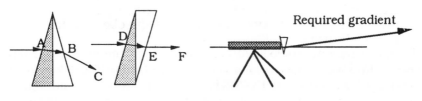

Figure 3.19

The wedge is used within and externally to instruments. Fitted to the telescope of a level it can establish an inclined line of sight needed to set out pipework etc.

Corner cube prism

The prism formed by cutting away the corner of a perfect glass cube, to form a prism of equal side length d = AD (Figure 3.20), is of great importance as a reflector in electro-optical distance measurement.

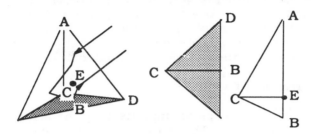

Figure 3.20

The prism has the property of returning an incident ray parallel to the original direction. This can be demonstrated with a pocket laser or proved mathematically by considering the reflections of a vector, with direction cosines L, M an N, at three orthogonal surfaces. The returned vector has direction cosines – L, – M, – N, and also undergoes a displacement, unless it is the central ray. The path is illustrated in the Figure 3.20.

The *geometrical* path length p = CE of a ray normal to the front face is given by

$$p = CE = d / \sqrt{6}$$

and the *optical* path length of interest to EDM is given by

$$np = nd / \sqrt{6}$$

Figures for a typical Aga corner cube, and n = 1.5270, the refractive index for a carrier wavelength of 0.910 μm, are

$$p = 41.5\,mm \quad giving \quad np = 63.4\,mm$$

If the ray makes an angle θ with the normal, its optical path length is increase to p sec θ. These results can be obtained from Figure 3.20 almost by inspection, if the direct central ray is considered. The proof for a general ray is more lengthy. This information is required when considering the geometrical requirements of electro-optical distance measurers capable of 0.1 mm accuracy. For example this same corner cube misaligned by 5° increases the path length by

$$np\,(sec\,\theta - 1) = 0.24\,mm$$

For use in the field, the corner cube is mounted in a housing which fits into a tribrach, usually interchangeable with the total station or theodolite.

It can also be mounted on a telescopic rod provided with a bubble for detail surveying. This device is often called a *pogo stick.* Modern pogo sticks also incorporate a level sensor and a speech link to the total station, which in one case is motorised and telemetered to track the pogo stick as the surveyor moves around to fix detail points.

3.5 Instrument stands and centering

The most common instrument stand used by the field surveyor is a tripod with telescopic legs.The feet have to be pressed firmly into the ground, or are set in concrete blocks, after greasing, or they may be stuck down with resin glue to a steel surface. Stability is essential. Tripod screws and fittings must be tight. The tolerances of industrial work require the use of concrete pillars, or other special metal heavy stands, often fitted with wheels and jacks. Such stands usually cost ten times as much as a tripod.

Centering and levelling the instrument on an ordinary tripod are usually achieved together, with the aid of an optical plummet. This should be mounted on the alidade of the instrument, to allow for collimation error. Instruments with no plummet are centered by plumb line, or plumbing rod.

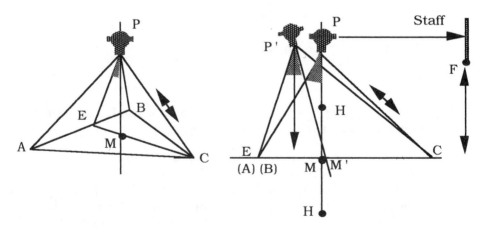

Figure 3.21

Some ingenuity is often needed to erect an instrument over or under a mark, close to a wall etc. Telescopic legs are essential in such circumstances. Sometimes a theodolite has to be placed directly on the ground for two observers to operate from a prone position, or has to be suspended upside down from a roof. Industrial applications employ special stands clamped to scaffolding.

Centering by optical plummet

The theodolite P, resting on the tripod with telescopic legs whose feet are at A, B and C, (Figure 3.21 left), has to be centred over the mark M, shown here at ground level, with the primary axis PM vertical. This is the final objective.

To set up the instrument begin by placing it in an approximate position P ' so that the observer can see the mark through the plummet. With the tripod feet firmly in place, the footscrews are used to centre M from P '. Next, one tripod leg, say PC, is shortened as in Figure 3.21 right, to level the theodolite approximately, with mark M still visible. Since this levelling is in only one direction, it has to be repeated using another leg. It is quite remarkable to the beginner to see how far P ' and P can be off centre whilst the line of sight of the plummet still passes through M. For level ground, the mismatching distance MM' is given by

$$2 \ EM \ \sin^2 \tfrac{1}{2} e$$

where the primary axis P' M' is off the vertical by an angle e.

Suppose $e = 5°$ and   EM = 0.8 m   then

MM ' = only 3 mm,   when PP ' = 1.2 sin e = 0.1 m.

Clearly the plummet alone gives no indication of centering accuracy, without a carefully levelled instrument. An instrument height of 1.2 m infers levelling to better than 3' for 1 mm centering accuracy. Industrial work to 0.1 mm requires greater attention (18" arc) to levelling.

When the mark M is not at ground level, being either in a hole at H or on a tall peg at H, the mismatch of the plummet line of sight gives a much better indication of dislevelment.

The final process is to level by the footscrews and shift the instrument laterally for centering, trying not to rotate its base relative to the tripod top, which is probably not level. Industrial stands are fitted with a precise xy stage for lateral movement to be finely tuned.

Height of instrument

It is not easy to measure the height of instrument directly to 0.1 mm. It can be done in two stages using an auxiliary point F whose height above M has been levelled before setting up the instrument. The height of collimation can be read on a small scale at F when the line of sight is horizontal.

Centering under a roof mark
Centering under a roof mark is carried out in in a similar manner; either by plumb line, or using the main telescope as a plummet to look vertically upwards, if this is possible and if a diagonal eyepiece is available. Collimation error is eliminated by rotating the alidade in azimuth so that the roof mark appears at the centre of the collimation circle. Some surveyors use a pentagonal prism fitted to the objective to achieve this centering, or to measure the amount the instrument is eccentric to the mark. Another technique is to read off an eccentric position from a local coordinate system established by graph paper about the roof point.

Centering from auxiliary point
When centering has to be assured in one direction, such as in EDM calibration, a theodolite placed approximately perpendicular to the line is used to sweep up to the target, as shown in Figure 3.22. To assure against collimation error, the centering is repeated on both faces and a mean taken.

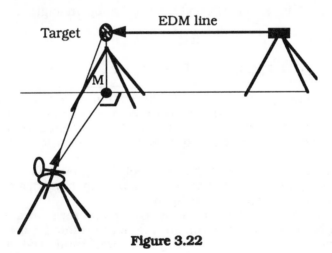

Figure 3.22

Forced centering systems
Once the tripod has been centred over the mark, the mechanical relationship between theodolite, target, EDM, and lamp or helio mirror, can be preserved to a high order (about 0.2 mm) by a forced centering system of locating pins and matching holes. All components are manufactured to the same height and all fit above

the levelling system, or tribrach containing the footscrews. The system is illustrated in Figure 3.23.

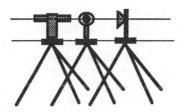

Figure 3.23

3.6 Simple distance measurement

Before the advent of EDM, all distances had to be measured by tape or wire or optical methods. Any distance less than a 30 m tape length is still usually measured by tape of that length, although longer tapes are still in use. Optical tachymeters and stadia methods are still very much in use. We outline these systems very briefly.

Tapes rods and bars
Tapes, wires, rods and bars are very varied. Each has its use according to need. Attainable accuracy varies from 1 cm with a cloth tape to 0.02 mm with a tooling bar. Long 50 m steel and invar tapes are seldom used today.

Rods used in levelling generally conform to the "E" pattern of marking. Bar-coded staves for automatic reading by pattern recognition are also common.

Figure 3.24

Tapes
Tapes are used in a straight line along a surface such as a railway line, or suspended in catenary, see Figure 3.24. In the latter case care has to be taken to apply the correct tension, noting that the correction dL to catenary length L is given by

$$dL = -\frac{w^2 L^3}{24 T^2} \cos^2 \theta$$

The tension T can be applied by a spring balance, or a special tension handle, or by a weight suspended over a pulley, or by a tensioning lever device. The weight per unit length w has to be determined by weighing the tape; the slope θ is measured by clinometer or obtained from difference in height ΔH.

Corrections to catenary and line measurements are also applied for changes from standard temperature t, from the standard tension T, and for slope, as summarised below.

(a)   Temperature t : $L_t = L_0 + \alpha L (t_F - t_0)$
where the coefficient of linear expansion of the tape material is α.

(b)   Tension T : $L' = L + L(T_F - T_0)/AE$
where A is the tape cross-sectional area and E Youngs modulus of elasticity of the tape.

(c)   Slope: $L_S = L - \Delta H^2/2L$

Good practice is to standardise the equipment empirically against a laser interferometer, or high precision EDM.

Special industrial systems using wires cut to size, automatic tensioning devices and laser interferometer calibration, can readily achieve accuracies of 0.025 mm up to 60 m in length.

Subtense bars
Bars made of invar or carbon fibre with end targets for use in subtense methods have applications in industrial surveying (see Figure 3.24 right). They also need to be calibrated against a standard before use.

Distance is obtained from the subtense bar by a miniature triangulation method. The bar, usually 2 m long, forms one side of a triangle, from which the other sides are computed. To simplify the process, the bar can be fitted with a sighting vane enabling it to be aligned normal to a theodolite, at which a horizontal angle is read. The observed angle B subtended by the bar b, enables the horizontal distance S to be computed from

$$S = \frac{b}{2} \cot \frac{1}{2} B$$

At close range up to 20 m with a two metre bar, the method is still much used in precise industrial surveying. Modern levels use the technique by pattern matching a bar coded staff to give the distance at which a precise difference of level is also read off.

Stadia tachymetry
This subtense idea is incorporated into a theodolite telescope by arranging the stadia crosshairs to be set at twice the angle whose cotangent is 200: in other words the angle B is fixed at a value of about 34' arc. Thus a variable staff intercept b can be converted to the distance S from

$$S = 200\,b/\,2 = 100\,b$$

This expression is valid when the line of sight is normal to the staff. If the line of sight is inclined at vertical angle V, the horizontal distance D is given, with ample accuracy, by

$$D = 100\,b\cos^2 V$$

and the difference in height to the mid point of b by

$$\Delta h = 100\,b\sin V\cos V = 50\,b\sin 2V$$

The attainable accuracy naturally falls off rapidly as the distance increases, both due to the poor geometry of the triangle and because the staff becomes very difficult to read 100 m away. Air turbulence in warm climates makes the system almost useless.

Optical tachymeters and range finders
The subtense principle has been utilised in many ingenious instruments to make calculations by analogue methods. Range finders of various designs were also much in use before EDM. Although still of limited value, their importance has almost evaporated except as fascinating historical curiosities.

Plane table surveying
In the past, a small portable *plane table* was much used to draw maps directly in the field. The device still has its professional uses and is much valued in teaching. The surveyor sights through a telescope or open sight to which a parallel rule is attached. The direction is drawn directly on to the plotting surface: paper or plastic. The *alidade* may also incorporate a small EDM and clinometer to measure distances and slopes.

The system is very versatile, and able to be used for any ground survey technique: intersection, resection, contouring, radiation etc. It suffers from the disadvantage of simplicity and lacks a technological image. In the hands of an expert it is a most efficient and cost-effective tool with which to carry out small surveys.

3.7 Future vision and automated systems

The most obvious vision system used in surveying is the human eye together with its linkage to the human brain. This system has the ability to *interpret* what is seen as well as to resolve an image. Much effort is being made to copy this process by electronic detectors coupled to a computer as part of automated measurement systems to be used by automated levels, motorised total stations and totally robotic survey instruments. Consideration of the human vision system tells us *what automated systems must do*. The automated systems have the advantage of direct links to computer processing, and the potential to achieve better accuracy.

The total automation of the surveying process is unlikely if only on the grounds of costs. For many applications however, such as in the de-commissioning of nuclear reactors, it is the only way ahead.

4 Geometrical geodesy

Chapter Summary

This chapter deals with the computation of surveying data on the surface of the Earth as a whole, where simple plane reference systems are inadequate. The geometry of reference ellipsoids, their relation to the geoid, and their projection on a plane, are also discussed. The computation of geodetic data in curvilinear and three dimensional cartesian coordinate systems is described. The chapter concludes with a discussion of the computation of surveys on the transverse Mercator and Lambert's orthomorphic projections.

Geodesy

The word *geodesy* is derived from the Greek word "ge" = the Earth. Geodesy therefore is about a treatment of the Earth as a whole for surveying purposes. With the widespread use of artificial Earth satellites for general surveying, geodesy has become an essential part of every day surveying, providing positions for control and detail work. In this chapter we treat the geometrical aspects of geodesy. Geophysical aspects are only treated briefly where they relate directly to geometrical aspects.

Geodetic coordinate systems

As in all forms of surveying, a suitable framework of reference has to be adopted. This framework has greatly altered in response to modern computing power. Because of the very large ten-digit numbers involved in geodesy, past methods were geared to computational simplicity and convenience, rather than to theoretical elegance or effectiveness. The solid Earth was described with reference to departures from a reference surface, for which a sphere or ellipsoid was chosen. Thus a point in space was described in a two-plus-one *geographical* or *curvilinear* system of (latitude ϕ, longitude λ, and height h). Whilst this system is still used, most of the actual computations are done in a *rectangular* or *cartesian* system (x, y, z), being transformed into the geographical and *map projection* (E, N) systems to present the final data.

Part One: the geometry of the ellipsoid

For many purposes in land surveys it is possible to neglect the Earth's curvature, and to treat its surface as plane over a limited area. This greatly simplifies computation. Sometimes it is sufficient to consider the Earth as a sphere and work in terms of spherical trigonometry. In the most refined work, a spheroid or ellipsoid of reference has to be used. The ellipsoid of geodesy is the figure described by the rotation of an ellipse about its minor axis, often called an oblate spheroid.

The geoid
If the land masses were covered by a network of canals through which the waters of the oceans were permitted to flow freely under gravity, neglecting tidal effects, the surface of water on the canals and the oceans would form an equipotential surface called the *geoid*, which simply means 'Earth-shaped'. Note this is a surface based on a force field which has no simple *geometrical shape* other than that it approximates to a ellipsoid of revolution.

As a result of various measurements carried out since the 18th century in different parts of the world a number of geodesists have produced estimates of the size of ellipsoid which best fits the geoid for the part of the Earth considered in their calculations. These ellipsoids were adopted in various countries for geodetic computations: e.g. the United Kingdom is computed on Airy's ellipsoid, much of Africa is computed on the Clarke 1880 ellipsoid.

It is often sufficiently accurate to consider the surface of a sphere which closely approximates to the ellipsoid at the particular place in question, and whose radius equals the radius of curvature of the ellipsoid at that point. This approximate treatment is sufficient to give the provisional coordinates needed for precise computation by the least squares method.

4.1 The meridian ellipse
The ellipse which defines an ellipsoid is called the meridian ellipse. The parallels of latitude are small circles in planes parallel to the equator, which is a great circle. An ellipse is defined in many ways and has a multiplicity of geometrical properties. We shall consider it defined with respect to its semi-major axis a and semi-minor axis b by the equation

$$\frac{x^2}{a^2} + \frac{y^2}{b^2} = 1 \tag{1}$$

The coordinates of a point on the ellipse with respect to the origin at its centre are (x, y). The following properties will also be used:

$$b^2 = a^2 (1 - e^2) \tag{2}$$

where e is the eccentricity of the ellipse, which is about 1/12 for a terrestrial ellipsoid. Also the flattening f given by

$$f = (a - b) / a$$

is about 1/300 for the Earth. The area of an ellipse is πab.

Given either of the size parameters a or b, and a shape parameter e or f, the others may be derived. It is usual to define a meridian ellipse, and therefore an ellipsoid, in terms of a and f. For historical reasons, there are many reference ellipsoids recommended for use in different parts of the world, because all previous surveys and maps were based on them.

The parameters of early ellipsoids were calculated from terrestrial arc measurements in specific parts of the Earth. With the advent of artificial Earth satellites, geodesists have been able to use the Earth as a whole to define an ellipsoid. The basic parameters, obtained from orbital analyses, are the semi-major axis a, and the first harmonic of gravitational potential J_2. The latter can be used to derive an equivalent flattening according to a selected mathematical model. Datums, such as the world geodetic reference system WGS 84, are used for satellite work. The subject is one of some complexity, and much influences position fixing, especially off-shore in oil exploration, where geodetic cadastral problems arise. Table 4.1 gives approximate values of a few of these system parameters.

Table 4.1

Ellipsoid		a	1/f
Everest	1830	6 377 304	300.9
Bessel	1841	6 377 397	299.2
Clarke	1858	6 378 293	294.3
Clarke	1866	6 378 206	295.0
Clarke	1880	6 378 249	293.5
Helmert	1906	6 378 200	298.3
Hayford	1910	6 378 388	297.0
WGS	1984	6 378 137	298.3

It will be clear that, because the flattening is not listed to the same precision as the semi-major axis, the semi-minor axis cannot be calculated to the same precision. Early measurements were unable to establish ellipsoidal shape sufficiently accurately. However, the WGS 1984 flattening reciprocal has been quoted with sufficient precision as

298.257 222 101

Consultation with geodetic sources, to obtain the latest information on parameters, should be made before embarking on important computations involving ellipsoids and datums.

Types of latitude

Since the reference ellipsoid will not fit the geoid exactly at all points, two main types of latitude may be distinguished:

(i) *geodetic latitude*, ϕ_G, is the angle between the normal to the ellipsoid at a point and the plane of the equator (see Figure 4.1). The plane of the equator is perpendicular to the spin axis of the Earth.

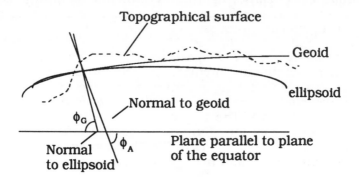

Figure 4.1

(ii) *astronomical latitude*, ϕ_A, is the angle between the equator and the meridian component of the normal to the geoid at that point. The normal to the geoid is the direction of the vertical and will not usually lie in the plane of the meridian ellipse. The angle between the direction of the plumb line and the normal to the ellipsoid is the deviation of the vertical. This depends on the ellipsoid chosen and the point or points at which the ellipsoid and geoid are related to each other. Latitude is positive north of the Equator.

Other types of latitude, such as *geocentric*, *reduced* and *isometric* latitude used in projection theory, will not concern us here.

Types of longitude

In a like manner to latitude, we distinguish two types of longitude which arise out of the lack of coincidence between the direction of the plumb line, or the vertical, and the normal to the ellipsoid.

(i) Geodetic longitude, λ_G, is the angle measured along the equator between the meridian of Greenwich and the meridian ellipse of the place.

(ii) Astronomical longitude, λ_A, is the angle measured along the equator between the astronomical meridian of Greenwich and the astronomical meridian of the place: the astronomical meridian of a place being defined as the plane containing the Earth's axis and the astronomical zenith of the place.

Longitude is reckoned positive East of Greenwich.

Types of azimuth: Laplace azimuth equation

Since there are two meridians at any place, the astronomical and the geodetic, directions referred to these meridians, azimuths, will be different, i.e. we distinguish astronomical azimuth α_A and geodetic azimuth α_G. Since computation is carried out in terms of geodetic azimuth, and only astronomical azimuth can be observed to control a survey, the geodetic value corresponding to an observed astronomical azimuth must be obtained. The connection between the two is called the *Laplace azimuth equation*.

Azimuth is positive clockwise from north.

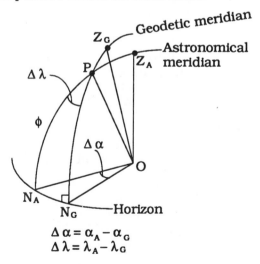

$$\Delta \alpha = \alpha_A - \alpha_G$$
$$\Delta \lambda = \lambda_A - \lambda_G$$

Figure 4.2

Let the geodetic and astronomical longitudes and azimuths be respectively

$$\lambda_G, \lambda_A, \alpha_G, \text{ and } \alpha_A$$

Then the Laplace azimuth equation connecting them is

$$\alpha_A - \alpha_G = (\lambda_A - \lambda_G) \sin \phi \qquad (3)$$

This may be proved by reference to Figure 4.2 in which the following are shown: P the pole, O the centre of the Earth; Z_A and Z_G the respective astronomical and geodetic zeniths, N_A and N_G the astronomical and geodetic north points on the horizon; and astronomical and geodetic azimuth and longitude differences $\Delta\alpha$ and $\Delta\lambda$.

It should be noted that, because there is only one pole, the effect is to align the minor axis of the ellipsoid parallel to the spin axis of the Earth.

In the triangle $P\ N_A\ N_G$, arc $N_A\ N_G = \Delta\alpha$, and angle $PN_G\ N_A$ is a right angle if N_AN_G O is the geodetic horizon. Then

$$\sin \Delta\alpha = \sin \Delta\lambda \sin \phi$$

and, since both $\Delta\alpha$ and $\Delta\lambda$ are small angles,

$$\Delta\alpha = \Delta\lambda \sin \phi \qquad (4)$$

It makes no difference which latitude is used in this equation.

Traditionally the Laplace equation was used to align the ellipsoid parallel to the Earth's spin axis and to control the swing of traverses and networks. It required a very precise and tedious determination of astronomical longitude.

Black has shown that a *geodetic* azimuth may be determined directly from *astronomical* observations without the need to observe astronomical longitude. Black's method is outlined in Chapter 6.

Radii of curvature of the ellipsoid

The double curvature of the surface of the ellipsoid is usually resolved into two components; one in the plane of the meridian ellipse, and the other, in a plane at right angles to the meridian: the plane of the *prime vertical*. The respective radii of curvature in these two directions, denoted by the Greek letters ρ and ν, are given by

$$\rho = a^2 b^2 / p^3$$
$$\nu = a^2 / p$$

where

$$p^2 = a^2 \cos^2 \phi + b^2 \sin^2 \phi \tag{5}$$

where φ is the latitude of the place. For any given ellipsoid for which the values of a and e are defined, both radii are functions of latitude alone. See Appendix 4 for derivations of these results. The distance p is the pedal distance to the ellipse.

The radius of curvature R at latitude φ at any azimuth α is given by

$$\frac{1}{R} = \frac{\cos^2 \alpha}{\rho} + \frac{\sin^2 \alpha}{\nu} \tag{6}$$

This is Euler's theorem, which is used to calculate R for computations involving survey lines and their reduction to the ellipsoid. See Chapter 7.

The mean value of the radius of curvature at a point is the geometrical mean of the principal radii of curvature, i.e.

$$\sqrt{\rho\nu}$$

This expression is used when areas are considered, e.g. in the computation of spherical excess. The following section gives outline proofs of these expressions.

Principal radii of curvature of the ellipsoid
Consider Figure 4.3 showing the meridian ellipse.

In Appendix 4 it is shown that

$$\nu = PS = a^2 / p$$

and that

$$x = a^2 \cos \phi / p$$

therefore

$$x = \frac{a \cos \phi}{(1 - e^2 \sin^2 \phi)^{\frac{1}{2}}}$$

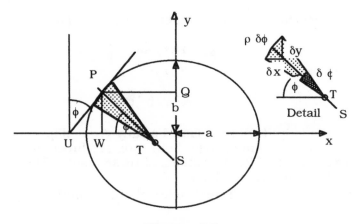

Figure 4.3

To derive the expression for ρ we use the fact that

$$\rho \, \delta\phi = \delta x \, \text{cosec} \, \phi$$

or

$$\rho = \frac{\delta x}{\delta \phi} \text{cosec} \, \phi \tag{7}$$

which on differentiation of x with respect to ϕ and substitution in (7) gives

$$\rho = \frac{a \, (1 - e^2)}{(1 - e^2 \sin^2 \phi)^{\frac{3}{2}}}$$

or

$$\rho = a^2 b^2 / p^3 \tag{8}$$

Euler's Theorem

Consider a point P on the surface of the ellipsoid touched by a tangent plane AA' (see Figures 4.4). If BB' is a plane parallel to AA' at a small distance z below it, it will cut the ellipsoid whose surface will describe an ellipse in plane BB'. This ellipse is called the Tissot indicatrix.

Now consider the normal section through P at azimuth α and let P' be a point on the indicatrix at this azimuth a distance s from P. Let the semi-major and semi-minor axes of the indicatrix be m and n respectively. Then from Figure 4.4

$$z = s^2 / 2R \qquad z = m^2 / 2v \qquad z = n^2 / 2\rho$$

And the equation of the indicatrix is

$$\frac{x^2}{m^2} + \frac{y^2}{n^2} = 1$$

or

$$s^2 \sin^2\alpha / m^2 + s^2 \cos^2\alpha / n^2 = 1$$

giving, from the equations in z,

$$\frac{1}{R} = \frac{\cos^2\alpha}{\rho} + \frac{\sin^2\alpha}{v}$$

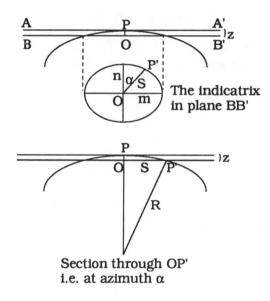

The indicatrix
in plane BB'

Section through OP'
i.e. at azimuth α

Figure 4.4

The mean radius of curvature

The mean value of R is given by

$$\frac{1}{2\pi} \int_0^{2\pi} R \, d\alpha$$

and from the indicatrix this is

$$\frac{1}{2\pi} \int_0^{2\pi} \frac{s^2}{n^2} \rho \, d\alpha$$

But

$$\frac{1}{2}\int_0^{2\pi} s^2\, d\alpha$$

is the area of the ellipse, i.e. πmn whence the

$$\text{mean radius} = m\, \rho/n = \rho\sqrt{\frac{\rho}{v}} = \sqrt{\rho v}$$

Distances along a meridian

The length of a small arc of the meridian in the vicinity of the point P whose latitude is ϕ is given by

$$ds = \rho\, d\phi$$

where ds is the arc length subtended by $d\phi$. This expression is often used in practice for approximate computations when tables of meridional distances are not available. For longer arcs, e.g. between latitudes ϕ_1 and ϕ_2, the length of the arc is given by

$$s = \int_{\phi_1}^{\phi_2} \rho\, d\phi$$

$$s = a\,(1 - e^2) \int_{\phi_1}^{\phi_2} \frac{d\phi}{(1 - e^2 \sin^2 \phi)^{\frac{3}{2}}}$$

To evaluate this elliptical integral we expand the elliptical part as a power series and integrate term by term to the accuracy required, obtaining a result of the form

$$s = a\,(1 - e^2)[\, A\,(\phi_2 - \phi_1) - \tfrac{1}{2} B\,(\sin 2\phi_2 - \sin 2\phi_1) + \tfrac{1}{4} C\,(\sin 4\phi_2 - \sin 4\phi_1)]$$

where A, B, and C are constants for the particular ellipsoid, being themselves power series of the eccentricity of the meridian ellipse. For the Clarke 1866 ellipsoid these coefficients have the following values: A= 1.005 1093; B= 0.005 1202; C= 0.000 0108.

An alternative is to carry out the integration by numerical addition of many small arcs, evaluating ρ for each latitude, which is ideal for a computer program.

Length of a parallel of latitude

The length of a parallel is simply found, because the parallel is a circular arc of radius $v \cos \phi$. Hence for a difference in longitude of $\Delta\lambda''$ the corresponding arc length Δs is given by

$$\Delta s = v \cos \phi \, \Delta\lambda'' \sin 1''$$

Geodetic computations

The complex formulae of geodesy were most arduous to handle before the advent of computers. Much use was made of special tables laboriously calculated by hand methods. Although tables still have some uses for cartographic work they have been entirely superseded by computer programs. Because of the very large numbers involved, care is needed in retaining the necessary accuracy throughout. It is important to remember that one second of arc subtends about 30 m on the Earth's surface.

4.2 The computation of geodetic position

Again as a result of computational facilities, the work of geodetic computation has changed from one of extreme tedium and some approximation, using long series-formulae, to one of comparative ease and complete rigour. Generally the approach is to work in cartesian coordinates and by least squares iterative methods. Approximate formulae for ellipsoidal calculations suffice to produce provisional coordinates for the least squares estimation process.The accurate conversion from geographical coordinates to cartesian coordinates is fundamental to the whole process.

Geodetic cartesian coordinates

After the theory has been developed from Figure 4.5, an example is given.

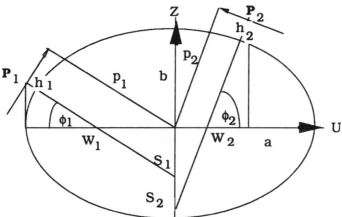

Figure 4.5

The meridian ellipse shows two points P_1 and P_2 at heights h_1 and h_2 above the ellipsoid. Since the length PS in each case is $(h + v)$ and the distance WS is $e^2 v$ (see Appendix 4) we have the transformations

$$z = [(1 - e^2) v + h] \sin \phi \qquad (9)$$
$$u = (v + h) \cos \phi \qquad (10)$$

These equations are easy to solve in the forward case, i.e. given ϕ and h, because v is itself a function of ϕ. The ellipsoid parameters a and e are generally given.

The reverse computation is not so simple because v is a function of the unknown ϕ. An iterative approach may be adopted using a first approximation

$$\tan \phi \approx z/u$$

to give a first value of v and then h, and so on. However Bowring and Vincenty, reference (15), have shown that the following nearly-closed formula gives adequate accuracy.

If we put $w^2 = u^2 + z^2$ and $\tan k = (z/w)(a/b)$ then we have the working formula

$$\tan \phi = \frac{z + \varepsilon \sin^3 k}{w - a e^2 \cos^3 k} \qquad (11)$$

where the square of the second eccentricity is given by

$$\varepsilon = \frac{a^2 - b^2}{b^2}$$

Once ϕ has been found, the height h is calculated directly from

$$h = u \sec \phi - v$$

Having found u and z from ϕ and h, the coordinates with respect to another meridian at an angle of longitude λ are given by

$$x = u \cos \lambda \qquad y = u \sin \lambda \qquad (12)$$

In this case the reverse calculation poses no problems, for

$$\tan \lambda = y/x \qquad u = x \sec \lambda \qquad (13)$$

Practical example

Many of the concepts of this chapter will be calculated using the data for a triangle ABC, situated in the southern hemisphere, given in Table 4.2. It should be noted that calculation to mm precision ensures that no errors arise as a result of the computational process itself, even though the observations may not warrant this. It will also be noted that longitude is quoted before latitude to preserve the other convention of listing eastings before northings. The triangle is also computed on the transverse Mercator projection later in the chapter.

Table 4.2

	Adjusted angles	Side lengths : metres
A	30° 48' 03".96	a = BC = 13 327.628
B	61° 05' 49".07	b = AC = 22 785.523
C	88° 06' 07".74	c = AB = 26 013.282

spherical excess 0".77

	Longitude	Latitude
A	35° 18' 16".7559	– 15° 52' 29".9096
B	35° 23' 37".4540	– 15° 39' 22".5964
C	35° 28' 22".7070	– 15° 44' 56".7567

Forward and back azimuths

AB	21° 32' 28".16	BA	201° 31' 01".02
AC	52° 20' 32".11	CA	232° 17' 47".00
BC	140° 25' 11".95	CB	320° 23' 54".74

Geographical coordinates to cartesian coordinates

Table 4.3 outlines the forward calculations of the cartesian coordinates of point B, and lists the results for two other points A and C, which will be used later throughout this chapter. All three points lie on the ellipsoid, i.e. at zero height, because the original observations were reduced to the ellipsoid. This also simplifies the calculations, using the formulae:

$$z = [(1 - e^2) \, v + h] \sin \phi$$
$$u = (v + h) \cos \phi$$
$$x = u \cos \lambda \qquad y = u \sin \lambda$$

Table 4.3

	major axis	Clarke 1880 Spheroid flattening	$e^2 = f(2 - f)$
	a = 637 8249	1/293.5	0.006 8027 013
		v_B = 637 9829.554	
	x	y	z
B	500 7820.801	355 8049.226	−170 9985.126
A	500 7981.152	354 6459.113	−173 3273.197
C	500 0631.698	356 3358.275	−171 9872.296

Distance between two points

The chord distance s between two points is given from the theorem of Pythagoras from

$$s^2 = \Delta x^2 + \Delta y^2 + \Delta z^2$$

In the case of the line AB we have

$$s = 26\ 013.279$$

Normal section azimuths

The normal section azimuth A between two points on the ellipsoid is given by the formula

$$\tan A = P/Q \tag{14}$$

where

$$P = -\Delta x\ \sin \lambda_1 + \Delta y \cos \lambda_1$$

and

$$Q = -\sin \phi_1\ (\Delta x\ \cos \lambda_1 + \Delta y \sin \lambda_1) + \Delta z \cos\ \phi_1$$

For the line AB of the example we obtain the result

$$A_{12} = 21° 32' 28".09$$

We will explain the slight differences from those of Table 4.2 after the derivation of the formulae.

Consider Figure 4.6 which is obtained from folding Figure 4.5 about the Earth's axis to create a difference of longitude $\Delta\lambda$ between the two meridians of points P_1 and P_2. The reader is encouraged to make a paper model of this to gain an immediate understanding of the three dimensional model represented in Figure 4.6. All points marked by black dots lie in the plane of the meridian of P_1. while all points marked by circles lie in the plane of the meridian of P_2.

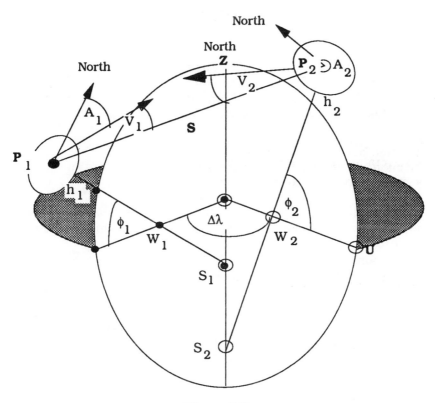

Figure 4.6

To obtain the forward and back *normal-section* azimuths A_1 and A_2. and also the respective slope angles V_1 and V_2. we transform the coordinates of each point into a local system based on the ellipsoid normals and the direction of north. The essence of the problem is expressed by Figure 4.7 which is extracted from 4.6. Consider the coordinate system based on P_1 shown in Figure 4.7.

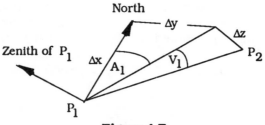

Figure 4.7

We have

$$\tan A_1 = \Delta y / \Delta x \qquad \text{and} \qquad s \sin V_1 = \Delta z$$

To find Δx, Δy and Δz in this system, we transform from the original coordinates of the Earth-centered system as follows:

    (a) a translation of origin to the point P_1

    (b) a rotation about the new z axis of $+ \lambda_1$

    (c) a rotation about the new y axis of $+ (90 - \phi_1)$.

These operations are depicted in Figures 4.8, but we repeat that a paper model shows very clearly the correctness of this mathematical procedure.

Translation of the origin to P_1 gives Δx, Δy, and Δz.

The first rotation is achieved by the matrix operation

$$\begin{bmatrix} x' \\ y' \\ z' \end{bmatrix} = \begin{bmatrix} \cos \lambda_1 & \sin \lambda_1 & 0 \\ -\sin \lambda_1 & \cos \lambda_1 & 0 \\ 0 & 0 & 1 \end{bmatrix} \begin{bmatrix} \Delta x \\ \Delta y \\ \Delta z \end{bmatrix}$$

The second rotation is achieved by the matrix operation

$$\begin{bmatrix} x'' \\ y'' \\ z'' \end{bmatrix} = \begin{bmatrix} \sin \phi_1 & 0 & -\cos \phi_1 \\ 0 & 1 & 0 \\ \cos \phi_1 & 0 & \sin \phi_1 \end{bmatrix} \begin{bmatrix} x' \\ y' \\ z' \end{bmatrix}$$

Then

$$\tan A_1 = y'' / -x'' = P/Q \qquad\qquad (14)$$

Also

$$s \sin V_1 = z''$$

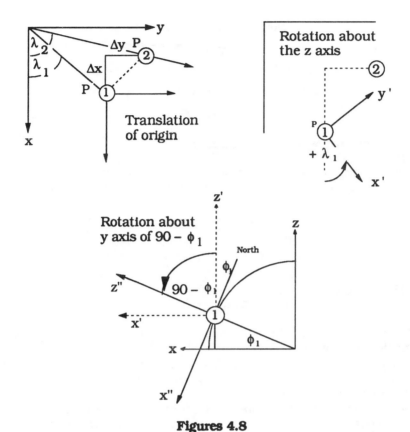

Figures 4.8

These formulae are generally used for the calculation of provisional azimuths, distances and slope angles from provisional values of the geographical coordinates, which are compared with their measured counterparts to give the l vector of the observation equations.

The azimuth and distance of the example do not agree exactly with the quoted values for the geodetic triangle, because the latter are geodetic quantities for curves traced out along the ellipsoid surface. The curves traced out along the ellipsoid surface by the normal sections from P_1 and P_2 and the shortest line along the surface, called the *geodesic*, are shown in Figure 4.9. The very small angles between them are divided in the ratio of 1/3 to 2/3. Little use is made of the geodesic today. Work is generally carried out in terms of the formulae for normal sections and chords.

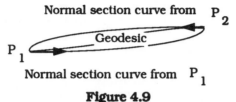

Figure 4.9

Variation of geographical coordinates

By differentiation of formulae (9), (10), and (12), the equations relating changes in the three-dimensional cartesian coordinates (x,y,z) to changes in northings, eastings and heights, (N,E,h), are

$$
\begin{bmatrix} dN \\ dE \\ dh \end{bmatrix} = \begin{bmatrix} -\sin\phi\cos\lambda & -\sin\phi\sin\lambda & \cos\phi \\ -\sin\lambda & \cos\lambda & 0 \\ \cos\phi\cos\lambda & \cos\phi\sin\lambda & \sin\phi \end{bmatrix} \begin{bmatrix} dx \\ dy \\ dz \end{bmatrix} \quad (15)
$$

where

$$ dN = (\rho + h)\,d\phi \qquad \text{and} \qquad dE = (\nu + h)\cos\phi\,d\lambda \quad (16) $$

The various differential equations of Chapter 2, linking measured quantities to coordinate changes dx, dy, and dz, may be extended via equations (15), to express $d\phi$, $d\lambda$ and dh, as the parameters to be estimated. Notice that the values for the radii of curvature at every point will differ slightly.

These expressions (16) follow by inspection of Figure 4.10 in which H is the centre of curvature in the meridian plane and S the centre of curvature in the plane perpendicular to the meridian, the *prime vertical* plane.

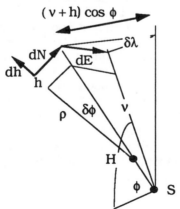

Figure 4.10

The orthogonal rotation matrix represents the rotations between the local axis system and the original cartesian system. Because the matrix is orthogonal, its inverse is equal to its transpose, therefore the reverse transformation is simple.

Calculation of provisional coordinates
Because all geodetic coordinates are now computed as part of a least squares process the need for precise ellipsoidal formulae is almost nil. Only approximate formulae are required to obtain provisional coordinates. These formulae, Legendre's theorem and the Gauss Mid-latitude formulae, are methods which may be derived by analogy from plane calculations.

Legendre's theorem
This theorem states that the side lengths of a spherical triangle can be computed by plane trigonometry, if the angles of the triangle first have one third of the spherical excess of the triangle subtracted from each. The *spherical excess* is amount by which the sum of the spherical angles exceeds 180°. For example the triangle ABC of Table 4.2 has a spherical excess of 0".77 which may be checked from the formula

$$\varepsilon'' = (\Delta / \rho \nu) \ 206 \ 265$$

The area of the triangle Δ is obtained from

$$\Delta = \frac{1}{2} \ ab \ \sin C$$

In the example, we deduct 0.253 " from two angles and 0.254" from one to obtain the "Legendre angles" which are then used in the plane formula
$$a = b \sin A / \sin B$$

to calculate the two unknown sides from one other known side such as AB.

Note: a rigorous proof of this theorem, given on page 157 of the first edition of this book, consists of expanding terms in the cosine formula of spherical trigonometry, and comparing an approximate plane triangle with its spherical counterpart.

The mid latitude formulae
Consider Figure 4.12, showing the following:

(a) The pole P; x and y are points on the surface of a sphere approximating to the ellipsoid, whose respective latitudes and longitudes are ϕ_x, ϕ_y, λ_x, and λ_y.

(b) α_x is the forward azimuth of y from x.

(c) α_y is the back azimuth of x from y:　　$\alpha_y{}'' = \alpha_y - 180°$

Due to the convergence of the meridians, α_x will not as a rule equal $\alpha_y - 180°$. The forward and back azimuths will differ by 180° only if x and y lie on the same meridian, or if both are on the equator.

The length of xy = s. If the position of x, the forward azimuth α_x, and the distance s are given, we require to compute the position of y and the reverse azimuth from y.

Whilst it is possible to solve the triangle by spherical trigonometry, it is inconvenient to do so because great precision is needed in the angles subtended at the centre of the Earth to yield the side lengths. One second of arc subtends about 30 m on the surface of the Earth.

It is easier and sufficient to work in terms of the surface triangle itself.

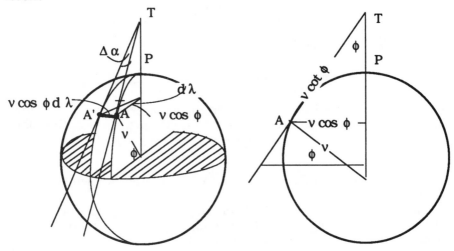

Figure 4.11

Meridian convergence

The convergence of the meridians is zero at the equator, and increases towards the poles until it equals the difference in longitude between the two points considered. At any intermediate latitude the convergence is given by

$$\tan \tfrac{1}{2} \Delta\alpha = \tan \tfrac{1}{2} \Delta\lambda \, \sin \tfrac{1}{2}(\phi_x + \phi_y) \, \sec\tfrac{1}{2} \Delta\phi$$

where

$$\Delta\alpha = \alpha_y\text{''} - \alpha_x$$

However because of the approximate nature of the task, just to obtain provisional coordinates, it is adequate to use the approximation

$$\Delta\alpha \;=\; \Delta\lambda \, \sin \phi_m \tag{17}$$

The differences in latitude and longitude between the points are given by the formulae

$$\Delta\phi\text{''} = s \cos\alpha_m \, / \, \rho_m \sin 1\text{''} \tag{18}$$

$$\Delta\lambda\text{''} = s \sin\alpha_m \, / \, v_m \cos\phi_m \sin 1\text{''} \tag{19}$$

Justification of formulae
By inspection of the equivalent plane triangle in Figure 4.12 we see that

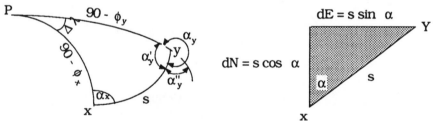

Figure 4.12

$$dN = s \cos\alpha \qquad dE = s \sin\alpha$$

Also by inspection of Figure 4.10 we see that

$$dN = (\rho + h)d\phi \qquad \text{and} \qquad dE = (v + h) \cos\phi \, d\lambda$$

If h = 0 the mid latitude formulae follow by equating dE and dN.
   Again from Figure 4.11 equating two expressions for the arc of the parallel AA' gives

$$v \cot\phi \, \Delta\alpha = v \cos\phi \, \Delta\lambda$$

therefore

$$\Delta\alpha \;=\; \Delta\lambda \cos\phi$$

Table 4.4

	Data	Formulae
ϕ_A	$- 15° 52' 29".9096$	$\Delta\phi" = s \cos \alpha_m / \rho_m \sin 1"$
λ_A	$+35° 18' 16".7559$	$\Delta\lambda" = s \sin \alpha_m / v_m \cos \phi_m \sin 1"$
α_A	$21° 32' 28".16$	$\Delta\alpha = \Delta\lambda \cos \phi$
s_{AB}	$26\ 013.282$ m	
$1/ \rho_1 \sin 1"$	325.354×10^{-4}	Computation
$\Delta \phi$	$+ 787".2$	of approx
$\frac{1}{2}\Delta \phi$	$+393".6$	mid -latitude
	$+ 6' 33".6$	
ϕ_A	$- 15° 52' 29".9$	
ϕ_m	$- 15° 45' 56".3$	
$1/ \rho_m \sin 1"$	325.35779×10^{-4}	Computation
$\Delta \phi$	$+ 787".2474$	of
	$13' 07". 2474$	latitude of B
ϕ_A	$- 15° 52' 29".9096$	
ϕ_B	$- 15° 39' 22".6622$	
$1/ v_m \sin 1"$	323.30660×10^{-4}	Computation
$\Delta \lambda$	$+ 320".8701$	of
	$5' 20". 3701$	longitude of B
λ_A	$+35° 18' 16".7559$	
λ_B	$+35° 23' 37".1260$	
$\Delta \alpha$	$- 87.1817$	Computation
	$1' 27". 18$	of
α_{AB}	$21° 32' 28".16$	back azimuth
α_{BA}	$201° 31' 00".98$	

The sequence of computation is first to find ϕ_y, then use this with ϕ_x to obtain the mean latitude and so evaluate the radii of curvature and the convergence to find α_m. Table 4.4 outlines the key stages in sequence of the first iteration of the computation of the line AB. A second iteration was needed for the result to converge to that of Table 4.1.

Reverse computation
The reverse computation occurs when the two points are given and their distance apart and respective azimuths are required. Since the mid-latitude is obtainable at once, no successive approximation is necessary. However, the Vincenty-Bowring approach, described above, is accurate and also straightforward.

Geodetic datums
Perhaps the most serious problems that can arise in modern geodetic surveying are due to uncertainty about the relationships between different reference systems. These differences arise because of historical and other reasons.Various ellipsoids have been adopted for computations, and even the same ellipsoid can be aligned with respect to the Earth's axis and centre of gravity in non unique way.

The outcome is that reference systems will be incompatible until their relationships are known. Once this is so, all data can be related by a transformation. Information to evaluate the parameters of the transformation is obtained by examining the discrepancies between the positions of well spaced points on two systems.

The problem takes many forms depending on the information available, not all of which will necessarily be what it might seem. For example hybrid computations are not unknown in which part of a survey was computed on one datum with the incorrect ellipsoid parameters.

At its simplest, a datum needs to have an initial reference point requiring three pieces of information referred to an axis system say (x, y, z) or (ϕ, λ, h). It is wise to align the z axis parallel to the spin axis of the Earth by the Laplace condition, and to adopt some reference for longitude, usually Greenwich or Greenwich related. If a reference ellipsoid is used, two further pieces of information are essential to describe its shape and its size. In all this makes seven parameters to define a datum. Additionally because the scale of a survey network must also be determined an eighth parameter is needed although this is strictly not part of the datum.

The identification of these parameters is obtained from the general least squares transformation of three-dimensional coordinates by the method of centroids, already described in detail in Chapter 2 for a two-dimensional problem.

4.3 Location of geographical boundaries

Sometimes the surveyor has to set out or relocate a geographical boundary defined by a meridian, or a great circle between two points, or a parallel of latitude. The first two cases are comparatively simple because the theodolite can be made to point along a great circle, for all practical purposes. The parallel of latitude however has to be set out from a great circle by offsets, as with a railway curve.

In all three operations a point close to the required line has to be located on the correct datum. This may be carried out in many possible ways using any recognised survey method. The basic principle is to locate a point as close to the required position as possible by reconnaissance surveys or aerial photography, then establish its exact position and finally to compute an offset to the required boundary.

Suppose that x is the provisional position whose coordinates can be calculated through the survey. Then assuming that the point y lies on the required meridian or parallel, the azimuth and distance of y from x are computed by the mid-latitude formulae and the required point is located on the ground, making due allowance for height above sea-level, slope, etc.

Points are then set out along the required route, with checks applied from time to time by tying-in to control surveys or making astronomical determinations. Such surveys can often involve the

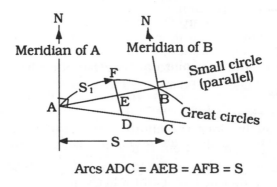

Arcs ADC = AEB = AFB = S

Figure 4.13

greatest of ingenuity on the part of the surveyor to overcome obstacles, though EDM and satellite sytems have made the work easier. The main problem will be to establish that the work is on the correct datum.

Setting out a parallel of latitude
This operation may be carried out from key points by offsets from a tangent to the parallel, or by offsets from a chord. Refer to Figure 4.13 in which AC is the tangent to the parallel AEB at A, and AFB is the chord of this parallel. Both arcs AFB and AC are great circles, and AEB is a small circle. The small angles of Figure 4.13 have been exaggerated. The length CB is the offset from the tangent at a distance ADC equal to s. Since BC is small it is sufficiently accurate to take the lengths AFB, AEB, and ADC to be equal to s. The diagram is grossly exaggerated for clarity.

Offset from the tangent AC
Let $\Delta\alpha$ be the difference between the forward and back azimuth from A to C, neglecting 180°. The angle ACB is $90 - \Delta\alpha$ and the offset BC is given by

$$BC = \frac{s^2 \tan\phi}{2v}$$

For, remembering that $\alpha = 90°$,

$$BC = s\cos(90 + \tfrac{1}{2}\Delta\alpha)$$
$$= s\sin\tfrac{1}{2}\Delta\alpha = s\tfrac{1}{2}\Delta\alpha = s\tfrac{1}{2}\Delta\lambda\sin\phi$$
$$= s\tfrac{1}{2}\sin\phi\, s / v\cos\phi$$
$$BC = \frac{s^2 \tan\phi}{2v}$$

After positioning the theodolite at C by radiation from A, the angle ACB = $90° - \Delta\alpha$ and distance BC are used to set out B.

Offset from the chord AFB
To locate B directly along the chord from A, sight along AFB at an azimuth of $90 - \tfrac{1}{2}\Delta\alpha$, and set out B at a distance s. Intermediate points along the parallel are often required, say at intervals of one km. The offset FE from an intermediate point F is given by

$$FE = \frac{s_1\,(s - s_1)\tan\phi}{2v}$$

For
$$FE = FD - ED = \frac{s_1 \Delta\alpha}{2} - \frac{s_1^2 \tan\phi}{2v}$$

$$= \frac{s_1 \, BC}{2 \, s} - \frac{s_1^2 \tan \phi}{2 \, v}$$

$$FE = \frac{s_1 \, (s - s_1) \tan \phi}{2 \, v}$$

When point B is reached, the next stage of the setting out is repeated as from A, and so on. In thickly wooded country the method of offsetting from a chord is preferable since less cutting is required for the shorter offsets. The survey points are often marked by huge concrete pillars so that warring factions cannot plead ignorance of the boundary location.

Part Two : Survey Projections

Although the positions of points may be computed on the surface of a reference ellipsoid or in three-dimensional cartesian coordinates a common end product of surveying, the map, has to be drawn on a plane surface. Hence some system of map projection is required.

In the past, hand computation on the curved surface of an ellipsoid or a sphere was more tedious than on a plane surface so preference was given to projection methods. Current practice is still to work on a projection for limited surveys which have to be tied into a national coordinate system. Large survey organizations dealing with several countries compute on the ellipsoid and convert to map projections for plotting purposes.

4.4 Map projections

A *map projection* is any orderly system of representing points of the sphere or ellipsoid on a plane.

We shall confine our theoretical deliberations to the sphere since the main principles can be more easily understood by this approach. With the exception of the stereographic projection, the projections used by surveyors cannot easily be drawn by direct geometrical perspective, but are mathematical projections.

The two most commonly used are the *transverse Mercator* and *Lambert's conical orthomorphic*, both of which are conformal or orthomorphic projections, i.e. on these projections the scale factor at a point is independent of azimuth. Attention is further confined in this book to these two surveyor's projections, with brief mention of the Cassini projection as an interim stage in the development of the theory. The formulae derived are restricted to closed spherical trigonometrical formulae, except in a very few cases, notably the arc-to-chord or t–T formula, and the scale factor formula for the transverse Mercator projection. These latter form the basis of practical computation on the projection, whilst the spherical trigonometrical formulae are useful for cartographical purposes.

Although the conversion of geographical or ellipsoidal coordinates to projection coordinates is now carried out by computer software some use of tables has been retained to provide data for the examples and to verify computer software.

Basic concepts of projection

The following basic concepts are essential for a grasp of the subject.

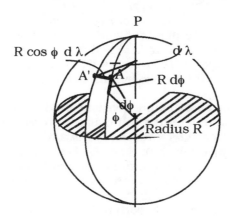

Figure 4.14

A *graticule* is the network of lines formed by the meridians and parallels either on the sphere or on the map, the latter being the projected positions of the former.

A *grid* is a system of squares drawn on a map, from which the map graticule and points of detail are plotted. It is unusual, though not impossible, to plot directly in terms of a graticule.

Transformation formulae

Let A be a point on a sphere of radius R representing the Earth, whose pole is P. The map projection system relates the latitude and longitude (ϕ, λ) of A to its grid coordinates (E,N) by some functional relationship which establishes a unique one-to-one correspondence between a point on the sphere and its plotted position on the map. Expressing this mathematically we write

$$E = F_1 (\phi, \lambda) \qquad\qquad N = F_2 (\phi, \lambda)$$

$$\phi = F_3 (E, N) \qquad\qquad \lambda = F_4 (E, N)$$

These transformation formulae have to be explicitly stated for each projection according to some rules which will preserve some properties of the original surface, but not all of them. *Conformal* or *orthomorphic* projections preserve local small shapes, *equal area* or *orthembadic* projections preserve areas, and so on. Much of the theory is developed around the *scale factor* of the system.

Scale factor

The scale factor K is the ratio of a straight line distance on the map to the corresponding distance on the sphere or ellipsoid. The scale factor is often considered under the headings *nominal scale* and *differential scale* or scale error. The nominal scale of a map is its representative fraction, say 1: 50 000, stating in a general way the magnitude of the scale factor. The scale factor over the whole map is not constant but varies from the nominal scale by small differential amounts, often called scale errors.

Surveyors usually work implicitly with a nominal scale of unity (at life size); hence the differential scale or scale error is the amount by which the scale factor differs from 1. For example if the scale factor at a point on the projection is 0.9996 the corresponding scale error is – 0.000 4. The subsequent reduction of coordinates from life size to some convenient map scale merely alters the nominal scale of the survey.

The use of the word 'error' is perhaps unfortunate, since it connotes the idea of a mistake, whereas the introduction of a scale factor is deliberate and its effects are predictable. In a sense it is a known *systematic* error.

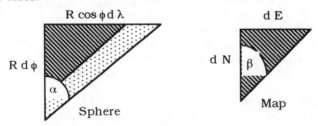

Figure 4.15

The concept of differential scale is explained as follows. Let a small linear element of the meridian of A in the vicinity of A be **R d ϕ** and a small element of the parallel of A be **R cos ϕ d λ**. Then if dN and dE are the elements on the map corresponding to R d ϕ and R cos ϕ d λ, the scale in a north-south direction is K_1 and in an east-west direction is K_2 given by

$$K_1 = dN / R d \phi$$

$$K_2 = dE / R \cos \phi \, d\lambda$$

It is sometimes convenient to define a new variable ψ, called the *isometric latitude*, by putting

$$d\psi = \cos \phi \, d\lambda$$

Although isometric latitude is not be employed in this book, it is important to know the meaning of the term.

The projection is conformal or orthomorphic if $K_1 = K_2$. A corollary to this is that *for a small element*, angles on the surface of the sphere are correctly represented on the map. Thus in Figure 4.15

$$\tan \alpha = R \cos \phi \, d\lambda / R \, d\phi$$

$$= (dE / K_2) \, (K_1 / dN)$$

$$= dE / dN = \tan \beta$$

thus

$$\alpha = \beta$$

In practice a small element will be the theodolite circle used for the angle measurement. For such a small part of the Earth's surface the angle equality of the conformal map can be considered exact.

It is important to note that these properties of a conformal projection hold only for small differential amounts. Since the surface of the sphere cannot be correctly represented on a plane over large areas, the scale factor for long lines will not be independent of azimuth, neither will angles be correctly preserved in conversion from the sphere to the map.

Just what is meant by 'long lines' depends on the ultimate accuracy desired. The angular distortion over a line of 8 kilometres is normally less than 1 second of arc on a conformal projection, hence the projection effect could be neglected for most work over this distance, and the survey computed in simple rectangular coordinates. Before computing a survey over a wide area it is usual to investigate the angular and scale distortion which will be accumulated as a result of neglecting these factors, and reach a decision on the method of computation on the grounds of economy, paying due regard to possible future requirements.

In non-conformal projections, angular and differential scale distortions seriously restrict the lengths of lines that may be computed without applying corrections.

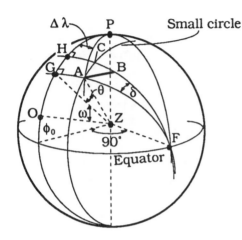

Radius of sphere is R
Angle GZO = $\omega = x/R$
Angle GZA = $\theta = y/R$
AC is arc of small circle
Angle PAC = Convergence c

Figure 4.16

4.5 Rectangular spherical coordinates or Cassini's projection

The simplest form of projection used by surveyors for early mapping
is the Cassini projection. Although many old cadastral maps were
plotted by this method, it is little used today. However, it does provide
a simple way to develop formulae for the more useful conformal
projection, the transverse Mercator.

Figure 4.16 shows the terrestrial sphere. P is the pole; OGHP is the
central meridian of the projection system; O is the origin of the
projection ; A and B are two points on the surface of the sphere; G and
H are the feet of the perpendiculars from A and B respectively to the
central meridian; arc AC is part of a small circle passing through A
lying in a plane which is parallel to that of the central meridian; F is
the pole of the central meridian; the radius of the sphere is R;
latitudes and longitudes of the various points are denoted by (ϕ_A, λ_A),
with the suffix referred to the point.

Let the angles subtended at Z, the centre of the sphere, by OG and GA
be ω and θ respectively. Let the arc length of OG be x_A and that of GA be
y_A ; to construct a projection draw a straight line O'G'H', mark off O'G'=
OG = x_A, and G'A' perpendicular to O'G' at G' equal to GA = y_A, the point A'
so obtained is the position of A on the the Cassini projection whose
origin is at O and whose central meridian is OGH (O'G'H') (see Figures
4.16 and 4.17).

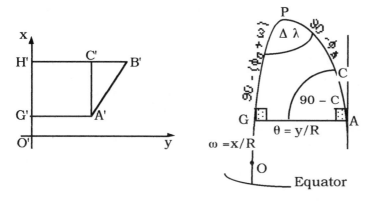

Figure 4.17   **Figure 4.18**

Transformation formulae for Cassini projection

Projection computation is concerned with the forward and reverse transformation of coordinates from geographicals to rectangulars and vice versa, and the calculation of convergence from either starting point. Also involved are the calculation of graticule-grid cutting points, point and line scale factors, and arc-to-chord corrections.

Forward transformation

Figure 4.18 shows the spherical triangle formed by A, P, and G. The angle at G is a right. Since CA is perpendicular to GA and if the angle PAC = C angle PAG is 90 – C.

The line AC is projected as A'C' on the map (Figure 4.17), which defines the direction of grid north. The angle C is therefore the difference between true and grid north. It is called the *map convergence* or simply *the convergence at* A. Notice this is strictly different from the meridian convergence $\Delta\alpha$ on the ellipsoid although they both are about the same magnitude.

Applying Napier's rules (See Appendix Two) for circular parts to Figure 4.18 we have

$$\sin(y/R) = \sin\theta = \sin\Delta\lambda \ \cos\phi_A \qquad (1)$$

$$\cot(\phi_0 + x/R) = \cot(\phi_0 + \omega) = \cos\Delta\lambda \cot\phi_A \qquad (2)$$

$$\tan C = \tan\Delta\lambda \sin\phi_A \qquad (3)$$

where

$$\Delta\lambda = \lambda_A - \lambda_o$$

Given the value of R for the sphere, the coordinates of A' on the map can be calculated from its given geographical coordinates (ϕ_A, λ_A).

Reverse transformation
For the reverse computations, given (x and y) and therefore ω and θ can be derived, the following expressions are used to obtain ϕ_A and λ_A

$$\sin \phi_A = \cos \theta \sin (\phi_0 + \omega) \qquad (4)$$

$$\cot \Delta\lambda = \cos (\phi_0 + \omega) \cot \theta \qquad (5)$$

Cutting points on a map sheet
Although the details of a map will be plotted from the grid, it is convenient to have some parts of the graticule plotted. These points will be plotted by computer software in just the same way as any others. The boundary of the map sheet may even be a part of the graticule.

If hand methods are still used, at most scales it is sufficient to plot three points on each graticule line and draw in the line by fitting a very slightly curved spline to these points. However, when the grid and graticule intersect at a very small angle, it is sometimes uncertain where the two cut. In some cases it is uncertain if a cut lies on a particular map sheet at all. Therefore these critical cutting points have to be computed. This is true even if computer plotting is used.

Case (a) : where an east-west grid line cuts a parallel. In this case it is required to compute the easting y of the cutting point, from a given northing x and latitude ϕ. The formula is

$$\cos (y/R) = \cos \theta = \sin \phi \, \text{cosec} \, (\phi_0 + \omega)$$

Case (b) : where a north-south line cuts a meridian. In this case ω is derived from

$$\cos (\phi_0 + \omega) = \cot \Delta\lambda \tan \theta$$

and so x can be found, since longitude and easting are known. Calculations using these formulae are usually sufficiently accurate for plotting purposes provided geodetic tables are used to obtain the values of the radii of curvature and the meridional distance from the origin.

Scale and angular distortions on the Cassini projection

Figure 4.17 shows points A and B plotted in positions A' and B' on the map. C' lies on H'B' such that H'C' = G'A' = y. Figure 4.19 shows an enlarged view of triangle A'B'C' with the original spherical triangle ABC superimposed upon it. The grid bearing of A'B' is denoted by T, and angle B'A'B by dT. Since B'C' = BC, the scale east-west is correct.

But the line AC, equal to A'C' = H'G' = HG = R δ, is plotted too long. The angle HFG = δ, or δ is the angle subtended by GH at the centre of the sphere (Figure 4.16).

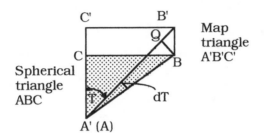

Figure 4.19

In Figure 4.19 we have

$$AC = R δ \cos θ = GH \cos θ = A'C' \cos θ$$

therefore the scale factor in a north-south direction at A is

$$A'C' / AC = \sec θ = K \text{ say}$$

then

$$BB' = CC' = AC' - AC = A'C' (1 - \cos θ)$$
$$= A'C' \tfrac{1}{2} θ^2 \text{ approx}$$
$$= \tfrac{1}{2} θ^2 s \cos T$$

where s is the length of the line AB and T is the grid bearing.

Angular distortion

The angular distortion dT is given approximately as follows:

$$dT'' \sin 1'' = BQ / s = BB' \sin T / s$$
$$= \tfrac{1}{2} θ^2 \sin T \cos T$$
$$= \tfrac{1}{4} θ^2 \sin 2T$$
$$= (y^2 / 4 R^2) \sin 2T \tag{6}$$

Hence dT is zero for a line north to south, and a maximum at azimuths of 45°, 135°, 225°, and 315°. Notice the angular distortion even for a short line is considerable at points away from the central meridian; for example its maximum value when y = 500 km is about 30'.

Scale distortion

The line AB becomes A'B' on the map, hence the scale is increased by

$$B'Q = BB' \cos T = \frac{1}{2} \theta^2 s \cos^2 T$$

Expressed as a fraction of the length s this scale error is

$$\frac{1}{2} \theta^2 \cos^2 T$$

and the scale factor K is

$$K = 1 + \frac{1}{2} \theta^2 \cos^2 T$$

When T = 90° or 270° K = 1 : when T = 0° or 180° K = 1 + $\frac{1}{2} \theta^2$.

Put another way, the scale factor is sec θ in a north-south direction, and unity in an east-west direction.

Because of these excessive angular and differential scale distortions the projection is not well suited to mapping, though it was widely used in the past on account of its computational and graphical simplicity.

4.6 The transverse Mercator or Gauss-Krüger projection

This projection can be thought of as a Cassini projection modified to be made conformal: i.e. so that the point scale factor is independent of bearing T, with no angular distortion over a very short line.

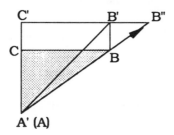

Figure 4.20

The angular distortion at a point, such as A in Figure 4.20, will be eliminated by increasing the projection easting y of point B' until B' lies in position B", i.e. until A'BB" is a straight line. This is achieved if the scale factor for the element BC is made the same as for the element AC, i.e. BC is increased by the value of sec (y/R) = sec θ.

This increase has to be made to all short lines out from the central meridian; that is each element of the arcs G'A' and H'B' of Figure 4.16 will be increased by its appropriate sec θ.

The easting thus obtained is the transverse Mercator easting E which is given by

$$E = \int_0^\theta R \sec \theta \, d\theta$$

or

$$E = R \ln \tan [\pi/4 + \theta/2] \tag{7}$$

The transverse Mercator (TM) northing N is the same as x on the Cassini projection.
Note: this is the transverse case of Mercator's projection. In the direct case, the latitude φ replaces the angle θ of this projection, and we have
$$N = R \ln \tan [\pi/4 + \phi/2]$$

Computation on the transverse Mercator projection
It is easy to apply this formula for E with modern calculators or computer software. Traditionally this was not so and an alternative approach using expansion series was preferred. Since this second method still has its counterpart in ellipsoidal treatment of the projection problem we will give a brief account of it here.

Sec θ is expanded as a power series of θ and integrated term by term. We have

$$E = \int_0^\theta R \sec \theta \, d\theta$$

$$E = R \int_0^\theta (1 + \frac{1}{2}\theta^2 + \frac{5}{24}\theta^4 ...) \, d\theta$$

$$= R (\theta + \frac{1}{6}\theta^3 + \frac{1}{24}\theta^5 ...)$$

The third term has a value of 0.1 m for θ = 3° which is normally the maximum distance from the central meridian used. When a country has a larger extent east-west than this, the area is divided into two or more projection 'belts' or 'zones' with a suitable overlap (see the Universal Transverse Mercator projection).

Reverse case

To obtain θ from E, the method of successive approximation is adopted, putting $\theta = E/R$ in the formula instead of the correct $\theta = y/R$. We have

$$E = R \left(\theta + \frac{1}{6} \theta^3 + \frac{1}{24} \theta^5 \ldots \right)$$

$$= R\theta \left(1 + \frac{1}{6} \theta^2 + \frac{1}{24} \theta^4 \ldots \right)$$

or

$$\theta = (E/R) \left(1 - \frac{1}{6} (E/R)^2 \ldots \right) \text{ approx}$$

The new value of θ is then used in the reverse series until convergence is achieved.

In practice the precise ellipsoidal formulae are programmed for computer computation without much difficulty.

Scale factors on the transverse Mercator projection

The scale factor used in the design of the projection was sec θ which was integrated to give the eastings. Thus for a short line the scale factor is sec θ. This means that all lines away from the central meridian are increased by an amount which can be quite large. To reduce the overall effect, a central scale factor K_0, less than unity, is applied to the whole map, making the scale factors at the central meridian and at the designed maximum distance from it to be equal and opposite in size. TM projection zones are usually set at $\pm 3°$ from the central meridian, i.e. θ max is 3°.

Without the application of K_0 the maximum scale factor sec 3° = 1.0014 approx. If we make $K_0 = 1 - 0.0007 = 0.9993$, then the combined new scale factor is $K_0 K = 0.9993$ sec θ for any point. At two eastings either side of the central meridian the net scale factor is 1. These eastings are at angles $\theta = 0.7 \times 3° = 2.1°$, i.e. whose secant is 1.0007.

Line scale factors

Because the scale factors apply only to points, the appropriate value to apply to a line is the average for the points on that line. Sometimes the mid-point value will suffice; for others Simpson's rule for approximate integration is required. An example is this is given later.

False eastings

To avoid the unnecessary nuisance of positive and ne~~gative castings~~ the origin of a projection is moved to the west to a *False Origin*. This is usually 500 000 metres for a 6° belt, or in the case of the United Kingdom, 400 000 metres. Before the projection formulae can be used, the listed eastings must be reduced to the true eastings with respect to the central meridian.

The Universal Transverse Mercator projection (UTM)

The UTM projection has been adopted for mapping by many countries and agencies. Before giving a worked example of its use, we describe its basic characteristics.

Zones

Because the scale error becomes excessive, even with a central scale factor, the practical width of a transverse Mercator projection is limited to about 3° on either side of the central meridian. This limitation also permits relatively simple formulae to be used for the calculation of scale and angular distortions with ample accuracy for most work. To map areas of greater east-west coverage than 6°, two or more belts or zones, consisting of identical transverse Mercator projections, are used.

The Universal Transverse Mercator projection system consists of sixty belts each 6° wide, beginning at 180° west longitude with Zone 1 as its western edge, i.e. its central meridian is 177° west longitude. The zones are then numbered eastwards, e.g. Zone 31 has the Greenwich meridian as its western edge.

UTM projections

Each zone consists of an identical transverse Mercator projection whose characteristics are as follows:

(a) The north-south extent is from 80° N to 80° S latitude.

(b) In the northern hemisphere the origin is on the equator at a point 500 000 m to the west of the central meridian, i.e. the false easting (FE) = 500 000 m.

(c) In the southern hemisphere the origin is 10×10^6 m to the south of the equator and 500 000 m to the west of the central meridian.

(d) The central scale factor K_o = 0.9996 exactly.

(e) Various ellipsoids are recommended for different parts of the world, e.g. the Clarke 1880 ellipsoid is suggested for Africa.

(f) Projection tables, for all the recommended ellipsoids, are available for the conversion of geographical coordinates to grid coordinates and vice versa.

(g) Tables are also available for zone-to-zone transformations.

(h) Tables are available giving the grid coordinates to 0.1 m of the intersections of the graticule at intervals of 5 minutes of arc, and at intervals of 7.5 minutes of arc. These are of great use to the cartographer since the edges of various map sheets may be plotted directly from the tables.

Projection tables
The closed formulae for the spherical model, which become more complex when the ellipsoid is treated rigorously, may be recast as power series of the arguments, latitude and longitude for the forward case, and northing and easting for the reverse case..
  The coefficients of these series are available in projection tables or as part of computer software. Full formulae are given in Snyder's admirable textbook, see reference (2).
  To enable the reader to follow the general method, a limited entry from projection tables is given here. Most tables are of the following form:

(a) The argument for the direct case is $p = \Delta\lambda\, 10^{-4}$. Where $\Delta\lambda$ is the true longitude difference from the central meridian.

(b) The argument for the reverse case is $q =$ true easting x 10^{-6}.

(c) The conversion from (ϕ, λ) to (E, N) and calculation of convergence c, are carried out by the polynomials :

Northings: $N = (I) + (II)\, p^2 + (III)\, p^4 + A_6$

Eastings: $E = (IV)\, p + (V)\, p^3 + B_5$

Convergence: $c = (XII)\, p + (XIII)\, p^3 + C_5$

(d) The reverse transformations are expressed as polynomials in terms of q and other coefficients.

Example

Consider the point A of Table 4.1 relative to a central meridian at 33°
east. The parameter p = 0.82967559. Coefficients for the latitude of A
are

(I) = 1 754 951.147   (II) = 1971.918   (III) = 1.778 A_6 = 0.000
(IV) = 297 389.355   (V) = 99.749 B_5 = 0.000
(XII) = 2735.391   (XIII) = 2.020 C_5 = 0.000

Which give the values for A

N = 1 756 309.372   E = 246 793.673   c = 2270.641" = 37' 50.64"

The northings and eastings are converted to the false origin to give
finally

N = 8 243 690.628   E = 746 793.673

The reader may care to verify the order of magnitude of these results
by computation using the spherical formulae (2), (7) and (3).

The various factors given in this book are for the UTM system
referred to the Clarke 1880 ellipsoid obtained from the tables of the
United States Army Map Service.

When a survey crosses a belt junction its coordinates are referred to
two new central meridians. It is possible to avoid going through the
geographical coordinates by direct zone-to-zone transformation for
which tables also exist.

These tables and formulae have now been cast into computer
programs by most organizations.

4.7 The computation of a survey on a map projection

As discussed above in section 4.2, geographical coordinates, latitude
and longitude, may be computed on the surface of a reference
ellipsoid. These geographical coordinates can then be projected on to
a plane surface by the formulae for a particular projection system.
Thus a set of plane rectangular coordinates representing the points of
a survey is obtained, which will be used as the basis for mapping, for
the calculation of areas etc.

As an alternative to computation on the surface of the ellipsoid, it is
possible to compute directly in terms of plane trigonometry, provided
the ellipsoidal lengths and the measured angles are distorted to suit
the particular projection used. This method of computation is still of
some relevance for surveys of limited areas. Computer software
written for plane coordinate computations by least squares is easily

adapted for use on the projection by suitably altering the absolute terms l of the matrix of observation equations.

The computation of a triangle on the UTM is considered in detail to show the general principles of projection computation.

Computation on the UTM projection

The data for triangle ABC used in the ellipsoidal computation will be computed on the Universal Transverse Mercator system Zone 36, central meridian 33 east longitude, origin at 10×10^6 m south, 500 000 m west. The ellipsoidal angles of the triangle are denoted by A', B', and C'; their corresponding projection or grid counterparts are denoted by A, B, and C. The ellipsoidal lengths are denoted by a', b', c'; the corresponding grid lengths are denoted by a, b, and c. Let the angular distortions be dA, dB, and dC and the side distortions be da, db, and dc such that;

$$A = A' + dA \quad B = B' + dB \quad C = C' + dC$$
$$a = a' + da \quad b = b' + db \quad c = c' + dc$$

If the spherical excess of triangle A'B'C' is ε, then since $A+B+C=180°$, $dA + dB + dC = -\varepsilon$. The grid angles and sides have been calculated from the grid coordinates of A, B, and C obtained from the geographical coordinates of A', B', and C' by the direct projection formulae (2) and (7). The corresponding ellipsoidal and grid data are given in Table 4.5.

Table 4.5

Angles	Ellipsoidal	Plane	Diff	"
A'	30° 48' 03".96	30° 48' 10".48	dA	+06.52
B'	61° 05' 49".07	61° 05' 57".90	dB	+08.72
C'	88° 05' 07".74	88° 05' 51".55	dC	−16.01
Sides	m	m		m
a'	13 327.628	13 333.512	da	+5.884
b'	22 785.523	22 794.865	db	+9.342
c'	26 013.282	26 023.251	dc	+9.969

Alternatively, from the measured ellipsoidal angles and sides, we could compute the grid coordinates entirely in plane trigonometry once we had the minimum of starting data on the projection, say one point A and a bearing to B. Often this information is given as part of the control from which a survey is begun. We need only compute the distortions given in the fourth column of Table 4.5, apply them to the observed data, and proceed with the ordinary formulae thereafter.

There is the difficulty that the formulae for the angular and scale distortions involve the known coordinates of the starting points and also those of the points required. Thus a process of successive approximation is required, though only one approximation is necessary in most work. The method is first to compute approximate coordinates, by plane trigonometry, to a metre precision using the observed data, or a mixture of observed and grid data, and then to repeat the process with ellipsoidal lengths altered to grid lengths, and ellipsoidal angles to grid angles. Whichever combination of fixation methods is used, a unique result is obtained. For example, an intersected point computed from grid angles will give the same result as the same fixation computed by a linear grid distance and a grid bearing using the same original data. The working formulae are for

(a) the line scale factor; and
(b) the arc-to-chord, or t–T, correction.

Line scale factors
Assuming a required accuracy of 0.01 m and distances greater than 1 km, the line scale factor is obtained from the point scale factors, evaluated at its terminals and mid-point, by Simpson's rule. Consider the line AB whose mid-eastings and northings are E_M and N_M, and whose point scale factors are K_A, K_M, K_B. The scale factor for AB, i.e. K_{AB}, is given by Simpson's rule.

$$K_{AB} = \frac{1}{6}(K_A + 4\ K_M + K_B)$$

Example: If the scale factors for AB at A, its mid-point, and B are found to be respectively

$$1.000\ 353\ 4, \quad 1.000\ 383\ 6, \text{ and } 1.000\ 414\ 5$$

then

$$K_{AB} = 1.000\ 383\ 7.$$

Since the ellipsoidal length of A'B' is 26 013.282, this gives the grid length

$$AB = 26\ 023.263$$

which agrees with that obtained from coordinates to the precision expected.

An alternative formula for the line scale factor K is

$$K_{AB} = 1 + \frac{E_A^2 + E_A E_B + E_B^2}{6 R^2} \qquad (8)$$

Arc-to-chord correction or t–T correction

To convert the ellipsoidal angle A' to a plane projection angle A we proceed as follows. Refer to Figure 4.21.

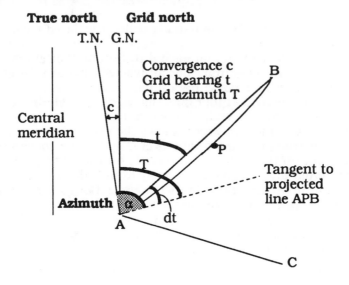

Figure 4.21

The grid angle A is derived from the grid bearings of AC and AB denoted by t_{AC} and t_{AB} respectively, i.e.

$$A = t_{AC} - t_{AB}$$

Let T_{AB} and T_{AC} be the grid *azimuths* of AB and AC respectively. The grid azimuth or spherical bearing is the azimuth α reduced by the convergence C. That is

$$T = \alpha - C$$

Because the projection is orthomorphic, the angle at A between the projected ellipsoidal arcs, or (which is the same thing) the angle between the projected tangents to these arcs, equals the ellipsoidal angle A', i.e. $A' = T_{AC} - T_{AB}$.

Let the differences between the grid azimuths and grid bearings be denoted by dt, then

$$dt_{AC} = t_{AC} - T_{AC}$$

$$dt_{AB} = t_{AB} - T_{AB}$$

Subtracting gives

$$dt_{AC} - dt_{AB} = t_{AC} - t_{AB} - (T_{AC} - T_{AB})$$

$$= A - A'$$

Then

$$A = A' + dt_{AC} - dt_{AB}$$

The distortion factor $dt = t - T$ is usually called the "$t - T$" correction, or the "arc-to-chord correction".

In practice the values of $t - T$ for each direction are found first by the formula given below. They may be subtracted in pairs to give the distortion to each angle, if computation uses the angles.

The sign of the correction to a direction is obtainable from a strict application of the formulae with due regard to signs; but in practice it is quicker to draw a diagram which shows the signs at once. In Figure 4.21, A and B are the projected positions of A' and B', c = convergence at A. If P' is a point on the ellipsoidal line A'B', on the projection it plots as P on the dotted line which curves away from the line of least scale, i.e. away from the central meridian of the projection. Thus the sign of $t-T_{AB}$ is at once evident from the diagram.

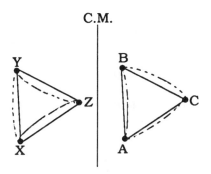

C.M.

Figure 4.22

Figure 4.22 shows the projected ellipsoidal arcs for the triangle ABC lying to the east of the central meridian, and the projected ellipsoidal arcs for the triangle XYZ lying to the west of the central meridian. Remembering that azimuths and bearings are reckoned clockwise from north, even in the southern hemisphere,

Grid bearing = Grid azimuth + (t -T) with due regard to signs.

Thus the (t-T) $_{ZY}$ is negative and (t-T) $_{YZ}$ is positive. The working formula for the arc-to-chord correction for the bearing AB is

$$(t - T)''_{AB} = -\frac{(2\,E_A + E_B)\,(N_B - N_A)}{6\,\rho\,v\,K_o^2\,\sin 1\,''} \tag{9}$$

and that for the bearing BA is

$$(t - T)''_{BA} = -\frac{(2\,E_B + E_A)\,(N_A - N_B)}{6\,\rho\,v\,K_o^2\,\sin 1\,''}$$

It will be evident that the (t–T)'s at either end of the same line are not identically equal in magnitude, though they will be of the same order. The size of the correction varies with easting, i.e. increases away from the central meridian, and it also increases with the difference in northing. A line running east-west will have no correction.

Complete formula for (t–T)
The approximate formula is the first and by far the largest term of a series from which the theoretically exact value of the t–T correction is derived. In practice the short formula will normally be quite sufficient, though in geodetic work involving very long lines the full formula should be used. For a discussion of this question see reference (1).

Computation of (t–T) corrections
Since the denominator of the formula changes very slowly with latitude it may be tabulated for particular ellipsoids and central scale factors, or programmed for computer.

Because the algebraic sum of the t–T corrections for the angles of a closed figure must equal the spherical excess of that figure, the spherical excess should always be calculated to check the arithmetic.

Example of solution of triangle ABC
Given the projection coordinates of A and B and the adjusted angles of triangle ABC, the coordinates of C will be derived using the cotangent formula for angles

$$E_P = \frac{E_A \cot B + E_B \cot B - N_A + N_B}{\cot A + \cot B} \tag{10}$$

$$N_P = \frac{N_A \cot B + N_B \cot B + E_A - E_B}{\cot A + \cot B} \tag{11}$$

Table 4.6 gives the complete computation. Provisional coordinates of C, E'_C and N'_C are obtained from the ellipsoidal angles, and the coordinates of A and B, using equations (10) and (11).

From these provisional coordinates and the coordinates of A and B, the six t–T corrections for the directions are derived, remembering to subtract the false easting of 500 000 m.

The factors tabulated in column three are

$$F = \frac{2\,E_1 + E_2}{6\,\rho v\,K_0^2 \sin 1''}\ 10^3$$

Hence the t–T correction is given by

$$-F\,(N_2 - N_2)\ 10^{-3}$$

The reader can easily verify the magnitude of F using R = 6 378 000 m and a central scale factor of 0.9996, and that the t–T is 8.81" by this means, and the spherical excess from

$$\varepsilon = \tfrac{1}{2}\,a\,b\,\sin C\ \sin 1''$$

The correct radii of curvature are used in the example.

Table 4.6

	True E	True N 8 000 000 +	Ellipsoidal angle	t – T	Grid angle
A	246 793.673	243 690.628	30° 48' 03".96	+ 6".52	10".48
B	256 612.703	267 790.343	61° 05' 49".07	+ 8".72	57".79
C'	264 990.5	257 418.5	88° 05' 67".74	–16".01	51".53
C	264 991.40	257 418.28	180° 00' 00".77	–00".77	00".00

Line	$2\,E_1 + E_2$	Factor	$N_2 - N_1$	t – T Direction	t – T Angle
AC	758 578	0.64558	– 13 727	– 8"86	+ 6".52
AB	750 201	0.63846	– 24 099	– 15".38	
BA	760 020	0.64681	+ 24 099	+ 15".59	+ 8".72
BC	778 216	0.66228	+ 10 372	+ 6".87	
CB	786 593	0.66940	– 10 372	– 6".94	– 16".01
CA	776 774	0.66105	+ 13 727	+ 9".07	
			Sph excess	– 0".77	– 0".77

The t–T correction for each angle is obtained by subtraction of corrections for pairs of directions. The spherical excess of the triangle is found to check the sum of the t–T's. The grid angles are then derived by application of the t–T corrections to the adjusted angles. The final coordinates of C are computed using the cotangents of the grid angles. In least squares and semi–graphic methods of fixation, the t–T corrections are incorporated into the O–C terms at the stage at which the coordinates are known to within a metre of their correct values.

Other fixations on the projection
This procedure of converting observed angles and measured sides into their grid counterparts is used in all methods of fixation, i.e. in traverse, trilateration, etc. The process is similar to that given for the triangle, namely, scale factors and/or t–T corrections are derived from provisional coordinates. The plane trigonometrical formulae are then applied to grid lengths and grid bearings or grid angles as the case may be.

Consistency between ellipsoidal and grid computations
The triangle A'B'C' on the ellipsoid of Table 4.2 and its grid counterpart in Table 4.6 agree within the precision of computation expected. For example the known azimuth of BA should agree with its value obtained from the azimuth of BA by calculation through the projection. The data for the triangle is given in Table 4.7.

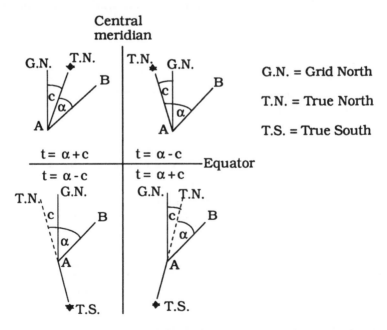

Figure 4.23

Figure 4.23 shows the signs of the convergence c in the four differing positions with respect to the central meridian and the equator.

As has been mentioned before, modern computational strategy is to work entirely in geographical or three dimensional cartesian terms usually by variation of coordinates and least squares estimation. Points are then transformed to a projection for mapping. Computation on the projection is now applied to surveys of limited scope only.

Table 4.7

Line	Azimuth	Convergence	t – T	Bearing
AB	21° 32' 28".16	37' 50".64	– 15".38	22° 10' 03".42
BA	201° 31' 01".02	38' 46".83	+ 15".59	202° 10' 03".44
BC	140° 25' 11".95	38' 46".83	+ 6".87	141° 04' 05".65
CB	320° 23' 54".74	40' 17".84	– 6".94	321° 04' 05".64
AC	52° 20' 32".11	37' 50".64	– 8"86	52° 58' 13".89
CA	232° 17' 47".00	40' 17".84	+ 9".07	232° 58' 13".91

Proof of formula for (t – T) correction

The following simplified derivation of the arc-to-chord correction on the transverse Mercator projection assumes a spherical model Earth. The reader who prefers a more rigorous proof should refer to the first edition of this book. Refer to Figure 4.24.

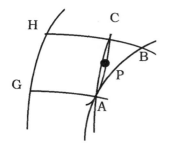
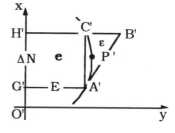

Figure 4.24

Figure ABCHG on the sphere is projected on the map as A'B'C'H'G'. Point C lies due north of A. The arc APC is a great circle whereas AC is a small circle. The spherical excess of Figure APCHG is e, given by

$$e = \text{area}/R^2 = \Delta N. E / R^2$$

By inspection and symmetry the small angle CAP = C'A'P' = e/2.

If the spherical excess of triangle ABC is ε then

$$\varepsilon = \Delta N . \Delta E / 2 R^2$$

In the spherical triangle we may use Legendre's theorem to compute sides, and therefore deduct $\varepsilon/3$ from angle A for calculations. Since the projection is conformal, the angle A = A' of the map. Hence the bearing of A'B' will be

$$A - \frac{e}{2} - \frac{\varepsilon}{3}$$

which on substitution for e and ε gives the required bearing as

$$A - \Delta N (2 E_A + E_A) / 6R^2$$

or

$$t - T "_{AB} = - \Delta N (2 E_A + E_A) / 6R^2 \sin 1"$$

If a central scale factor of K_o has been introduced to the projection, this amounts to a reduction of the radii of the ellipsoid model, and we have the final result

$$t - T "_{AB} = - \Delta N (2 E_A + E_A) / 6 \rho v K_o^2 \sin 1"$$

4.8 The Lambert conical orthomorphic projection

The Lambert conical orthomorphic projection is much used to map areas which have a great extent east-west and a small extent (about 6°) from north to south. For example, about one half of the states of the USA are mapped on this projection. It is easier to construct than the transverse Mercator projection, since its graticule is formed by meridians which project as straight lines, and parallels which plot as circular arcs. Interpolation of coordinates between tabulated portions of the graticule is relatively uncomplicated. One drawback is that the convergence becomes excessive away from the centre of the projection.

Simple conical projection
The Lambert projection is formed by adapting a simple conical projection, in a similar way to that in which the transverse Mercator projection is adapted from the Cassini projection.

Simple conical coordinates are produced by developing the surface of a cone which touches the sphere along a parallel of latitude passing through the middle of the area to be mapped. This parallel is called the *standard parallel* defined by the standard latitude ϕ_o. The apex of the cone lies on the prolongation of the axis of the sphere. In Figure 4.25 the cone is shown touching the sphere along the standard parallel.

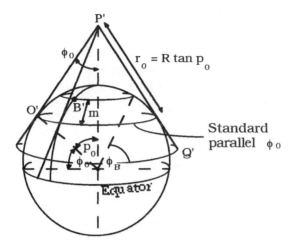

Figure 4.25

A point at the centre of the east-west dimension of the area to be mapped and lying on the standard parallel is chosen to be the origin of the coordinates. The meridian through this origin O is the central meridian. As will be seen later, the character of the projection depends much more on the standard parallel than on the central meridian. The latitudes of points to be mapped are reckoned from the standard parallel, and the longitudes from the central meridian.

If the radius of the sphere is R, the slant length of the cone from the standard parallel, P'O', is given by

$$P'O' = R \cot \phi_0 = R \tan p_0$$

where $p_0 = 90 - \phi_0$ is the co-latitude of the standard parallel.

The cone is developed by cutting it along a line such as P'O' producing the shape shown in Figure 4.26. The standard parallel on the map is plotted as a circular arc of radius r_0 = P'O' = PO centred on the projected apex of the cone, i.e. at P. The length of this arc equals the length of the standard parallel on the sphere. The length of the standard parallel on the sphere = $2 \pi R \cos \phi_0$, hence

$$\text{arc } Q_1 O Q_2 = 2 \pi R \cos \phi_0$$

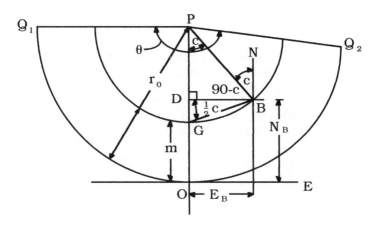

Figure 4.26

The angle $Q_1 P Q_2 = \theta$ is given by

$$\theta = \frac{2 \pi R \cos \phi_0}{R \tan \phi_0} = 2 \pi R \sin \phi_0 = 2 \pi n$$

Hence 2π radians of longitude on the sphere are represented by $2 \pi n$ radians on the map; in general, any longitude difference $\Delta\lambda$ on the sphere will be represented by an angle $n \Delta\lambda$ where n, called the *constant of the cone*, defines the spacing of the meridians on the map.

The other parallels are represented by circular arcs centred on P such that their spacing equals the meridian arc distance separating these parallels on the sphere. Consider a point B' on the sphere at latitude ϕ_B. The spherical arc distance $m = R (\phi_B - \phi_0)$ separates the projected parallel of B from the standard parallel. The radius of the projected parallel of B is given by

$$r_B = r_0 - m$$

The projected point B lies on a projected meridian which makes an angle c with the central meridian such that

$$c = n (\lambda_B - \lambda_0)$$

where c is the *map convergence* at B. The map coordinates of B are given from Figure 4.26 as follows.

$$E_B = DB - r_B \sin c = (r_0 - m) \sin c \qquad (12)$$

The arc-to-chord correction on the Lambert projection

By analogy with the transverse Mercator projection, the first term of
the formula for the t–T correction is

$$t - T''_{AB} = - \Delta E \, (2 N_A + N_A) / 6 \rho v^2 K_0^2 \, \sin 1''$$

Northings are interchanged with eastings by comparison with the
transverse Mercator projection.

4.9 Oblique or skew projections

In Figure 4.16 a projection system was derived from the point P and
the great circle PHGO. In the cases of the Cassini and transverse
Mercator projections, P is the Earth's geographical pole, and PHGO is
a meridian.

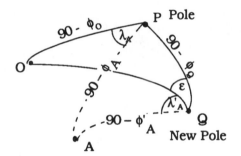

Figure 4.27

However, the same formulae can be used to obtain similar projection
systems, if P is any point on the surface of the sphere, and PHGO is
any great circle; provided the coordinates analogous to 'latitude' and
'longitude' with reference to this 'pole' and this 'meridian' can be
calculated.

In Figure 4.27 Q is the new 'pole' and the great circle QO is the new
'central meridian' of a projection whose origin is at O. A is any point
to be projected. The geographical latitudes and longitudes (ϕ, λ) of Q, O,
and A respectively can be used to derive new 'latitudes' and 'longitudes'
(ϕ', λ'), with respect to the pole at Q, from spherical triangle OPQ.

Also the new latitiudes and longitudes can be found on solution of
the spherical triangle PQA. Thence a Cassini type or transverse
Mercator type projection can be computed, based on pole Q and
meridian QO and new coordinates of points such as A.

The general case is an *oblique* or *skew* form of projection. The skew projection is suited to a country which is orientated at some azimuth intermediate between 0° and 90°. For example Malaysia and Borneo use oblique Mercator projections.

Although the projection character and formulae are based on a skew 'meridian', the coordinates obtained are finally rotated on to a north–south grid axis to accord with convention and reduce the convergence between the final grid and graticule.

The name transverse Mercator derives from the fact that this projection is the transverse case of the Mercator projection in the sense described here. The Mercator projection is based on the equator as the 'central meridian' and was developed historically before the transverse case and before the Gauss-Kruger projection as the latter is also called. Alternatively the Mercator projection could be called the 'transverse Gauss-Kruger' projection.

All map projections can be considered as conical projections. For example, if the standard parallel of the Lambert projection is the equator, Mercator's projection is produced, and thus the Gauss-Kruger projection is a special transverse case of Lambert's conical orthomorphic projection. If the standard parallel is at latitude 90°i.e. at the pole, an azimuthal projection is obtained. A full treatment of map projection is beyond the scope of this book. For more information the reader should refer to Schnyder or Maling, see references (2) and (3).

5 Theory of errors and quality control

Chapter Summary

In this chapter we discuss how to assess the quality of measurements and results derived from them.This involves an understanding of the theory of errors of measurement, statistical methods of selecting the best estimates of derived parameters, such as coordinates, ways in which to combine extra observations of various kinds into a consistent set of data, and whether or not to accept or reject values on the basis of statistical hypotheses.

Much information is now readily demanded by clients, as a means of assessing the quality of the product presented to them by the surveyor. *Quality control* forms part of a wider management system called *quality assurance.*

The comparative simplification of modern computer systems and software, makes the calculation of such sophisticated quality data a normal course of events, and what was formerly a somewhat esoteric branch of geodetic theory is now commonplace. It is therefore vital that the land surveyor is fully conversant with the fundamentals of error theory and its applications.

At first we establish mathematical ways to describe measurements and results derived from them; then we examine some basic statistical procedures used to handle data in a consistent manner, and finally we look closely at the major methods of reconciling mixed and inconsistent observational data into consistent figures for use in dimension control and for other purposes.

In our explanations we assume that the reader is familiar with the methods and purposes of surveying such as are described elsewhere in this book.

5.1 Introduction to errors of measurement

At school we were once asked to draw a straight line 300 mm long and then to measure it, using a 100 mm rule for both processes; and subsequently we were not a little surprised to discover that the exercise was not as ridiculous as it appeared at first, and that it embodied many basic concepts to do with the errors of measurement.

Some of the discoveries made were

(a) Most lines drawn by pupils were of different lengths, as could easily seen by direct comparison without the use of any rulers. This direct comparison is an example of an independent checking procedure, something always to be sought after in surveying, which gives a measure of *reliability*.

(b) When we measured with a ruler graduated in centimetres only, the same result was obtained each time, but when we used a ruler with small precise millimetre divisions, the results varied within small limits. This illustrates the need for sufficient *precision* in measurement.

(c) Different precise answers were obtained with different rulers, illustrating the need to calibrate the rulers by comparison with a known standard length. The calibration process reduces the *systematic* error of the rulers to within known limits.

(d) One pupil had miscounted, and gave the length as 400 mm. Mistakes are the most serious of all the errors.

(e) In the end we were uncertain as to the exact lengths of all the lines, which were each supposed to be 300 mm. This is an example of uncertainty in meeting the defined specification of 300 mm. No measurement *tolerance* was asked for nor was there a *statistical criterion* for this tolerance. Arguments about specifications are often the subject of legal disputes and often arise because these matters were not defined beforehand.

Thus an apparently simple exercise was full of problems; problems which arise in all attempts to set out measurements according to a specification, and to measure the exact value of a quantity. In addition it raised the issue of what is meant by 'the exact value'.

Since land and geodetic surveying is almost wholly concerned with measurements and setting out to specifications, these problems are of paramount importance; indeed in many instances it is almost as important to know the accuracy of a result as to know the result itself. Clearly there is a need to define the meaning of 'quality 'in unambiguous ways.

5.2 Errors in derived quantities

The real benefits from error analysis arise in the assessment of the likely quality of results derived from measurements whose errors are known or which are postulated in design problems.

Consider Figure 5.1 which shows a point P (x,y). The abscissa x and the ordinate y are measured and plotted at right angles. For the moment we assume that a perfect right angle is possible. Suppose that these measurements are in error by small amounts δx and δy respectively. If we plot the new position of P as P' we see that there are

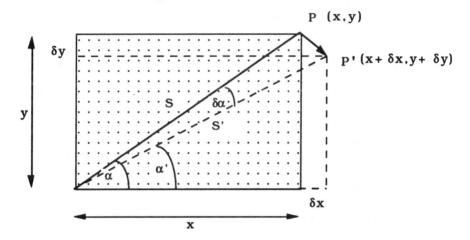

Figure 5.1

corresponding changes to the angle α, the distance s = OP and the area of the rectangle A= xy. Hence if we know the values of the errors δx and δy we can calculate their effects on the derived quantities α, s and A. Let these effects be δα, δs and δA.

A simple way of finding these changes is to recalculate them using new values derived from the originals altered by the known changes. Although inelegant, this procedure is not to be depised because it can give quite good quick rough estimates. Consider the following example.

> Let x = 300 mm, y = 200 mm, δx = 0.1 mm and δy = −0.1 mm.
> Thus s = 360.555 mm, α = 33° 41 ′ 24", and A = 6 x 10⁴ mm²
>
> Recalculation with x = 300.1 mm and y = 199.9 mm gives
> s = 360.583 mm, α' = 33° 40 ′ 05", and A' = 59 989.99 mm².
>
> Thus by subtraction we have the changes
> δs = 0.028 mm, δα = −01 ′ 19 ", and δA = −10.01 mm².

Fine as this may be, the procedure does not lend itself to treatment of random errors in very complex problems, nor the statistical assessment of results. To do this we need a different approach using the differential calculus and matrix algebra, which puts this simple procedure on a much more flexible footing.

The three formulae used in these calculations are

$$s^2 - x^2 - y^2 = 0 \quad \text{or} \quad F_1(s : x, y) = 0$$
$$\tan \alpha - y/x = 0 \quad \text{or} \quad F_2(\alpha : x, y) = 0$$
$$A - xy = 0 \quad \text{or} \quad F_3(A : x, y) = 0$$

A more general statement is to relate any derived parameter p to the measured variables x and y by the expression $F(p : x, y) = 0$.

We find it is helpful to separate the parameters from the observed quantities by a colon, to retain the full notation of the original problem, to carry out its partial differentiation, and only finally to convert the expression to the general form suited to matrices. For example the partial differentiation of this general expression is written

$$(\partial F/\partial p)\, \delta p + (\partial F/\partial x)\, \delta x + (\partial F/\partial y)\, \delta y = 0$$

which when applied to these three explicit functions gives

$$+2s\, \delta s \quad - \quad 2x\, \delta x \quad - \quad 2y\, \delta y \quad = 0$$
$$\sec^2 \alpha\, \delta \alpha \quad + \quad (y/x^2)\, \delta x \quad - \quad (1/x)\, \delta y \quad = 0$$
$$1.\delta A \quad - \quad y\, \delta x \quad - \quad x\, \delta y \quad = 0$$

Putting x = 300 mm and y = 200 mm, and ignoring the fact that the coefficients can be simplified, the equations become

$$721\, \delta s \quad - \quad 600\, \delta x \quad - \quad 400\, \delta y \quad = 0$$
$$1.444\, \delta \alpha \quad + \quad 0.00222\, \delta x \quad - \quad 0.00333\, \delta y \quad = 0$$
$$\delta A \quad - \quad 200\, \delta x \quad - \quad 300\, \delta y \quad = 0$$

From these equations, on putting $\delta x = 0.1$ mm and $\delta y = -0.1$ mm we obtain the results $\delta s = -0.028$ mm, $\delta \alpha = 3.8435 \times 10^{-4}$ radians $= 01'\ 19''''$, and $\delta A = -10.00$ mm^2.

The very slight disagreement in δA is due to the approximation in taking first differentials only.

It will be seen that the above equations may be assembled together in matrix form as follows

$$\begin{pmatrix} 721\, \delta s \\ 1.444\, \delta \alpha \\ 1\, \delta A \end{pmatrix} + \begin{pmatrix} -600 & -400 \\ +0.002\ 22 & -0.003\ 33 \\ -200 & -300 \end{pmatrix} \begin{pmatrix} \delta x \\ \delta y \end{pmatrix} = \begin{pmatrix} 0 \\ 0 \\ 0 \end{pmatrix}$$

or

$$\begin{pmatrix} 721 & 0 & 0 \\ 0 & 1.444 & 0 \\ 0 & 0 & 1 \end{pmatrix} \begin{pmatrix} ds \\ da \\ dA \end{pmatrix} + \begin{pmatrix} -600 & -400 \\ +0.002\,22 & -0.003\,33 \\ -200 & -300 \end{pmatrix} \begin{pmatrix} \delta x \\ \delta y \end{pmatrix} = \begin{pmatrix} 0 \\ 0 \\ 0 \end{pmatrix}$$

Which is of the form

$$\mathbf{A}\,\mathbf{x} + \mathbf{C}\,\mathbf{s} = 0$$

It should be remembered that the coefficients of the variables are the partial differentials of the various functions, assembled above as a matrix, for example

$$\mathbf{C} = \begin{pmatrix} \partial F_1 / \partial x & \partial F_1 / \partial y \\ \partial F_2 / \partial x & \partial F_2 / \partial y \\ \partial F_3 / \partial x & \partial F_3 / \partial y \end{pmatrix}$$

It is defined as the *Jacobian matrix* of the vector $\mathbf{F} = (F_1, F_2, F_3)^T$ with respect to the vector $\mathbf{x} = (x,y)$.

Very often the coefficients in this matrix can be obtained by a semi–graphic method with little difficulty. This semi–graphic method always seems to give the coefficients in their simplest form, and is particularly helpful in the spherical triangle as in Chapter 6, and in the treatment of the geometry of the ellipsoid as in Chapter 4.

Reconsider the problem illustrated in Figure 5.1. The error figure for this is given in Figure 5. 2. We see at once the effects of the changes δx and δy in producing s $\delta\alpha$ and δs which can obtained by inspection as

$$\delta s = \cos \alpha\, \delta x - \sin \alpha\, \delta y$$
$$s\, \delta\alpha = \sin \alpha\, \delta x + \cos \alpha\, \delta y$$

These are the same first two equations as before, only reduced to their simplest forms. It will also be noticed that these equations can be derived by a rotation of the error coordinates (δx, δy) through the angle α to give the new coordinates (δs, s $\delta\alpha$). This illustrates the power of the semi–graphic approach to simple geometrical problems. Complex mathematical models are best treated by conventional differentiation.

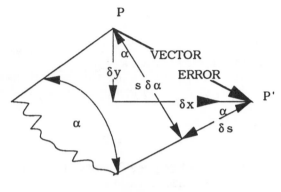

Figure 5.2

Summary to date
The above section has shown how errors in derived quantities can be found and expressed in terms of matrix algebra. The treatment assumed that errors of a known sign are present, as is often the case with systematic errors. As will be shown later in the chapter, errors with unpredictable signs and magnitudes can be treated in a similar manner.

5.3 Definition of concepts and terms

The following definitions and symbols are essential for an understanding of subsequent sections. Whilst, alas, there is no generally accepted form of notation for the various quantities used in error theory, we shall adopt forms which most nearly accord with the main texts and publications. But readers are warned of the great lack of consistency in the literature.

Although the following treatment may seem complicated to the beginner, we need to define mathematical and statistical formulae so that problems can be treated consistently and by computer. Most of the problems ultimately end up as the near intersection of two lines in space, or combinations of such intersections. It should never be forgotten that a good appreciation of graphic surveying allows the surveyor to apply common sense to check results.

True value T
Just as the concept of 'truth' is an abstract idea, so also in most cases is the concept of true value. In general, the true value of a quantity will never be found; or if it is found, we will never know that we have found it. For example we can only know the length of our ruler within specific limits set by calibration.This lack of knowledge of the truth applies to a single observed quantity only, for we often know the true value of a combination of quantities that are observed singly. For

example the sum of the three angles of a plane triangle is known to be 180°.

Population and sample means μ and x̄

By *population* we mean all the possible values that could exist of a particular kind, such as a measurement of the 300 mm line we described above, made by every 100 mm ruler in the world. This is a very large population indeed.

Our particular small set of measurements drawn from this population is called a *sample*. It is also a sample of another population made up of all the measurements that could be made of the line using one ruler alone.

The mean or average of all possible values is the population mean usually denoted by the Greek letter μ; the mean or average of the sample is denoted by x̄. Clearly these two means will not generally be equal. But it would be reasonable to think that, if we went on measuring more and more samples, the average of the sample averages would tend to the population mean.Thus we introduce the concept of mathematical expectation.

Expectation E

The *mathematical expectation* of a quantity is the value to which the average of such quantities will tend as the number of measured values increases to infinity. Thus the expectation of the sample mean x̄, written as E (x̄), is the population mean μ. If we were able to remove all systematic errors (see below) from the measurements, this *population mean* would be the *true value* (T). Some authors define the true value in this sense. We think that the assumption that all systematic errors have been removed is unlikely in practice, and prefer to keep the two concepts separate.

Observed value x̊

The observed value is the first numerical information you obtain about a result, for example a length from a tape or an angle from a theodolite. In fact these numerical values are very often derived within an instrument from a series of other measurements, such as time in an EDM instrument or satellite receiver, which the user generally does not see. Thus we must settle for the first set of numerical values we have as being the 'observations'.

The superscript ^{o} is used over a parameter, say x, to indicate that it is the observed value of x. For example we would write the observed values of a distance s and of an angle α as

$$\overset{o}{s} \text{ and } \overset{o}{\alpha}$$

True error Δ

Like true value, the true error Δ of a single observed quantity can never be found; hence it is merely an abstract idea defined to be the difference between the true value and the observed value, i.e.

$$\Delta = T - \overset{o}{x}$$

The best estimate $\hat{x}$

The best estimate derived from observations is that value which is more likely than any other to be the true value, judged on the evidence available. This was called the *most probable value* (m.p.v.) in older texts. We shall accept that the best estimate which can be derived from a series of observations of a quantity is the arithmetic mean of the set, provided that the observations are independent of each other and that they are of the same quality. Observations of differing quality will be treated later.

The superscript $^\wedge$ is used over a parameter x to indicate that it is the best estimate of x. For example we would write the best estimates of a distance s and of an angle α as

$$\hat{s} \text{ and } \hat{\alpha}$$

The use of the hat has widespread acceptance in literature to indicate that the best estimate is derived by invoking the principle of least squares or, which is the same thing, the principle of minimum variance (see below).

Residual v

A residual v is defined to be the difference between the best estimate and the observed value i.e.

$$\text{residual} = \text{best estimate} - \text{observed value}$$
$$v = \hat{x} - \overset{o}{x}$$

Note that in statistical literature deviations from the mean, or residuals, are defined in the opposite sense of

$$v = \overset{o}{x} - \hat{x}$$

This makes no difference to any results provided we are consistent. We use the convention common to survey.

Corrections

It should be noted that in deriving the best estimates from observed values, in no sense are these observations "corrected". The observations are simply used to derive best estimates. It is legitimate to use the word "correction" for the change made to an arbitrary provisional value to give the best estimate.

Weight w

The weight w of an observation is a measure of its quality relative to other observations, usually expressed as a number proportional to the inverse of the variance (see later). Weights are needed to reduce mixed mode types of observations e.g. angles and sides to a dimensionless number for combined work, and to allow for the differing quality of various instruments and observers (see 5.9 for further explanation).

Precision and accuracy

If a quantity is measured several times, the degree of agreement between the measures is the *precision* of the set. Thus if the residuals are small the observations are precise, and vice versa. The *accuracy* of the set is the difference between the best estimate and the true value. Hence in most cases the accuracy can never be found.

A high degree of precision is no indication of great accuracy. For example an expensive watch may be precise to a second, but could easily be two hours slow, i.e. inaccurate by two hours.

The efforts to obtain accuracy usually involve a high degree of precision but, more importantly, they require much effort to remove systematic errors by calibration, and mistakes by a quality control system. For example the watch should be compared with radio time signals at regular intervals, as for astronomical observations.

Systematic and random errors

In the classification of errors, two main types are distinguished according to the way they affect a result:

(a) Errors which have a cumulative or constant effect are *systematic errors*.

(b) Errors which have a tendency to compensate one another are *random* errors, sometimes called *accidental* errors.

In our example of length measurement, clearly if the ruler is either too long or too short, it will always give an incorrect answer. This information will have been found by *testing* it against a standard. The way to solve the problem is to *calibrate* the ruler, that is having compared it with a known standard of length, apply a calibration correction to all readings.

Even after this calibration has been applied, there will be small differences each time anyone uses the ruler, provided it is precisely graduated. If there are even chances that these small errors are either too large or too small their effects tend to compensate and are therefore called random errors. Notice that the calibration process itself is subject to these random errors, and the calibration figure which is applied is still burdened by random error.

It can be very dangerous to cite examples of these types of error out of context. For example, the statement that 'failure to record temperatures in taping is an example of systematic error', may either be correct or incorrect according to the circumstances of the measurement. If the standardization temperature of the tape is say 10°C, and if all the measurements are taken at a temperature of 30°C, and no corrections for temperature are applied, the errors introduced are systematic. If on the other hand, the field temperatures vary between 8° C and 12° C in a random manner, the errors introduced will tend to compensate each other. For this and other reasons the temperature at which a tape is standardized should be as close to the expected field temperatures as possible. This principle of changing a systematic into a random error is applied widely in surveying, where possible.

Mistakes or blunders
A third type of error is sometimes distinguished on the basis of the mere size. This is the *gross error*, or *mistake*, or *blunder*. Since mistakes may be systematic or random this is not a genuine third type of error. It is however true to say that they are treated in a different manner by most surveyors simply because they may be detected by a self-checking procedure built into the particular survey process involved.

Mistakes are the most serious of all the errors. This fact can easily be forgotten in the light of the attention paid to small errors by the method of least squares, and to the space taken up by the latter in survey literature, including in this book. The importance of a topic is not proportional to the number of words written on the subject nor to the effort required to understand it. Quality assurance techniques are therefore of paramount importance if reliable work is to be performed.

Mistakes are very serious simply because of their size, and care must be taken to avoid them. The following list of common mistakes should serve to indicate how serious they can be.

(a) Reading a micrometer or altimeter in the wrong direction.

(b) Misreading scale divisions.

(c) Transposition of numbers, e.g. writing 3457 for 3475.

(d) A mistake between observer and booker due to a faulty procedure, e.g. saying 'oh' which is mistaken for 'four', instead of calling 'zero'.

(e) Pressing the wrong trigonometrical function key on a hand calculator, e.g. tan instead of sin.

(f) Alteration of a bearing through 180°, because the wrong face is booked, and two zeros are not observed.

(g) Misnumbering a station in a field book so that the survey is swung through the angle between the station named and that actually observed.

(h) Using a wrong foot/metre conversion factor, e.g. using the factor 1 metre = 3.048 feet. (1 metre is 3.2808 ft but 1 foot is 0.3048 metres.) The result looks about correct in either case.

(i) Failing to look at a calculator display when keying in data.

With the advent of automated production line work, for example when observations made from total stations are recorded by data loggers which give direct output for processing, there is less likelihood of mistakes in booking, but at the same time, there are fewer opportunities to check work by common sense methods. It is therefore necessary to incorporate into computer software various statistical tests to locate blunders and to incorporate self checking observational procedures to locate any mishaps. This means that statistical theory has to be used more than ever it was.

5.4 Systematic errors and calibration

Systematic errors arise from some physical phenomenon or psychological tendency on the part of the observer, and may only be eradicated if the laws governing these contributory factors are known, albeit empirically.

A troublesome source of systematic error in levelling is atmospheric refraction, and an example of a personal tendency is personal visual bias when observing a vertical angle. Systematic error is often difficult to assess and equally troublesome to remove, although it is often practicable to reduce its effect considerably by calibration, or by balancing observations, or even to remove it completely by the application of the important principle of instrument reversal such as changing face in theodolite work.

In dealing with an overdetermined set of observations and parameters by the method of least squares, efforts can be made to locate systematic errors by incorporating a suitable mathematical model, such as a scale factor in a tape. In this case the systematic effect becomes another unknown parameter. Sometimes, but not always, this error can then be determined and eradicated.

Processes most likely to be affected significantly are those involving long repetitive techniques, such as levelling and traversing, where a chain of small errors can accumulate with significant effect.

In some cases, although the magnitude of a systematic error may be assessed, its sign may be unknown. For example, the standardization of a tape is itself subject to error. If the length of the tape is quoted as 100.025 m, all field measurements will be corrected by the 0.025/100 x length measured. The standardization itself could be in error by say ± 0.0005, i.e. either plus or minus. Whatever its sign it will affect the measurements in a systematic manner. Generally this sign is unknown. If ± 0.0005 is the *standard error* of the calibration measurements, it is a *standard systematic error* when affecting the field measurement. Of course the standard error of the calibration could be worse than 0.0005. This figure is taken only if no other information is available.

The propagation of systematic errors

As an example of the propagation of systematic errors, consider a base line of n tape lengths each with a systematic error e_1, e_2,.... e_n. Then the resultant systematic error of the whole base is given by

$$E_s = \Sigma\ e_i$$

If these values, e_1, e_2,.... e_n, are standard systematic errors, they are *either all positive or all negative*.

Hence E_s the standard systematic error of the base is either + E_s or $-E_s$. If the e_1..., e_n are all numerically equal to σ, then we have

$$E_s = n\,\sigma$$

It will be seen later that this contrasts with the propagation law for random errors which gives σ√n for n equal standard errors σ.

The proportional error for the base as a whole does not improve with its length but remains σ/L where L is the tape length. Again this contrasts with the effect of accidental error, because the proportional error of the base improves with distance measured.

5.5 Random or accidental errors

The theory of random errors is now considered at length. However, attention is redrawn to the importance of mistakes and systematic error, neither of which conform to exactly the same laws. They must be studied in each particular case, and the treatment of discrepancies carried out accordingly. In theory random errors remain after mistakes and systematic errors have been removed. In practice some systematic errors will remain and, for want of an alternative, they are treated along with random errors.

The concepts of error theory are really quite simple; it is the necessary notation and algebra used to develop ideas which are confusing to the beginner. Although it is quite possible to develop these ideas without employing matrix algebra (as for instance in the first edition of this book), it is very inconvenient not to use this very effective tool of mathematics. The reader is therefore encouraged to study the elements of matrix algebra before attempting to master the subject matter of this chapter.

However, we will develop many basic concepts through the medium of simple examples, both to assist the beginner and to define terms. A more direct treatment is given in Appendix 5.

5.6 Statistical analysis of results

The study of random errors is concerned with probability, with the presentation of results as histograms and probability density functions, with the calculation of sample statistics and population parameters, and with the detection of outliers which do not seem to accord with the theoretical distributions expected. Typical problems to be handled are:

(a) When should we reject what appears to be a bad observation or, more likely, what rejection criteria do we write into our computer software to do this?

(b) How can we tell if an instrument is performing to the manufacturer's specification?

(c) How can we assess the reliability and precision of work computed by the method of least squares, so that we may inform the client of this, and if necessary defend our results in a court of law?

Probability
Most of our findings are based on probabilities, and very few results are certain. Because we shall be dealing with levels of statistical

significance and probabilities which are themselves subject to error, it is essential to have a good grasp of some fundamental concepts. Thus we shall deal with the idea of probability in basic terms.

If there are two balls in a hat, one white and one red, the probability that we shall draw out a red ball, without looking into the bag of course, is one in two, or 1/2. This does not mean that we shall draw out a red ball in one of two chances. We might pick a white ball in the first six attempts. It means that if we make a very large number of attempts, say 1000, fifty per cent of the time we shall draw out a red ball, and fifty per cent a white ball. Notice that the probability of drawing any colour of ball is a certainty of one hundred per cent, and that the two probabilities of drawing either colour add up also to one hundred per cent.

Or again, if there are two white and ten red balls in the hat, the probability of drawing out a white ball is two in twelve or 2/12, and that of drawing out a red ball is ten in twelve, again understood that a very large number of chances have to be involved. Hence we obtain the rule that, the probability of a result,

$$P(x) = \frac{\text{the number of ways of obtaining a result}}{\text{the total number of possible results}}$$

In this way, the probability of the occurrence of an error is related to the number of times this particular error can occur, and the total possible number of all errors that could occur. This involves analysing and counting errors of various magnitudes. For such an analysis to be valid, a very large number of errors should be *sampled*. A sample of more than thirty observations can often be taken as large, when it comes to statistical testing.

However, it is seldom practicable to make a very large number of observations, nor is it necessary. Usually we make samples of ten or so unless the measurement process is highly automatic.

The theory below will therefore deal mainly with small *samples* and their *statistics,* the mean and the standard deviation, and estimates from these samples of the *population parameters* which they represent.

Histograms
The histogram is a diagrammatic way of displaying sets of data which have a statistical or *stochastic* distribution about a central value.

This method of analysing results will be explained in terms of the game of 'shove halfpenny'. This traditional game, once played in English country pubs with a small coin, the halfpenny, is easy to make up for oneself, and is a good way to obtain statistical data. Figure 5.3 shows a smooth board on which a number of parallel lines are ruled at a distance of just over an inch, i.e. slightly larger than the diameter of the coin. The player places his coin at a position just protruding over the edge of the board. The idea is to strike the coin with the wrist and get it to land as close to the middle as possible. The centre of the coin is taken as the reference point. If this lies within the range marked 0 to –1 the score is recorded for that group and so on.

The figure shows the results of two players A and B, each of whom took 157 shots. It is obvious that A was a much better player than B, because his observations are clustered closer to the mean, or we say they show less *dispersion* that B's.

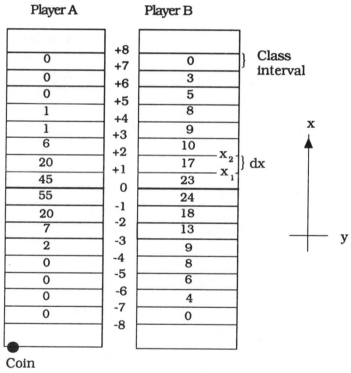

Figure 5.3

The bands between the lines are called the *class intervals* of the analysis. Their width has to be carefully chosen with respect to the skill of the players if a difference between them is to be detected. For instance, if the class interval had been the complete depth of the board no difference in the respective skills would have been found. Again, if the intervals had been very small, an abnormally large number of shots would have to be played to land more than one shot in each space. The selection of class intervals always needs care and experiment.

The number of times that a shot lands in each space, or class, is plotted on a block diagram called a *histogram* (see Figure 5.4). The histogram for A is bounded by the solid line and that for B by the broken line. The area of each histogram is the same since it represents the same number of shots. For a fair comparison to be made this should always be the case.

It is possible to analyse observational data just in the manner of the game. A decision has to be made about the size of the class interval,

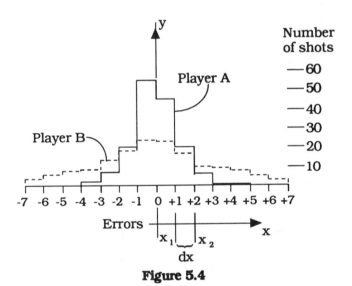

Figure 5.4

selecting about one fifth of the range of measurements for this. If the interval proves either too large ar too small, it has to be reselected.

For instance the data of Figure 5.3 could have been the triangle closures obtained by two separate observers A and B, the class intervals being in seconds of arc.

Mathematical treatment
The statistical information may also be expressed mathematically as follows:

(a) An error is denoted by x_i and the class interval by dx. Hence the $x_2 = x_1 + dx$, and so on.

(b) The height of the column of the histogram can be expressed as y_i to correspond with x_i.

Thus we convert the numbers in each class interval into probabilities by dividing each by the total number of observations made e.g. 55/157 = 0.35.

Notice that the probability of any result at all is 157/157=1, a certainty. The graph of all these probabilities is called a probability density function graph, which usually has a bell shape as shown in Figure 5.5. The equations of these bell shaped curves are considered later.

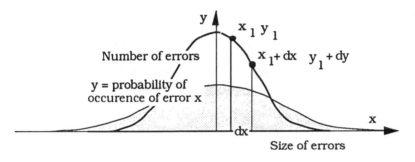

Figure 5.5

Random error and central tendency

Continuing with the shove halfpenny example, if the number of shots is increased to a very great number, i.e. towards infinity, while at the same time the class interval is decreased, we can think of a continuous curve towards which the histograms will tend.

The respective curves representing A's and B's performances are given in Figure 5.5. The area contained between each curve and the x-axis represents the total number of all possible errors in the system, an infinite number in theory. This is true for both curves, therefore neither curve will meet the x-axis but approach it asymptotically. For any point on the curve, the ordinate y can be considered as either the number of errors of abscissa x that occur, or the probability of the occurrence of an error of magnitude x. Hence the area under the curve represents either (a) the total number of errors n, or (b) the probability of all errors occurring, i.e. certainty, with a probability of 1, or 100%.

From the shape of the curve, of which shows *central tendency* about the mean, the distribution of random errors is described as follows:

(a) Small errors are more frequent than large ones.
(b) Positive and negative errors are equally likely to occur.
(c) Very large errors seldom occur.

These statements are reasonable if systematic errors have been removed and the observer, or player of the game, has a good degree of skill. In the halfpenny game an unskilled player would produce widely scattered results, as would an inexperienced observer of triangles.

It can be established theoretically by the *central limit theorem*, that the distribution of a combination of four or more error distributions approximates to the Normal. Since most survey "observations" are the result of at least four error sources, the Normal distribution is a reasonable statistical model to accept.

Sample deviation curve

More usually than in the case of triangle closures, where we are dealing with true errors, we obtain a set of observations distributed about a mean value. We can plot a histogram of deviations (residuals) about this mean value in just the same way, and also show their distribution on a smooth curve, which illustrates the degree of dispersion of the observations.

In this case the abscissae x_i become the residuals v_i, and the ordinates y_i the number of such residuals in each class.

Such a curve is defined by its mean $\bar{x}$ and its general shape, which shows the *dispersion* about the mean. A good indicator of this dispersion is the *variance* s^2. This is defined to be the average of the sum of squares of the residuals about the mean. For n observations we have

$$s^2 = \frac{\Sigma v_i^2}{n} = \frac{v_1^2 + v_2^2 + v_3^2 + v_4^2 + \ldots + v_n^2}{n}$$

or, in matrix terminology,

$$s^2 = \frac{\mathbf{v}^T \mathbf{v}}{n} \quad \text{where}$$

$$\mathbf{v}^T \mathbf{v} = \begin{bmatrix} v_1 & v_2 & v_3 \cdots v_n \end{bmatrix} \begin{bmatrix} v_1 \\ v_2 \\ v_3 \\ \cdot \\ \cdot \\ \cdot \\ v_n \end{bmatrix}$$

It will be obvious that the column vector $\mathbf{v}$ in matrix form takes up too much space on the page. For this reason it is usually written as the transpose of a row, that is as

$$(v_1 \ v_2 \ v_3 .. \ v_n)^T$$

We must stress that the definition of variance implies nothing at all about the nature of the distribution of the observations, and none is inferred. Variance is a convenient statistic which gives a measure of the spread or dispersion of results about a mean value. Provided there is more than one observation, a variance is calculable.

Practical example of variance
Before continuing further with theory, and to show how easy these calculations are in practice, we will work through a typical example.

An angle was observed nine times obtaining the results tabulated in column one of Table 5.1. The average

$$\bar{x} = \frac{\Sigma x_i}{n} = 109^\circ \, 25' \, 07.7''$$

Subtracting each observed value $\overset{o}{x}$ from $\bar{x}$ gives the positive and negative residuals v_i in columns two and three of the table. The sum of

Table 5.1

Obs	+ v	− v	v²
06.3	1.4		1.96
07.2	0.5		0.25
10.4		2.7	7.29
04.3	3.4		11.56
09.6		1.9	3.61
05.8	1.9		3.61
07.9		0.2	0.04
08.3		0.6	0.36
09.1		1.4	1.96
sum 68.9	sum + 7.2	sum −6.8	sum = 30.64
mean = 07.7	rem 0.4	checks	Var = 30.64/9
			s^2 = 3.40
			s = 1.8

the residuals should be zero, or nearly so if their has been some rounding error. The sum of squares of these residuals calculated from column four is found to be 30.64.

The variance s^2, defined to be the average of the sum of squares of these residuals, is therefore given by

$$s^2 = \frac{\Sigma v_i^2}{n} = \frac{30.64}{9} = 3.40.$$

And the *standard deviation*

$$s = \sqrt{3.40} = 1.8".$$

This *sample statistic* can be easily calculated by a direct function key in a hand calculator. However it is wise to check which formula is being used by the calculator. As we shall see later, another similar formula, with a denominator of n–1, is used to calculate an unbiased estimate of the population variance.

We shall return to this example later and recalculate the result a different way to illustrate the general method of observation equations used in the least squares process.

5.7 Statistical models

We now give a formal treatment which will incorporate all the above ideas and summarize them in mathematical form. We begin with the Normal or Gaussian distribution which is most generally applied to randomly observed data. This distribution, which applies to a population, will be treated in rather more detail than the others which follow, and which apply to samples: namely the Student's t, the Fisher ratio, and the Chi squared distributions.

The normal distribution function

Although many of our results and techniques, such as calculation of variance, have nothing to do with the normal distribution, this distribution is widely used, and rightly so, to represent the kind of error dispersion met with in practice. This is because most errors result from the combination of others which are centrally distributed. It can be shown by the Central Limit theorem that the combination of only a few such errors leads to a set of normally distributed errors. The bell shaped curve which represents the frequency of residuals about a mean derived from randomly observed variables may be expressed by the *probability density function* or PDF given by the formula

$$\phi(x) = y = \frac{h}{\sqrt{\pi}} e^{-h^2 x^2} \tag{1}$$

To accord with previous definitions and notation, the variable x in this formula is equivalent to the residual v defined by

$$x = v = \mu - \overset{\circ}{\overline{x}}$$

Such statistical, or stochastic, models are derived from probability theory in advanced statistical textbooks such as Aitken's *Advanced Theory of Statistics* to which the reader may refer. See reference (16). However, to give understanding of this most important distribution, we will trace this function and show that a bell shaped curve results.

$$\text{When } x = 0 \quad y = \frac{h}{\sqrt{\pi}}$$

giving the maximum height of the curve. A glance at Figures 5.3 and 5.4 will show that the less dispersed observations made by observer A have a greater height than those of B. The quantity h is therefore known as the *index of precision*.

Again, differentiating we obtain

$$\frac{dy}{dx} = -2h^2xy = 0 \quad \text{when } y = 0 \text{ and } x = \pm\infty$$

therefore $$\frac{d^2y}{dx^2} = -2h^2y\,(1 - 2h^2x^2) = 0 \quad \text{when } x = \pm\infty$$

or at the points of inflexion, when $x = \pm \dfrac{1}{h\sqrt{2}}$

A rough sketch of the curve can now be made, which clearly follows the general pattern shown in Figure 5.5.

We will now show that the index of precision h is related to the standard error σ by the expression

$$\sigma = \pm \frac{1}{h\sqrt{2}}$$

Remembering that

$$x = v$$

the variance is defined as *the expectation* of v^2

$$\text{or } E\,(v^2) = \sigma^2$$

Thus the population variance

$$\sigma^2 = \frac{\text{sum of squares of residuals}}{\text{total number of residuals}}$$

or in mathematical terms

$$\sigma^2 = \int_{-\infty}^{+\infty} x^2 y\, dx \Big/ \int_{-\infty}^{+\infty} y\, dx$$

(It is to be remembered that there are y values of each residual x)

But $\int_{-\infty}^{+\infty} y\, dx$ is the probability of all residuals being selected, namely a certainty, therefore

$$\int_{-\infty}^{+\infty} y\, dx = 1$$

but

$$\frac{d^2 y}{dx^2} = -2h^2 y\ (1 - 2h^2 x^2)$$

whence

$$\int_{-\infty}^{+\infty} \frac{d^2 y}{dx^2}\, dx = 4h^4 \int_{-\infty}^{+\infty} x^2 y\, dx - 2 \int_{-\infty}^{+\infty} h^2 y\, dx$$

therefore

$$0 = 4h^4 \sigma^2 - 2h^2$$

giving the result

$$\sigma = \pm \frac{1}{h\sqrt{2}}$$

We may now express the normal probability distribution function $\phi(x)$ in terms of the standard error σ, the population mean μ and the observed parameters $\hat{x}$ as follows

$$\phi(x) = y = \frac{h}{\sqrt{\pi}} e^{-h^2 x^2}$$

$$= \frac{1}{\sigma\sqrt{(2\pi)}} e^{-v^2/2\sigma^2}$$

$$= \frac{1}{\sigma\sqrt{(2\pi)}} e^{-(\mu - \hat{x})^2/2\sigma^2} \tag{2}$$

This function is completely defined if the population mean μ and the stardard error σ are known. However it is very convenient to *standardize* this formula by dividing the x and y by σ, and moving the origin to the mean μ. The new simpler equation is then of the form

$$\phi(t) = \frac{1}{\sqrt{(2\pi)}} e^{-\frac{1}{2}t^2} \tag{3}$$

where $t = v/\sigma = x/\sigma$. Hence the variance of t is 1.

This reduction to a standard form has two distinct advantages.

(a) It enables observed quantities of all different kinds to be treated in a general way, reduced to a common weight.

(b) It allows standard sets of tables to be used for all problems.

For example a residual of 15" (seconds of arc) belonging to a distribution whose standard error is 5", is converted to a dimensionless number $t = 15"/5" = 3$, which is conveniently tabulated. A residual of 15 mm and standard error of 5 mm is identically tabulated as 3. A special case is a residual equal to the standard error giving a ratio of 1.

Although the population standard errors σ can be used to bring about this reduction to a common dimensionless standard, in practice we must allow for errors in estimating the standard errors themselves, and if necessary reconsider our estimates in the light of more data becoming available.

5.8 Statistical tests

In statistical literature the standardized normal distribution function is written as N(0,1), meaning "a normal distribution with zero mean and variance 1". The distribution in its original general form of equation (2) would be written

$$N(\mu, \sigma^2)$$

This and other probability distribution functions are used to help decide whether we have made a mistake or not, or whether a systematic error has gone undetected, or to give statistical limits within which to expect a result to lie.

Any decision making is based on the probabilty of obtaining a result within selected bounds or limits. The area of the figure bounded by by the curve and two ordinates, with selected abscissae x=a and x=b, or in the standardised table $t = t_a$ and $t = t_b$, gives the probability that a value will lie in this region. This information is written by the equation

$$P(a < x < b) = \int_a^b y\, dx = \int_a^b \phi(x)\, dx = \int_{a/\sigma}^{b/\sigma} \phi(t)\, \sigma\, dt$$

or

$$P(t_a < t < t_b) = \frac{1}{\sqrt{(2\pi)}} \int_{t_a}^{t_b} e^{-\frac{1}{2}t^2} dt \tag{4}$$

These probabilities will always be less than one or 100%.
Equation (4) is used in two ways.

(a) From given values of a and b, we have to find P(x): for example, the probability that a value will lie within selected limits, such as ±σ about the mean.

(b) Or, we are given P(x) and have to find a and b. For example we select a probability such as 95% as acceptable for the retention of an observed value and thus find the range of acceptable limits within which to accept or reject values.

Although it is quite practicable to evaluate the integrals by computer, it is traditional to tabulate the area under the curve, the *cumulative probability function*, from minus infinity to successive values of x, and to interpolate subsequently within these tables. A required integral is obtained as the difference of two as follows:

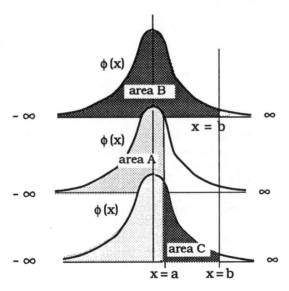

Figure 5.6

Figure 5.6 shows $P(a < x < b)$ in graphical terms, where

$$\text{area C} = \text{area B} - \text{area A}$$

$$P(a < x < b) = \int_a^b y\,dx = \int_{-\infty}^b y\,dx - \int_{-\infty}^a y\,dx$$

$$P(t_a < t < t_b) = \frac{1}{\sqrt{(2\pi)}}\int_{-\infty}^{t_b} e^{-\frac{1}{2}t^2} - \frac{1}{\sqrt{(2\pi)}}\int_{-\infty}^{t_a} e^{-\frac{1}{2}t^2}$$

For numerical work, these areas or integrals are tabulated in a standardised cumulative probability table as shown in Table 5.2. The specific entries of 0.750, 0.950, and 0.975 have been added for use in the examples that follow.

Table 5.2

Table of cumulative Normal probability

t	0	1	2	3	4	5	6	7	8	9
0.0	.500	.504	.508	.512	.516	.520	.524	.528	.532	.536
0.6								.750		
1.0	.841	.844	.846	.848	.851	.853	.855	.858	.860	.862
1.6					.950					
1.7						.960				
1.9							.975			
2.0	.977	.978	.978	.979	.979	.980	.981	.981	.981	.982
3.0	.999	.999	.999	.999	.999	.999	.999	.999	.999	.999

It will be noticed that when $x = 0$ the table gives half the total area covered by the integral from $-\infty$ to $+\infty$, or a cumulative probability of 0.5, or 50%. When x is less than 0 the required cumulative probability is (1 – the tabular entry).

Thus if $x = -2.05$ the cumulative probability from $-\infty$ to -2.05 is

$$1 - 0.980 = 0.020 \text{ or } 2\%.$$

This is shown by the dark area in Figure 5.7 (a).

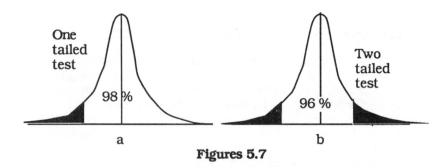

Figures 5.7

We interpret this result in this way: the probability of x being *less* than –2.05 is 2%. Likewise we can also say that the probability of x being *greater* than +2.05 is also 2%, because the function is symmetric. We write these ideas down more formally as

$$P(x < -2.05) = 2\% \text{ or } P(x > +2.05) = 2\%$$

One-sided statements or tests are called *one-tailed tests*. They arise in dealing with dimensional problems such as the length of a competition swimming pool, which must never be shorter than a given length, if swimming records are to be recognised. If the pool is too long, it is a serious disadvantage to the competitors, but does not invalidate a record time.

On the other hand we can say that the probability of x being *both* less than –2.05 *and* greater than + 2.05 is 4%. This is *a two-tailed test*, as shown by the two dark areas in Figure 5.7(b).

Another way of expressing this last result is to say the there is a 96% probability of x lying in the range –2.05 to + 2.05, which is written formally as

$$P(-2.05 < x < +2.05) = 96\%$$

This two–tailed test is common when a dimension, such as steelwork, has to be neither longer nor shorter than a given specification, to within acceptable statistical limits.

Critical tables

Another way to express these probabilities is by critical values only. These Critical Tables are much shorter and often give sufficient information by themselves. Table 5.3 gives similar information to Table 5.2 for selected critical values only. The most common values to be used are the 5 % or 2.5 % cumulative probabilities.

The argument $1 - f(x)$ of the second row is used in one–tailed tests: for example to answer such questions as :

"What is the value of x such that there is only a 5 % probability that it will be *exceeded*? " (a one–tailed test)

Look up the cumulative probability Table 5.2, choosing the entry at 0.950 and see that its argument is 1.64. Alternatively, from the critical percentage Table 5.3, select the 5.0 % entry from row two for a one-tailed test, and read off the critical percentage point x = 1.645.

A traditional percentage point in a two-tailed test was 50%, i.e. tabular entry of 75%, giving x = 0.6745. This relates a "probable" or "50% error e" to the standard error σ, by e = 0.6745 σ. The fifty percent probability is of value when considering the rounding errors in arithmetic processes.

Example
A typical example of the application of some of these ideas and tables, is as follows;

An angle of a primary triangulation is 63° 47' 12.6". During a third order extension, it is remeasured, with a standard error of 5", obtaining a value of 63° 47' 22". Is this value significantly different from the original?

Table 5.3

Critical percentage points for Normal probability

1	f(x)	95	97.5	99	99.5	99.9	99.95
2	1– f(x)	5.0	2.5	1.0	0.5	0.1	0.05
3	2 (1– f(x))	10.0	5.0	2.0	1.0	0.2	0.1
4	x	1.6459	1.960	2.3263	2.5758	3.0902	3.2905

There are two populations to consider.

(a) The primary angle with its given population mean, say μ_o.

(b) The observed third order population, say with unknown mean μ, from which the known sample mean $\hat{x}$ is drawn. The standard error σ of this population is known or it can be estimated from the sample itself.

We must now select a *level of significance* for the test.

A level of 5% is considered *significant.*
A 2.5% level is considered *very significant.*
A 1% level is considered *highly significant.*

It is common practice to deal with the 5% significance level or the 95% *confidence level.* We must also decide on the nature of the test; one-tailed or two? The question posed is "Is this value significantly different from the original". In other words we wish to know whether the two means are likely to be equal.

Because we are not asking whether one is bigger or smaller, we will apply a two-tailed test with confidence limits of ±1.96 derived from Table 5.3, selecting 5% from row three and the limit of 1.96 from row four. The test statistic is

$$\frac{\hat{x} - \mu_0}{\sigma} = \frac{9.4}{5} = 1.88$$

which falls inside the rejection limit of 1.96. We therefore accept that this sample could belong to a population which has the same mean as the primary population.

Scaling back to the dimensioned data gives the limits for the mean of

$$\mu = \hat{x} \pm (5 \times 1.96) = \hat{x} \pm 9.8$$

However, because the test statistic is only just within the acceptance criterionwe are a little concerned that we might have made a mistake in accepting the result of the test, and that this sample mean really belongs to another population with a different mean, say μ_1. If we know what this other mean is we can make a similar, or alternative, test on it.

If we assume that the alternative population mean is $5 \times \sigma$ greater than the actual value i.e. that it is 63° 47' 37.6", the test statistic is now

$$\frac{\bar{x} - \mu_1}{\sigma} = \frac{15.6}{5} = 3.1$$

which is well outside the 1.96 confidence limit and would be rejected from the alternative hypothesis.

Type I and type II errors
Remembering that acceptance or rejection depends upon arbitrary probability limits, it is possible that a decision is wrong.

Suppose in the first test, the test statistic

$$t = \frac{\bar{x} - \mu_0}{\sigma} \quad \text{had been 2.0, i.e. } >1.96$$

We would have rejected the value of $\hat{x}$. But there is a 2.5% probability that it does really belong to this population. Thus in rejecting it when it was actually correct, we would have made what is called a *type I error* with a known probability of 2.5%.

Conversely if we accept a value that actually belongs to another population, we have made a *type II error*.

The probability of making this error depends on the value of the alternative population mean chosen for the alternative hypothesis. Suppose the two means are separated by an amount given by

$$\Delta = \mu_1 - \mu_0 = (t + t_1) \sigma$$

The probabilities associated with t from the first mean and t_1 from the second are α and β. In Figure 5.8 the probability of making a type II error is β, the shaded area to the left of the line marked A.

The alternative hypothesis is often selected, as in our example, to differ from the sample mean by five times the standard error of the sample i.e.

$$\mu_1 = \mu_0 + \Delta = \mu_0 + 5 \sigma$$

The line marked A of Figure 5.8 is selected by the choice of 0.5 α in a two-tailed test on the null hypothesis. Thus the probability of accepting a value to the left of A, i.e. outside the limit of t_1, and that should rightly belong to the alternative sample, is known to be β. It will be seen that the values of (α, t), (β, t_1) and Δ, are linked together. Given any two of these groups, we may calculate the remaining one.

In accepting values within the range 0 to t we could be wrong by β. Thus we can say that we are β % sure that no value greater than 5 σ from the mean μ_0 has been accepted.

Identical tests can be applied to a residual, scaled by its standard error, to decide whether to accept or reject the observation from which it was derived, or to assess the reliability of an observation against an alternative hypothesis. The variance of a residual is obtained from the dispersion matrix $\mathbf{D}_v$, calculated as part of the least squares process. See Appendix 5.

Formal statement of ideas
Most of the concepts and ideas used in the above arguments have been formalised into statistical notation which at first sight is rather formidable.

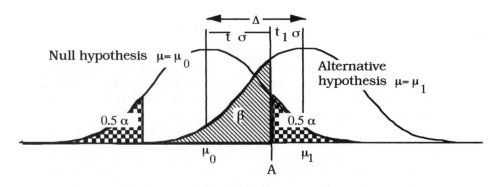

Figure 5.8

Refer to Figure 5.8 in the following discussion. The test statistic t is first standardised, as described above, to enable us to use normal tables.The outcome of the test is called a *null hypothesis,* written H_0. In this case is the null hypothesis states:

" does μ equal μ_0?", or simply, $\mu = \mu_0$?

The *alternative hypothesis* is called H_1: in this case is:

$$\mu \neq \mu_0 \; ?$$

The significance level, such as 5%, is denoted by α in one-tailed tests and by $0.5\,\alpha$ in two-tailed tests; whilst the corresponding confidence limits are denoted by t_α in one-tailed tests and by $t_{0.5\,\alpha}$ in two-tailed tests. In this case, $t_{0.5\alpha} = 1.96$.

All this information is brought together into the one grand statement

$$[\, H_0 : \mu = \mu_0 \,; H_1 : \mu \neq \mu_0 \,]$$
$$\text{accept } H_0 \text{ if } (-t_{0.5\,\alpha} < t < t_{0.5\,\alpha})$$
$$\text{because } P\,(t_{0.5\,\alpha} < t < t_{0.5\,\alpha}) = 1 - \alpha.$$
$$\text{where in this case, } t = \frac{\hat{x} - \mu_0}{\sigma}, \text{ and } 1 - \alpha = 95\%.$$

This is the general form used to express most statistical tests although the distributions and tests statistics vary according to the nature of the problem.

Degrees of freedom
In a mathematical problem, the essential amount of information required for a solution is called the *necessary information*. For instance, two distinct equations are needed to solve for two unknowns. If more than sufficient information is available, say we have five distinct equations in these two variables, we have a *degree of freedom* to select the two necessary equations.

If we have m equations in n unknowns (m > n) there are m – n degrees of freedom in the problem.

To obtain a best estimate of one observed quantity, we obviously need at least one observation. With only one observation there is no redundant information to yield residuals or statistical analysis. Therefore the necessary number of observations is one. Hence if there is more than one observation, say n, there will be n – 1 degrees of freedom. In statistics the number of degrees of freedom in a problem enhances the validity of predictions and calculations. The number of degrees of freedom is usually denoted by the greek letter ν.

Unbiased estimators of σ
We have already calculated a sample variance s^2, from nine observations. The question now arises, whether this sample statistic can be used to estimate the population parameter σ. Is, for example, the expectation, or average of more and more sample variances likely to be σ^2 ? The straight answer is "no", because each sample is biased by its mean. However, as will be shown later, the statistic

$$\frac{n}{n-1} s^2$$

does tend, on average, to equal the population variance σ^2. Or, put another way, we say the expectation

$$\text{written Exp} \left[\frac{n}{n-1} s^2 \right]$$

is σ^2. It is common to use the symbol $\hat{\sigma}^2$ for this expected value.

In statistical language we say that $\hat{\sigma}^2$ is an *unbiased estimator* of the population variance σ^2. Substituting for s^2 leads to the well known formula

$$\hat{\sigma}^2 = \frac{n}{n-1} s^2 = \frac{\Sigma v^2}{n-1} \tag{5}$$

If the population variance is unknown, we use this formula to estimate it. Hence from a calculated sample variance of 3.4, we obtain

$$\hat{\sigma}^2 = (9/8) \times 3.4 = 3.825 \text{ and } \hat{\sigma} = 1.96$$

Pooled variance

If we have several samples and their respective variances, such as from rounds of angles observed under similar conditions at several stations, may we not pool this information to give a better estimate of $\hat{\sigma}^2$? The answer is a qualified "yes".

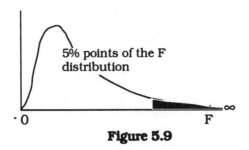

Figure 5.9

The formula for the *pooled variance* from two samples of n_1 and n_2 observations with variances s_1^2 and s_2^2 respectively, illustrates the general form to be adopted for any number of samples

$$\tilde{\sigma}^2 = \frac{n_1 s_1^2 + n_2 s_2^2}{n_1 + n_2 - 2} \tag{6}$$

Table 5.4

5% points of Fisher Table

$n_1 - 1$		5	6	7
$n_2 - 1$	7	3.97	3.87	3.79
	8	3.69	3.58	3.50
	9	3.48	3.37	3.29
	10	3.33	3.22	3.14

Before using this formula for the pooled variance, we should test that each of the samples could have belonged to populations with equal variances. The Fisher F test on the variances is applied. Figure 5.9 shows the approximate shape of the F statistic for degrees of freedom greater than three. The actual shape varies for each degree of freedom. It is also tabulated in Table 5.4.

Suppose we calculated the sample variance of nine observations to be 3.4, and suppose we have a set of similar results from another station giving a variance of 5.6 from six observations. To test whether

Suppose we calculated the sample variance of nine observations to be 3.4, and suppose we have a set of similar results from another station giving a variance of 5.6 from six observations. To test whether it is legitimate to derive a pooled variance to estimate the population variance, we test the ratio of the larger variance to the smaller, i.e. the test statistic is:

$$t = \frac{s_2^2}{s_1^2} = \frac{5.6}{3.4} = 1.64$$

The respective degrees of freedom are 5 and 8, giving a rejection limit of 3.69 from the Fisher Table 5.4, at the 5% confidence level.
Since 1.64 is well within the rejection limit of 3.69, we can accept these samples and pool them, giving the pooled unbiased estimate of the population variance:

$$\hat{\sigma}^2 = \frac{6 \times 5.6 + 9 \times 3.4}{9 + 6 - 2} = 4.9$$

Testing of sample values
When dealing with samples less than thirty, as is generally the case with surveying, the normal distribution is not strictly applicable.

Each value in a sample affects its mean, and therefore the residuals derived from it. Obviously at least one value is needed and any subsequent ones are not essential. The larger the sample, the better we are able to estimate the variance derived from it.

Table 5.5

Table of critical points in the Student's " t " distribution

Probability– two-tail test one-tail test	10 % 5 %	5 % 2.5 %
$\upsilon = n{-}1$ 1	6.31	12.7
9	1.83	2.26
10	1.81	2.23
20	1.72	2.09
30	1.70	2.04
∞	1.64	1.96

Instead of the normal distribution, we use a separate distribution for each sample size. This distribution is called the Student or "t" be noted that when the sample size is very large, the Student distribution is the same as the normal distribution and gives identical results for the critical values.

This table is used in the same way as for the normal distribution, entering the row with the appropriate *degree of freedom*, or $\upsilon = n-1$, where n is the size of the sample.

Using the data of the example worked earlier in which the mean is:

$$\bar{x} = 07.7,$$

and the sample variance is:

$$s^2 = 3.4, \quad s = 1.84$$

we can check the largest residual (3.4) for acceptability at the 5% level in a two-tailed test. The test statistic is

$$\frac{3.4}{1.84} = 1.85$$

The degrees of freedom are

$$v = n-1 = 9$$

This gives a rejection limit of 2.26 from Table 5.6. Clearly this residual is quite acceptable. A residual of 2.26 x 1.84 = 4.2 would be on the rejection limit.

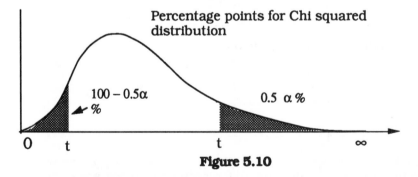

Figure 5.10

Distribution of sample variance; χ^2 (chi squared)distribution
Just as we wish to know the quality of a value drawn from a sample, we also wish to know how good is the estimate of the sample variance. There is no point in quoting a standard error to a second of arc, if it is uncertain to five seconds. To reach a conclusion about the

quality an estimated variance we use the probability distribution of the standardised residuals, i.e. the distribution of the function

$$\Sigma \left(\frac{v}{\sigma}\right)^2$$

One example of this distribution is illustrated in Figure 5.10, which, being a function of squares, has no negative values.

If the statistic (v/σ) has a normal distribution about a zero mean i.e. is $N(0,1)$, then the statistic $\Sigma(v/\sigma)^2$ has a "chi squared" or χ^2 distribution, which takes the approximate shape shown in Figure 5.10 for samples with more than three degrees of freedom.

Remembering that the sample variance

$$s^2 = \frac{\Sigma v^2}{n}$$

it follows that

$$\frac{n s^2}{\sigma^2}$$

has a χ^2 distribution.

This χ^2 probability distribution varies for each value of n, and is

Table 5.6

	χ^2 probability table critical percentage points			
P	97.5 %	95 %	5 %	2.5 %
n − 1				
5	0.83	1.15	11.07	12.83
8	2.18	2.73	15.15	17.53
9	2.70	3.33	16.92	19.02
10	3.25	3.94	18.31	20.48
20	9.59	10.85	31.41	34.17
60	40.48	43.19	79.08	83.30

tabulated for each degree of freedom v in the problem. Since we use the sample mean to calculate the residuals v, there are only n − 1 degrees of freedom in the simple problem. To save space, it is usual to tabulate the critical percentage points only, and for confidence limits set by a 95% probability limit, for one or two-tailed tests, as in Table 5.6.

Suppose we wish to know the confidence limits for σ for the example of section 5.1, where

$$s^2 = 3.40, \quad n = 9, \quad v = 8$$

From Table 5.7, applying a two-tailed test at 97.5% and 2.5% probabilities, yields the confidence limits of 2.18 and 17.53, which is equivalent to limits of 3.74 and 1.32 for the standard error. This can be seen from:

$$2.18 \quad < \quad n\,s^2 / \sigma^2 \quad < \quad 17.58$$

$$2.18 \quad < \quad 9 \times 3.4 / \sigma^2 \quad < \quad 17.58$$

$$\frac{9 \times 3.4}{2.18} \quad > \quad \sigma^2 \quad > \quad \frac{9 \times 3.4}{17.58}$$

$$14.0 \quad > \quad \sigma^2 \quad > \quad 1.74$$

$$3.74 \quad > \quad \sigma \quad > \quad 1.32$$

The calculated standard deviation of the sample is 1.8", or the estimated population standard error is $\sqrt{n/n-1} \times 1.8 = 1.9"$,

Table 5.7

Random Sample of Residuals

Random number	To 1.000 base	Value from N (0,1)	Scaled by sd of 5 mm	Rounded residuals
89	.895	1.25	6.25	6.2
90	.905	1.31	6.55	6.6
26	.265	−0.63	−3.15	−3.2
36	.365	−0.35	−1.75	−1.8
22	.225	−0.75	−3.75	−3.7
74	.745	0.66	3.30	3.3
71	.715	0.57	2.85	2.8
13	.135	−1.10	−5.50	−5.5
74	.745	0.66	3.30	3.3
05	.055	−1.60	−8.00	−8.0
		check sum = 0.1		sum = 0.0

From the above analysis this value of 1.9 could be in error by
3.74 − 1.9 = +1.84" or by 1.90 − 1.32 = − 0.68"

at the 95% confidence level. There is little point therefore in quoting a figure of better than 2" for the standard error, because the extra decimal would be misleading.

Creating a random sample from a normal distribution

Suppose we wish to simulate a sample of ten residuals drawn from a normal distribution. From a random number table, we selected ten values, entering the table anywhere at random. These two–digit random numbers are listed in column 1 of Table 5.7.

To convert these two digit random numbers to a symmetrical set, add 0.5 to all and multiply by 0.01, thus reducing them to random fractions of 1.00. We then select values of residuals corresponding to these fractional probabilities taken from a standardized normal distribution table. These values are given in column 3. Since the distribution set is chosen to have a standard deviation of 5 mm, these residuals are scaled up to give the final rounded values in column 5.The set of residuals is slightly adjusted to make their sum exactly zero as it should be from theory.

Readers may care to verify that the standard deviation computed from the created residuals is 5.1 mm, i.e. close to the expected 5 mm.

It is not always necessary to adopt this manual procedure, because there are many computer software packages which carry out the operation on demand.

5.9 Random errors in derived quantities

In section 5.2 above, we gave a simple treatment of errors in derived quantities. We now extend this theory to include random errors, expressed in terms of variances, and a new concept, covariance.

Covariance

Suppose we have two measured quantities x and y, which form the sides of a rectangle whose area is $A = x\,y$. The effects of small changes in x, y and A are related by the equation

$$\delta A = x\ \delta y + y\ \delta x$$

If these small changes represent the residuals dispersed about the best estimates of the measured quantities, this expression becomes

$$v_A = x\ v_y + y\ v_x$$

Because the variance, σ^2, equals E (Σv^2) we are interested in the squares of residuals so we consider

$$v_A^2 = (x\ v_y + y\ v_x)^2$$

or

$$v_A^2 = x^2 v_y^2 + y^2 v_x^2 + 2xy\, v_x v_y$$

We now take expectations. By definition, the expectation $E(v_x^2)$ is the variance of x, and so on for A and y. Thus we have the important result

$$\sigma_A^2 = x^2 \sigma_y^2 + y^2 \sigma_x^2 + 2x\, y\, E(v_x\, v_y)$$

The expectation of $(v_x v_y)$ is called the *covariance* of x and y. The population covariance is written σ_{xy} and the above equation becomes

$$\sigma_A^2 = x^2 \sigma_y^2 + y^2 \sigma_x^2 + 2xy\, \sigma_{xy} \tag{7}$$

The covariance is the limit to which the average of all these products $(v_x v_y)$ tends as the number of products tends to infinity, i.e. to cover the populations of which x and y are representatives.

If the measured values x and y are *completely independent* of each other, so are their residuals. In this case, each residual v_x will be combined with every residual v_y from the other distribution, so that the products will be made up of terms like $v_x \Sigma v_y$, which are all zero because $\Sigma v_y = 0$. This means that the covariance between independent observations is zero.

On the other hand, if there is some causal connection between the observed values of x and y, their covariance will not be zero.

Sample covariance

We calculate the covariance of a sample, written as s_{xy}, in a manner similar to the variance, as follows

$$s_{xy} = \frac{\Sigma v_x\, v_y}{n}$$

The notation is somewhat unfortunate because the dimension of the covariance is second order (the same as variance).

For example, in Table 5.8, six measurements of two lines made by six different rulers, each in order, gave the residuals shown, from which we have calculated their variances and covariance.

$$s_x^2 = 0.20 / 6 = 0.033$$
$$s_y^2 = 0.38 / 6 = 0.063$$
$$s_{xy} = 0.24 / 6 = 0.04$$

Because the covariance is not zero, we look for some connection or *correlation* between the measured values of x and y. This correlation is reduced to a dimensionless weighted number called the *product moment coefficient of correlation*, ρ, by the expression

$$\rho = s_{xy} / s_x s_y = 0.04 / \sqrt{0.033 \times 0.063} = 0.88$$

The maximum value which ρ can take is one. This indicates total correlation between the measurements. In the example the very high degree of correlation caused us to inspect the circumstances of the measurements. This inspection revealed that six different rulers, each of slightly different length, were used in turn to measure each line. Systematic errors had thus been detected by the correlation analysis.

Table 5.8

Ruler no	Line one residual v_x	Line two residual v_y	v_x^2	v_y^2	$v_x v_y$
1	− 0.1	0.0	0.01	0.0	0.0
2	+0.3	+0.3	0.09	0.09	+0.09
3	− 0.2	−0.4	0.04	0.16	+0.08
4	+0.1	0.0	0.01	0.0	0.0
5	− 0.2	−0.2	0.04	0.04	+0.04
6	+0.1	+0.3	0.01	0.09	+0.03
Σ	0.0	0.0	0.20	0.38	0.24
Σ/6			0.033	0.063	0.04

Correlation and statistical models
Correlations are of two types:

(a) Correlation which arises from some unknown connection between measurements. It is often identified by the type of analysis just given, in which the covariance and correlation coefficient are calculated. It is not always possible to find the exact reason for the correlation, and therefore to identify a systematic error. As will be seen later, the variances and covariances of derived quantities are obtained as part of the least squares estimation process.

(b) Correlation which is predictable from theory and can be modelled completely.

We shall now illustrate these two correlation concepts with reference to a simple example.

Consider Figure 5.11. The three sides x, y, z of two adjacent rectangles, whose areas are A and B respectively, are measured. In calculating the areas we use the following equations as a *mathematical or functional model*

$$A = xy$$
$$B = yz$$

Clearly, since we are using the same value of y in each area calculation, the areas A and B are correlated. Hence if we wish to known the error in the calculated area C = A + B it will be wrong to ignore this correlation.

If however the side y has been independently measured each time, we really have two quite independent quantities, called y and p say, giving as the *statistical or stochastic model* the equations

$$A = xy$$
$$B = pz$$

Although numerically equal values for y and p and for the derived areas A, B and C are quite possible, in each case the error analysis gives different results. We now derive the area covariance σ_{AB} in both cases to illustrate these differences.

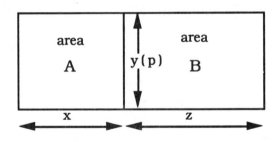

Figure 5.11

Case one
When the common side y is measured only once we have

$$\delta A = x\ \delta y\ +\ y\ \delta x$$
$$\delta B = y\ \delta z\ +\ z\ \delta y$$
$$\delta A\ \delta B = (x\ \delta y\ +\ y\ \delta x)(y\ \delta z\ +\ z\ \delta y)$$
$$= xy\,\delta y\,\delta z\ +\ xz\ \delta y\,\delta y\ +\ y^2\ \delta x\ \delta z\ +\ yz\ \delta x\ \delta y$$

Taking expectations gives

$$\sigma_{AB} = xy\sigma_{yz} + xz\,\sigma_y^2 + y^2\,\sigma_{xz} + yz\,\sigma_{xy}$$

Since x, y and z are independently measured, their covariances are zero and we have simply

$$\sigma_{AB} = xz\,\sigma_y^2$$

Thus the covariance has been *derived from theory*, unlike in the above example where it had to be *estimated from the sample measurements*.

Case two
In this case

$$\delta B = p\ \delta z + z\ \delta p$$

The argument in case two is similar to the above except that we obtain the result

$$\sigma_{AB} = xz\,\sigma_{py}$$

Since the values p and y are independent $\sigma_{py} = 0$ therefore $\sigma_{AB} = 0$.
The importance of these two examples can be seen when we consider the area

$$C = A + B$$

and we wish to calculate the variance σ_C^2.
We can obtain the result from very first principles because

$$C = (x + z)\,y = xy + yz$$

giving

$$\delta C = y\,\delta x + (x + z)\,\delta y + y\,\delta z$$

and

$$\sigma_C^2 = y^2\sigma_x^2 + (x + z)^2\sigma_y^2 + y^2\sigma_z^2$$

Suppose we do not have the original measurements of the sides, nor do we know whether the side y was common to both. We know only the values of A and B, their variances and covariance. We proceed as follows:

$$\delta C = \delta B + \delta A$$

Taking expectations gives

$$\sigma^2_C = \sigma^2_A + \sigma^2_B + 2\ \sigma_{AB}$$

The case in which y is measured twice (i.e. y and p are independent measurements) is of little interest since the covariance of areas is zero, and we have the result

$$\sigma^2_C = \sigma^2_A + \sigma^2_B$$

If A and B are uncorrelated we obtain the result

$$\sigma^2_C = y^2\ \sigma^2_x + (x^2 + z^2)\sigma^2_y + y^2\ \sigma^2_z$$

If A and B are uncorrelated because y and p were observed independently, this result is correct; but if the common value of y had been used in each case, it is wrong because the covariance term $2\,xz\,\sigma^2_y$ is missing.

Matrix treatment of the variance covariance problem

We now consider the treatment of several errors at once. In section 5.2 we derived the following equations linking errors in derived quantities, δs, δα and δA with the errors of measurement δx and δy.

$$\begin{pmatrix} 721\ \delta s \\ 1.444\ \delta\alpha \\ 1\ \delta A \end{pmatrix} + \begin{pmatrix} -600 & -400 \\ -0.00333 & -0.00222 \\ -200 & -300 \end{pmatrix} \begin{pmatrix} \delta x \\ \delta y \end{pmatrix} = \begin{pmatrix} 0 \\ 0 \\ 0 \end{pmatrix}$$

If these small errors now represent residuals we have the similar matrix

$$\begin{pmatrix} 721\ v_s \\ 1.444\ v_\alpha \\ 1\ v_A \end{pmatrix} + \begin{pmatrix} -600 & -400 \\ -0.00333 & -0.00222 \\ -200 & -300 \end{pmatrix} \begin{pmatrix} v_x \\ v_y \end{pmatrix} = \begin{pmatrix} 0 \\ 0 \\ 0 \end{pmatrix}$$

Dividing by the coefficients in the first column gives

$$\begin{pmatrix} v_s \\ v_\alpha \\ v_A \end{pmatrix} + \begin{pmatrix} -0.832 & -0.554 \\ -0.000228 & -0.000153 \\ -200 & -300 \end{pmatrix} \begin{pmatrix} v_x \\ v_y \end{pmatrix} = \begin{pmatrix} 0 \\ 0 \\ 0 \end{pmatrix}$$

We may express such equations in matrix form as:

$$\mathbf{e} + \mathbf{A}\varepsilon = \mathbf{0}$$
$$\mathbf{e} = -\mathbf{A}\varepsilon \tag{8}$$

where

$$\mathbf{e} = \begin{pmatrix} v_s \\ v_\alpha \\ v_A \end{pmatrix}, \qquad \mathbf{A} = \begin{pmatrix} -0.832 & -0.554 \\ -0.000228 & -0.000153 \\ -200 & -300 \end{pmatrix}, \qquad \varepsilon = \begin{pmatrix} v_x \\ v_y \end{pmatrix}.$$

Taking expectations of $\mathbf{e}\,\mathbf{e}^T$ we obtain

$$E(\mathbf{e}\,\mathbf{e}^T) = \mathrm{Exp} \begin{pmatrix} v_s \\ v_\alpha \\ v_A \end{pmatrix} \begin{pmatrix} v_s & v_\alpha & v_A \end{pmatrix}$$

$$= \mathrm{Exp} \begin{pmatrix} v_s v_s & v_s v_\alpha & v_s v_A \\ v_\alpha v_s & v_\alpha v_\alpha & v_\alpha v_A \\ v_A v_s & v_A v_\alpha & v_A v_A \end{pmatrix}$$

$$= \begin{pmatrix} \sigma_s^2 & \sigma_{s\alpha} & \sigma_{sA} \\ \sigma_{s\alpha} & \sigma_\alpha^2 & \sigma_{\alpha A} \\ \sigma_{sA} & \sigma_{\alpha A} & \sigma_A^2 \end{pmatrix}$$

This symmetric matrix contains the variances along the diagonal and the covariances as off-diagonal terms. It is called the *variance-covariance* or *dispersion matrix* of the derived parameters. We can now evaluate this matrix in terms of the dispersion matrix of the observed parameters, as follows. Reconsider the equation

$$\mathbf{e} = -\mathbf{A}\varepsilon$$

then

$$\mathbf{e}\,\mathbf{e}^T = -\mathbf{A}\varepsilon\,(-\mathbf{A}\varepsilon)^T$$
$$= \mathbf{A}\varepsilon\,\varepsilon^T\mathbf{A}^T$$

The matrix $\varepsilon \, \varepsilon^T$ is symmetric and of dimensions (2 x 2), and on taking expectations gives, by an identical argument to the above, the dispersion matrix of the observed parameters x and y

$$\varepsilon \, \varepsilon^T \; = \; \begin{pmatrix} \sigma_x^2 & \sigma_{xy} \\ \\ \sigma_{xy} & \sigma_y^2 \end{pmatrix}$$

The final stage is to premultiply this dispersion matrix by **A** and postmultiply it by **A**^{T} giving

$$\mathbf{e} \, \mathbf{e}^T \; = \; \begin{pmatrix} -0.832 & -0.554 \\ -0.000228 & -0.000153 \\ -200 & -300 \end{pmatrix} \begin{pmatrix} \sigma_x^2 & \sigma_{xy} \\ \\ \sigma_{xy} & \sigma_y^2 \end{pmatrix} \begin{pmatrix} -0.832 & -0.000228 & -200 \\ -0.554 & -0.000153 & -300 \end{pmatrix}$$

Suppose $\sigma_x^2 = 9$ mm², $\sigma_y^2 = 4$ mm² and the covariance is zero, the final dispersion matrix is

$$\begin{pmatrix} \sigma_s^2 & \sigma_{s\,\alpha} & \sigma_{s\,A} \\ \\ \sigma_{s\,\alpha} & \sigma_\alpha^2 & \sigma_{\alpha A} \\ \\ \sigma_{s\,A} & \sigma_{\alpha A} & \sigma_A^2 \end{pmatrix} \; = \; \begin{pmatrix} 7.464 & 0.0207 & 2163.7 \\ \\ 0.0207 & 5.732\mathrm{E}5 & 5.996\mathrm{E}5 \\ \\ 2163.7 & 5.996\mathrm{E}5 & 720000 \end{pmatrix}$$

Giving the standard errors

$$\sigma_s \; = \; \sqrt{7.464} \; = \; 2.73 \text{ mm.}$$
$$\sigma_\alpha \; = \sqrt{5.732\mathrm{E}5} \; = 0.00757 \quad \text{radians} \; = \; 26'.$$
$$\sigma_A \; = \; (720000) = \quad 848 \text{ mm }^2.$$

Although it is not essential to work in matrix terms, to do so greatly simplifies and formalizes the arithmetic operations when many variances and covariances are being calculated.

Summary
Covariance reflects the degree of correlation between quantities. It can be evaluated for derived quantities, but is much more difficult to estimate for observed quantities.

In the measurement of the areas of land plots, wrong error estimates could be important. In many measurement systems such as GPS and photogrammetry, useful results can be derived from known combinations of measurements. Some correlations can be predicted from theory, but others arising from measurement systems and propagation parameters, such as atmospheric refraction, are less well known. Seriously wrong error estimates can result if these matters are not modelled properly.

We shall see in subsequent sections how the analysis of these problems can be reduced to an automatic matrix process which forms part of least squares estimations.

Variance of the mean
So far we have dealt only with variance of a sample of observations. If a parameter is measured several times, the average or mean of these values will be used in further calculations. We thus need to have an estimate of the variance of this mean.

We shall show that the variance $\sigma_{\bar{x}}^2$ of the mean of n independent observations with variance σ^2 is given by

$$\sigma_{\bar{x}}^2 = \frac{\sigma^2}{n} \tag{9}$$

Consider the mean of n observations given by

$$\bar{x} = \frac{\Sigma x_i}{n} = (x_1 + x_2 + ... + x_n)/n$$

$$\delta\bar{x} = \frac{\Sigma \delta x_i}{n} = (\delta x_1 + \delta x_2 + ... + \delta x_n)/n$$

If these δ's represent residuals we have

$$v_{\bar{x}} = (v_1 + v_2 + ... + v_n)/n$$

Squaring and taking expectations we have

$$E(v_{\bar{x}}^2) = E(v_1 + v_2 + ... + v_n)^2/n^2$$

$$= (E(v_1^2) + E(v_2^2) + E(v_3^2) + E(v_4^2) + ... + E(v_n^2))/n^2$$

plus terms in expectations of products of residuals, or covariances, which are all zero if the observations are independent of each other. Thus

$$\sigma_{\bar{x}}^2 = (\sigma^2 + \sigma^2 + \sigma^2 + \sigma^2 + ... + \sigma^2)/n^2$$

$$= n \ \sigma^2 / \ n^2$$

$$= \frac{\sigma^2}{n}$$

And the standard error of the mean is given by

$$\sigma_{\bar{x}} = \sigma / \sqrt{n}$$

This relationship is used with the sample variance s^2 to determine an unbiased estimate of the variance of the population mean from the expression:

$$\hat{\sigma}_{\bar{x}}^2 = \hat{\sigma}^2 / \ n \ = \frac{s^2}{n-1} = \frac{v^2}{n(n-1)}$$

Statistical tests on the *mean* require the use of this standard error $\sigma_{\bar{x}}$ in arriving at the reduced test statistic t.

For example, suppose we wish to find the 5% confidence limits for the population mean μ estimated from a sample mean $\bar{x}$ of nine observations, whose sample variance $s^2 = 3.4$.

$$\hat{\sigma}_{\bar{x}}^2 \ = \ 3.4 / \ 8 \ = \ 0.425$$

$$\sigma_{\bar{x}} \ = \ 0.65$$

The test statistic is

$$t = \mu - \bar{x} \ / \ 0.65$$

From Table 5.5 with $v = 8$ we have, for a two tailed test confidence limit at 5% probability, the value of 2.31. Thus the range within which μ lies with respect to the sample mean is $\pm 0.65 \times 2.31 = \pm 1.50$.

Variance of the sum of observations
A result of special interest is the variance of the the sum of observed values, such as height differences obtained by levelling. Provided the observations are independent, each with the same variance σ^2, the variance of the sum σ_{Σ}^2 is given by

$$\sigma_{\Sigma}^2 \ = \ n \ \sigma^2$$

This result follows in a similar manner to the above result for the mean.

The sum Σx_i is given in terms of the n observations by

$$\Sigma x_i = x_1 + x_2 + x_3 + x_4 + ... + x_n$$

Differentiation with respect to each variable gives

$$\Sigma \, \delta x_i = \delta x_1 + \delta x_2 + \dots + \delta x_n$$

If these now represent residuals we have

$$v_\Sigma = v_1 + v_2 + \dots + v_n$$

On squaring, and taking expectations, and remembering that the covariances of the observations are zero, we have the result

$$\sigma_\Sigma^2 = n \, \sigma^2$$

Thus we find that the standard error σ_Σ of the sum of n independent observations is given by

$$\sigma_\Sigma = \sigma \sqrt{n} \tag{10}$$

Notice that this contrasts with the result for systematic errors given in section 5.4.

Variance of the difference of two observations
If Δ is the difference between two observations x_1 and x_2, whose respective variances are σ_1^2 and σ_2^2, the variance of this difference σ_Δ^2 is given by

$$\sigma_\Delta^2 = \sigma_1^2 + \sigma_2^2$$

This follows by taking the expectation of the square of the difference between the residuals v_1 and v_2, i.e.

$$E \, (v_2 - v_1)^2 = E \, (v_1^2) + E \, (v_2^2) - E \, (v_2 \, v_1)$$

Since the covariance of the independent observations is zero, we have the result

$$\sigma_\Delta^2 = \sigma_1^2 + \sigma_2^2$$

If the variances are equal to σ^2 we obtain

$$\sigma_\Delta^2 = 2\sigma^2$$

Notice this result is the same as the variance of the *sum* of two independent observations.

5.10 The theory of least squares or minimum variance

Since the nineteenth century, applied scientists have adopted the *principle of least squares* to select the best estimates of parameters from a series of observations. In fact when we accept the arithmetic mean of a number of observations as the best estimate of an observed parameter, we are making use of this principle. Most of the theory and procedures, which were developed by Legendre and Gauss, have been applied for over a century, but it has not been until the development of electronic computers that least squares has become commonplace in everyday applications. The principle gives a way of making the best use of observations when more than the minimum number has been observed, and also a way of combining different types of observations to yield best estimates of parameters.

Application to a single observed parameter

To define terms and explain basic ideas, consider the case of the nine observations of the angle whose sample variance was computed in Table 5.1. Firstly we relate each observation, denoted by

$$\overset{o}{x}_i$$

to the best estimate of the angle,

$$\hat{x}$$

by a typical *observation equation*

$$\hat{x} = \overset{o}{x}_i + v_i$$

The equation for the first observation is

$$\hat{x} = 109°25'\,06.3'' + v_1$$

This is only a statement of how much the observed value $109°25'\,06.3''$ differs by v_1 from the final best estimate $\hat{x}$. The procedure of least squares is to find this best estimate $\hat{x}$. Also if we wish some information about the quality of the work, we can substitute the best estimate in the observation equations to obtain the residuals. From these residuals statistical information, such as the variance, can be derived.

Returning to the example, we now introduce a device to reduce the size of the figures in calculation as follows: let

$$\overset{o}{x}_i = \overset{*}{x} + l_i$$

and

$$\hat{x} = \overset{*}{x} + \delta$$

where $\overset{*}{x}$ is some provisional, or base value, close to $\hat{x}$. Here the value of $\overset{*}{x}$ is chosen to be $109°25'$ simply for convenience in working the arithmetic by hand: but in a computer program, it is usual to adopt the first observation as the base value.

In linear problems, that is if the mathematical model only involves linear functions of parameters such as in levelling, there is no *need* to adopt this device. However it is always a good idea to do so, because the sizes of the numbers used in calculation are thereby greatly reduced. In non-linear problems, as in distance measurement, it is *essential* to adopt this procedure.

Substitution for $\hat{x}$ reduces the observation equation to

$$\delta = \overset{o}{x}_1 - \overset{*}{x} + v_1$$
$$= l_1 + v_1$$

e.g. for the first observation

$$\delta = 109°25' \ 06.3'' - 109°25' + v_1$$
$$= 06.3 + v_1$$

The absolute term l is always expressed in the sense: 'observed minus provisional' value, provided it is written to the right of the equality sign in the equations.

Casting all nine observations of Table 5.1 in this way gives equations in matrix form

$$
\delta =
\begin{pmatrix}
1 \\
1 \\
1 \\
1 \\
1 \\
1 \\
1 \\
1 \\
1
\end{pmatrix}
\quad
\begin{pmatrix}
06.3 \\
07.2 \\
10.4 \\
04.3 \\
09.6 \\
05.8 \\
07.9 \\
08.3 \\
09.1
\end{pmatrix}
+
\begin{pmatrix}
v_1 \\
v_2 \\
v_3 \\
v_4 \\
v_5 \\
v_6 \\
v_7 \\
v_8 \\
v_9
\end{pmatrix}
$$

These equations are of the form

$$\mathbf{A}\mathbf{x} = \mathbf{l} + \mathbf{v} \tag{11}$$

which is the standard form adopted for this type of least squares problem. In this simple case there is only one unknown, therefore

$$\mathbf{x} = \delta$$

Because there are more equations than unknowns, in this example nine equations to find one parameter, we have to adopt some principle to give a unique result. The most common principle adopted is the least squares, or minimum variance principle. This principle finds that value of δ which makes the sum of the squares of the residuals a minimum. Let the sum of squares of residuals be Ω. Then we have to make

$$\Omega = \Sigma v^2$$

to be a minimum. This will be the case when

$$\frac{\partial \Omega}{\partial \delta} = 0$$

We shall now explain the mechanism of applying the least squares principle to this simple problem using two mathematical methods side by side : simple algebra and matrix algebra. The reader should follow the left hand treatment to its conclusion, before considering the matrix equivalents to the right, which are merely restatements of the results derived by ordinary algebra on the left. There are m observations of one parameter.

Ordinary algebra			Matrix Algebra		
Ω	$=$	Σv^2	Ω	$=$	$\mathbf{v}^T \mathbf{v}$
	$=$	$v_1^2 + v_2^2 + ... + v_m^2$			
	$=$	$(\delta - l_1)^2 + (\delta - l_2)^2 + .. + (\delta - l_m)^2$			
$\partial\Omega/\partial\delta$	$=$	$2(\delta - l_1) + 2(\delta - l_2) + .. + 2(\delta - l_m)$			
	$=$	$2\Sigma v$	$\partial\Omega/\partial\mathbf{x}$	$=$	$2\mathbf{v}^T \mathbf{A}$
but					
$\partial\Omega/\partial\delta$	$=$	0	$\therefore \mathbf{v}^T \mathbf{A}$	$=$	0
	$\therefore$	$\Sigma v = 0$			

Thus we have the important result: $\Sigma v = 0$

or its equivalent in matrix terms

$$\mathbf{v}^T \mathbf{A} = \mathbf{0} \qquad\qquad (12)$$

The very important result in equation (12) will be derived for the more general case later. Note also that $A^T v = 0$.

We now proceed to the solution, using the two parallel mathematical models.

Adding up the columns of the nine equations we have:

$$m \delta = \Sigma 1 + \Sigma v$$
$$= \Sigma 1$$
$$\delta = \Sigma 1 / m$$

$$\boxed{\begin{aligned} A^T A x &= A^T 1 + A^T v \\ &= A^T 1 \\ x &= (A^T A)^{-1} A^T 1 \end{aligned}}$$

Although the matrix treatment seems more complex when applied to this simple example, it is important to study these two mathematical solutions side by side, and thus gain understanding of the matrix method.

The inverse $(A^T A)^{-1}$ is merely the reciprocal of m, and $A^T 1$ is $\Sigma 1$.

Finally we obtain the best estimate of x to be

$$\begin{aligned} \hat{x} &= \overset{*}{x} + \delta \\ &= \overset{*}{x} + \Sigma 1/m \\ &= \overset{*}{x} + (\Sigma \overset{o}{x}_1 - \Sigma \overset{*}{x}) / m \\ &= \overset{*}{x} + \Sigma \overset{o}{x}_1 / m - \overset{*}{x} \\ &= \bar{x} \end{aligned}$$

Thus we have proved that the best estimate is the arithmetic mean of the observations. Completing the arithmetic as in Table 5.1, we find again that $\delta = 07.7''$, giving the final result

$$\hat{x} = \bar{x} = 109° 25' 07.7''$$

Thus we obtain the same residuals as in Table 5.1 and the variance as before.

The beginner should be completely satisfied that only the arithmetic mean gives the minimum sum of squares of residuals. A simple exercise to demonstrate this is to select values slightly greater and slightly less that the mean, find the "residuals" from these means, and their sums of squares which both prove to be greater than Ω.

5.11 Weighted observations

Suppose that the observed angles of Table 5.1 differ in quality. This could arise for several reasons. Not all of them might have been observed with the same instrument, nor all by the same person, nor under the same conditions. We could carry out test measurements under these different circumstances to find the variances of each angle and thus have an estimate of their relative quality. Just as we depicted the results of the shove halfpenny game we can assign *weights* to the work of an observer or an instrument from test data, or from previous observations. The idea is to make allowances for the different quality of the residuals in a least squares process by scaling each residual by its standard error, thus producing new "residuals" each of the same quality. Let the new residuals **u** be given by

$$u = \frac{v}{\sigma} \tag{13}$$

Clearly the better observations have smaller standard errors and thus their scaled residuals are given more weight in the solution. To allow for the fact that we can only *estimate* the standard errors σ we introduce an unknown constant σ_0 into the equation as follows

$$\frac{u}{\sigma_0} = \frac{v}{\sigma}$$

or

$$u = \frac{\sigma_0 \, v}{\sigma}$$

If the standard errors have been well estimated we expect this constant σ_0, usually known as the *standard error of unit weight*, to be equal to 1. However in practice it is evaluated as a routine part of each problem and acts as an indicator of how well the observational variances have been estimated in the first place, or *a priori*.

The effect of this modification on the least squares procedure is that we now minimise the sum of squares of the reduced residuals u ; that is we minimise

$$\Omega = \mathbf{u}^T \mathbf{u}$$

where

$$\mathbf{u}^T = \sigma_0 (\frac{v_1}{\sigma_1} \quad \frac{v_2}{\sigma_2} \quad .... \quad \frac{v_9}{\sigma_9})$$

A neater way to introduce these standard errors and the unknown constant, is to create a special matrix containing their squares or variances. This is called weight matrix **W**. It is a diagonal matrix containing the reciprocals of the respective variances, scaled by the variance of unit weight. By the introduction of this weight matrix we convert

$$\Omega = \mathbf{u}^T \mathbf{u}$$

into an equivalent expression

$$\Omega = \mathbf{v}^T \mathbf{W} \mathbf{v} \tag{14}$$

where the weight matrix $\mathbf{W}$ is diagonal of size m x m, if there are m observations. To save space we write out only a typical 3 x 3 weight matrix in full to illustrate its structure as follows

$$\mathbf{W} = \sigma_0^2 \begin{pmatrix} 1/\sigma_1^2 & 0 & 0 \\ 0 & 1/\sigma_2^2 & 0 \\ 0 & 0 & 1/\sigma_3^2 \end{pmatrix}$$

Note that this weight matrix is the scaled inverse of the dispersion matrix. Also, for independent observations with zero covariances

$$\mathbf{W}^{-1} = \frac{1}{\sigma_0^2}\mathbf{D}$$

because

$$\mathbf{W}^{-1} = \frac{1}{\sigma_0^2} \begin{pmatrix} \sigma_1^2 & 0 & 0 \\ 0 & \sigma_2^2 & 0 \\ 0 & 0 & \sigma_3^2 \end{pmatrix} = \frac{1}{\sigma_0^2}\mathbf{D}$$

Example of weighted observations
Reconsider the example of Table 5.1. We now assign different weights to the twelve observations. Let us assume that each of the first six observations was observed with a standard error of 2", and the last three, with standard errors of 5". This could happen if two different theodolites had been used.

The diagonal dispersion matrix $\mathbf{D}$ of the observations is:

$$\begin{bmatrix} 4 & & & & & & & & \\ & 4 & & & & & & & \\ & & 4 & & & & & & \\ & & & 4 & & & & & \\ & & & & 4 & & & & \\ & & & & & 4 & & & \\ & & & & & & 25 & & \\ & & & & & & & 25 & \\ & & & & & & & & 25 \end{bmatrix}$$

The scaled inverse of **D** is the weight matrix **W** i.e.

$$\mathbf{W} = \sigma_0^2 \mathbf{D}^{-1} \tag{14}$$

$$\mathbf{W} = \sigma_0^2 \begin{bmatrix} \frac{1}{4} & & & & \text{All off diagonal terms} \\ & \frac{1}{4} & & & \text{are zero} \\ & & \frac{1}{4} & & & & & \\ & & & \frac{1}{4} & & & & \\ & & & & \frac{1}{4} & & & \\ & & & & & \frac{1}{4} & & \\ & & & & & & \frac{1}{25} & \\ & & & & & & & \frac{1}{25} & \\ & & & & & & & & \frac{1}{25} \end{bmatrix}$$

For correlated observations, the off-diagonal terms, the covariances, are generally non zero. In this case the inverse **W**, though not so easily found, is treated in just the same way. It is difficult to call such a matrix a "weight" matrix. Generally the term is retained for a diagonal matrix of uncorrelated observations, and we use the term "inverse of the dispersion matrix " for the general case. In many texts this matrix **W** is denoted by **P**.

Practical example

In practice we start by making $\sigma_0^2 = 1$. In our our simple problem converting to ordinary algebra we have

$$\mathbf{A}^\mathbf{T} \mathbf{W} \mathbf{A} = \Sigma w \quad \text{and}$$

giving the result
$$\mathbf{A^T W 1} \;=\; \Sigma w1$$
$$\hat{x} = \Sigma wl \,/\, \Sigma w \tag{15}$$

$$= \frac{\frac{1}{4}\,(06.3 + 07.2 + 10.4 + 04.3 + 09.6 + 05.8) + \frac{1}{25}\,(07.9 + 08.3 + 09.1)}{\frac{6}{4} + \frac{3}{25}}$$

$$= 07.35''$$

This result is called the *weighted mean.*
The residuals formed from the weighted mean are respectively

$$1.05,\; 0.15,\; 3.05,\; -3.05,\; 2.25,\; -1.55,\quad 0.55,\; 0.95,\; 1.75$$

giving

$$\mathbf{v^T W v} \;=\; \frac{1}{4}(\,1.05^2 + .. \;\; + 1.55^2) \;+\; \frac{1}{25}(\,0.55^2 + ... + 1.75^2)$$

$$=\;\; 27.195/4 \;+\; 4.2675/25$$

$$=\;\; 6.80 + 0.17$$

$$=\;\; 6.97$$

An estimate of the standard error of unit weight σ_0 is obtained from from the *weighted residuals* using the formula

$$\sigma_0^2 \;=\; \mathbf{v^T W v}\,/\,\text{m-1}$$

$$=\; 6.97/8 = 0.87$$

$$\sigma_0 \;=\; 0.93$$

That σ_0 is close to one indicates that the original weights were well estimated.

This very important concept of a weight matrix also enables observations of different types, e.g. directions and lengths, to be combined together in a least squares estimate of parameters.

Summary
The treatment of observations of differing quality is brought about by the use of a matrix **W**. It is formed from the inverse of the dispersion matrix of the observed parameters, scaled by the variance of an observation of unit weight, initially assumed equal to 1, and later estimated as part of the computation process. The expression to be minimised is:
$$\Omega \;=\; \mathbf{v^T W v}$$

We will show in the next section, that the effect of this is to give a solution in the form

$$x = (A^T W A)^{-1} A^T W 1$$

The variance of an observation of unit weight is given by:

$$\sigma_0^2 = v^T W v / m - n$$

where there are m equations in n variables, and $m \geq n$.

5.12 Observations to estimate more than one parameter

The extension of the above theory to deal with more parameters than one follows exactly similar lines except that the **A** matrix has no longer one column but one for each parameter to be estimated.

Consider the small levelling net shown in Figure 5.12. In it there are six lines of levels connecting four points A, B, C and D. Since only three such lines of levels would be sufficient to give relative heights between the points, three lines are extra to requirements. Therefore there are three degrees of freedom in the problem.

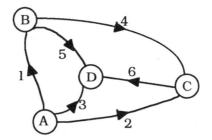

Figure 5.12

Since we only know *differences* in level, we can only find the relative heights of the points. We may select any one to be a datum. In this case choose the height of A to be fixed, hence $\delta h_A = 0$, leaving the three parameters h_B, h_C, and h_D to be estimated.

Consider Table 5.9 in which all the necessary data are presented. Although any three height differences could be selected as the provisional values, with the proviso that they are close to the observed values, it is neatest to select the first three observed values, thus making the first three values of the 1 vector zero.

In general, when there are m-n degrees of freedom in the problem, we select of the first n observed parameters to be the n provisional values, thus reducing the first n values of the 1 vector to zero.

The remaining provisional values are derived from these first n. In this case:

$$\overset{*}{h}_C - \overset{*}{h}_B = \overset{*}{h}_C - \overset{*}{h}_A - (\overset{*}{h}_B - \overset{*}{h}_A) = +69.36 - 12.00 = 57.36$$

and so on for the other lines.

Table 5.9

Line		Observation		obs	prov	1
AB	1	$h_B - h_A$	$=$	+12.00	+12.00	0.00
AC	2	$h_C - h_A$	$=$	+69.36	+69.36	0.00
AD	3	$h_D - h_A$	$=$	-1.55	-1.55	0.00
BC	4	$h_C - h_B$	$=$	+57.28	+57.36	-0.08
BD	5	$h_D - h_B$	$=$	-13.66	-13.55	-0.11
CD	6	$h_D - h_C$	$=$	-70.87	-70.91	+0.04

In this case of observation equations which are already linear, it is not essential but convenient to use the method of approximations. For example, the observation for the line CD is

$$\overset{*}{h}_D - \overset{*}{h}_C = -70.87$$
$$= -70.91 + 0.04$$

thus we have

$$\delta h_D - \delta h_C = +0.04$$

thereby reducing the sizes of the numbers in the problem. Care must be taken with signs. These follow strictly from the order in which we write down the parameters in the equations, as further indicated by the arrows on the lines of Figure 5.10. The observation equations for this problem can then be assembled in matrix form as follows;

$$
\begin{bmatrix}
1 & 0 & 0 \\
0 & 1 & 0 \\
0 & 0 & 1 \\
-1 & 1 & 0 \\
-1 & 0 & 1 \\
0 & -1 & 1
\end{bmatrix}
\begin{bmatrix}
\delta h_B \\
\delta h_C \\
\delta h_D
\end{bmatrix}
=
\begin{bmatrix}
0 \\
0 \\
0 \\
-0.08 \\
-0.11 \\
+0.04
\end{bmatrix}
+
\begin{bmatrix}
v_1 \\
v_2 \\
v_3 \\
v_4 \\
v_5 \\
v_6
\end{bmatrix}
$$

Once again these equations are in the standard form

$$\mathbf{A}\mathbf{x} = \mathbf{1} + \mathbf{v}$$

Minimising the the sum of squares of weighted residuals or

$$\Omega = \mathbf{v}^T \mathbf{W} \mathbf{v}$$

means that

$$\frac{\partial \Omega}{\partial \mathbf{x}} = \mathbf{0}$$

From the rules of the differentiation of matrices given in Appendix 3 this means that

$$\mathbf{v}^T \mathbf{W} \mathbf{A} = \mathbf{0}$$

and also

$$\mathbf{A}^T \mathbf{W} \mathbf{v} = \mathbf{0}$$

because the weight matrix **W** is symmetric.
Premultiplying

$$\mathbf{A} \mathbf{x} = \mathbf{l} + \mathbf{v}$$

by

$$\mathbf{A}^T \mathbf{W}$$

gives

$$\mathbf{A}^T \mathbf{W} \mathbf{A} \hat{\mathbf{x}} = \mathbf{A}^T \mathbf{W} \mathbf{l} + \mathbf{A}^T \mathbf{W} \mathbf{v}$$

$$\therefore \quad \mathbf{A}^T \mathbf{W} \mathbf{A} \hat{\mathbf{x}} = \mathbf{A}^T \mathbf{W} \mathbf{l}$$

or

$$\mathbf{N} \hat{\mathbf{x}} = \mathbf{b} \qquad (15)$$

This is a series of n linear equations in n unknowns. In older literature these n equations are known as the *normal equations*. As the name is useful it is retained in this book.

Here we take the weight matrix to be the unit matrix assuming the observations are all of the same weight. The normal equations become:

$$\mathbf{N} = \mathbf{A}^T \mathbf{A}$$

and the absolute term is given by

$$\mathbf{b} = \mathbf{A}^T \mathbf{l}$$

We now return to the example of the level net. The normal equations are formed to give

$$\begin{bmatrix} 3 & -1 & -1 \\ -1 & 3 & -1 \\ -1 & -1 & 3 \end{bmatrix} \begin{bmatrix} \delta h_B \\ \delta h_C \\ \delta h_D \end{bmatrix} = \begin{bmatrix} 0.19 \\ -0.12 \\ -0.07 \end{bmatrix}$$

giving the solution by any method described in Appendix 7

$$\hat{x} = \begin{bmatrix} \delta h_B \\ \delta h_C \\ \delta h_D \end{bmatrix} = \begin{bmatrix} 0.048 \\ -0.030 \\ -0.017 \end{bmatrix}$$

Substituting the solution in the original observation equations gives the residuals

$$v = \begin{bmatrix} 0.048 \\ -0.030 \\ -0.017 \\ 0.002 \\ 0.045 \\ -0.027 \end{bmatrix}$$

The arithmetic check on the work is that

$$A^T v = 0$$

In this case we find

$$A^T v = \begin{bmatrix} +0.001 \\ -0.001 \\ +0.001 \end{bmatrix}$$

the slight disagreement being due to rounding errors in the arithmetic processes. We can also compute the variance from

$$\frac{v^T v}{m - n} = 0.006251 / 3 = 0.002084$$

and the standard error

$$\sigma_0 = 0.0456$$

From this result we see that it would have been better to have assigned weights of $1/0.002084$ to each of the observation equations. Although this would not have changed the solution, because the weight matrix appears on each side of the equations, it would have altered the estimate of the variance. For this would have been calculated from

$$v^T W v \ / m\text{-}n$$

which would just have been

$$\frac{v^T v}{0.002084 \ (m\text{-}n)}$$

which is clearly 1.

Hence in problems where we assign different realistic weights to each observation we expect that the estimated variance will be 1.

5.13 Mathematical models with more than one observed parameter

In the two simple problems considered above, there was only one observed parameter in each equation, thus giving equations of the form

$$\mathbf{A}\mathbf{x} = \mathbf{1} + \mathbf{v}$$

To deal with problems with more than one observed parameter per equation we introduce equations of the form

$$\mathbf{A}\mathbf{x} + \mathbf{C}\mathbf{v} + \mathbf{C}\mathbf{1} = \mathbf{0}$$

These equations ultimately lead to the solution of a set of symmetric equations of the same form as before, i.e.

$$\mathbf{N}_1 \mathbf{x} = \mathbf{b}_1 \tag{17}$$

where

$$\mathbf{N}_1 = \mathbf{A}^T \mathbf{N}^{-1} \mathbf{A}$$
$$\mathbf{b}_1 = \mathbf{A}^T \mathbf{N}^{-1} \mathbf{b}$$
$$\mathbf{N} = \mathbf{C}\mathbf{W}^{-1}\mathbf{C}^T$$
$$\mathbf{b} = \mathbf{C}\mathbf{1}$$

$\mathbf{W}$ is the weight matrix of the observations.

We will derive these equations in the course of fitting a straight line to observed data points. See Appendix 5 for a purely formal derivation.

Example of fitting a straight line by least squares
In the following sections we use the example of fitting a straight line to data as a means of explaining the general principles connected with the least squares estimation of parameters. Not only does this particular problem occur in several fields, such as surveying, cartography, engineering surveying and photogrammetry, but the method of treatment is the same for all manner of curve fitting problems. Thus the following treatment of the straight line problem must be seen as typical of many.

Only two points are needed to define a straight line. For simplicity we confine our attention to the two dimensional problem of fitting a straight line in a plane. Three cases of fitting the corresponding values of x and y, listed in the first two columns of Table 5.10, will be treated.

Table 5.10

Line-Fitting data

x	y	Comp y	o − c = 1	v	ŷ
1	2.812	2.5	0.312	−0.023	2.789
2	3.066	3.0	0.066	−0.035	3.030
3	3.218	3.5	−0.282	0.053	3.271
4	3.482	4.0	−0.518	0.030	3.513
5	3.713	4.5	−0.787	0.041	3.754
6	4.033	5.0	−0.967	−0.037	3.996
7	4.278	5.5	−1.222	−0.040	4237
8	4.445	6.0	−1.555	0.034	4.478
9	4.783	6.5	−1.717	−0.063	4.720
10	4.920	7.0	−2.080	0.041	4.961

In two dimensions the equation of a straight line is

$$y = mx + c \tag{18}$$

$$F(m, c : x, y) = 0$$

Generally the problem is to find values for the coefficients m and c, gradient m and the intercept c on the y axis, from given values of x and y. In the formal functional statement, we adopt the convention of placing the unobserved parameters first, separated from the observed parameters by a colon.

There are two main cases of the line-fitting problem:

(a) when only one of the parameters x or y is observed,
(b) when both parameters x and y are observed.

In the first case, the least squares model follows the method already considered above.We shall deal with this briefly to set the scene for the more complex treatment which follows.

In both cases y is an observed quantity burdened by observational error. If a large scale graph is drawn of the values of y and x it will be seen that a straight line cannot be found to fit the data exactly. This graph is illustrated in Figure 5.13, showing the eye-balled line 1–10, from which the parameters can be measured as a check to give

$$c = 2.5 \quad \text{and} \quad m = \text{arc} \tan(13^\circ)$$

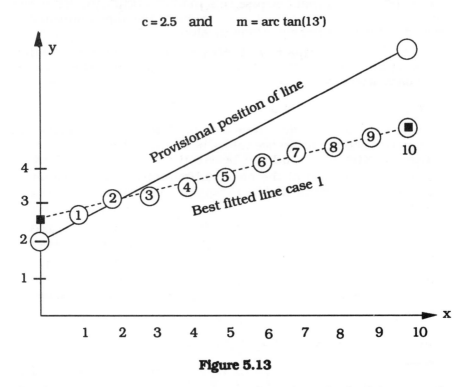

Figure 5.13

The least squares process is an analytical method of drawing this graph. It produces unique results and will not depend on the skill of the person drawing it. The criterion used is that the sum of squares of the observation residuals is minimum. The beginner is encouraged to draw this graph before continuing with the analytical method.

The first case is similar to a regression analysis of two variables, in which the small residuals are considered to be only in the direction of the y axis. The second requires that the residuals lie perpendicular to the line and thus needs to be treated by the general least squares method. In a later section we will deal with another case in which the solution is constrained in some way, for example to force the line to pass through a given fixed point. These examples illustrate the basic mechanisms for tackling a wide range of such problems.

The figures of the example have been selected from an initial line whose equation is

$$y - 0.5x - 2.0 = 0 \tag{19}$$

to which random residuals have been added.

Simple case with one observable per equation

It must be stressed that because this mathematical equation is non linear (because of the product m x), we must use the approximation method to linearize the equations involved.

For each observed value, $\overset{o}{y}$, there is an observation equation. Following the notation of section 5.10, we select provisional values of the parameters to be

$$\overset{*}{m} = 0.5, \qquad \overset{*}{c} = 2.0$$

Ten values of $\overset{*}{y}$, in of column three of Table 5.10, are calculated using equation (19) and the respective values of $x = 1$ to 10.

The objective is to find the best estimates of the unobserved parameters $\hat{m}$, $\hat{c}$, and of the observed parameters $\hat{y}$. As usual the various quantities are related by the equations:

$$\hat{m} = \overset{*}{m} + \delta m$$
$$\hat{c} = \overset{*}{c} + \delta c$$
$$\hat{y} = \overset{*}{y} + \delta y$$

Since there is only one observed parameter, $\overset{o}{y}$, in each equation the residuals v are written

$$v = \hat{y} - \overset{o}{y}$$
$$= \overset{*}{y} - \overset{o}{y} + \delta y$$

therefore

$$\delta y = \overset{o}{y} - \overset{*}{y} + v$$
$$= 1 + v$$

Because the best estimates and the provisional values satisfy equation (18) we may write:

$$F(\hat{m}, \hat{c} : x, \hat{y}) = 0$$
$$F(\overset{*}{m}, \overset{*}{c} : x, \overset{*}{y}) = 0 \qquad (20)$$

The connection between these equations and the small correction is given by the linear part of the expansion applied to equation (18) as follows

$$F(\hat{m}, \hat{c}: \hat{y}) = F(\overset{*}{m}, \overset{*}{c} : \overset{*}{y}) + \frac{\partial F}{\partial \overset{*}{m}} \delta m + \frac{\partial F}{\partial \overset{*}{c}} \delta c + \frac{\partial F}{\partial \overset{*}{y}} \delta y$$

Hence from equations (20) we have

$$\frac{\partial F}{\partial \overset{*}{m}} \delta m + \frac{\partial F}{\partial \overset{*}{c}} \delta c + \frac{\partial F}{\partial \overset{*}{y}} \delta y = 0 \qquad (21)$$

Applying (21) to equation (18) we have:

$$(x)\, \delta m + (1)\, \delta c = 1 + v \qquad (22)$$

It will be remembered that x is considered error free because it is not an observed parameter.

Evaluating equation (22) for each of the ten observation equations gives:

$$\mathbf{A x = 1 + v}$$

$$\begin{bmatrix} 1 & 1 \\ 2 & 1 \\ 3 & 1 \\ 4 & 1 \\ 5 & 1 \\ 6 & 1 \\ 7 & 1 \\ 8 & 1 \\ 9 & 1 \\ 10 & 1 \end{bmatrix} \begin{bmatrix} dm \\ dc \end{bmatrix} = \begin{bmatrix} 0.312 \\ 0.066 \\ -0.282 \\ -0.518 \\ -0.787 \\ -0.967 \\ -1.222 \\ -1.555 \\ -1.717 \\ -2.080 \end{bmatrix} + v$$

The normal equations, on the assumption that observations are all of equal weight, are:

$$\mathbf{N\ x = b}$$

$$\begin{bmatrix} 385 & 55 \\ 55 & 10 \end{bmatrix} \begin{bmatrix} dm \\ dc \end{bmatrix} = \begin{bmatrix} -69.458 \\ -8.75 \end{bmatrix}$$

Giving the solution

$$dm = -0.25858$$
$$dc = 0.5472$$

Back substitution in the observation equations gives the residuals of the fifth column in Table 5.10. Thus we can calculate the variance factor σ_0^2 from

$$\sigma_0^2 = \frac{v^T v}{m - n}$$

$$= 0.017\ 228/8$$
$$= 0.002\ 1535$$

and $\qquad\qquad\qquad \sigma_0 = 0.044$

Finally the best estimates of the derived parameters are obtained from

$$\hat{m} = \overset{*}{m} + dm = 0.2414 \quad \hat{\alpha} = 13'.5716$$
$$\hat{c} = \overset{*}{c} + dc = 2.5472$$

and the best estimates of the observed parameters from

$$\hat{y} = \overset{0}{y} + v_y$$

These are listed in the last column of Table 5.10.

The beginner is encouraged to work through this problem both by the graphical and computational methods before proceeding to the next section.

Statistical tests on data
Statistical tests for quality assessment can be made on the residuals and their variances, as in section 5.7. See also Appendix 6.

Case 2: both x and y are observed parameters
We now extend the treatment to the more general case in which both x and y are observed. The principles involved are also common to problems in astronomy and coordinate transformation.

The treatment is the same as before with the addition of the extra fact that we have to estimate

$$\hat{x} = \overset{0}{x} + v_x$$

The functional models to be satisfied are

$$F\,(\hat{m},\ \hat{c}:\hat{x},\hat{y}) = 0 \quad \text{and} \quad F(\overset{*}{m},\ \overset{*}{c}:\overset{*}{x},\overset{*}{y}\,) = 0$$

$$\frac{\partial F}{\partial \overset{*}{m}}\,dm + \frac{\partial F}{\partial \overset{*}{c}}\,dc + \frac{\partial F}{\partial \overset{*}{x}}\,dx + \frac{\partial F}{\partial \overset{*}{y}}\,dy = 0$$

$$(\overset{*}{x})\,dm + (1)\,dc + (\overset{*}{m})\,dx - dy = 0$$

The notation indicates that the coefficients of the variables are evaluated using the provisional values adopted for the problem. More casually the equation may be written without the superscript dots as:

$$(x)\,dm + (1)\,dc + (m)\,dx - dy = 0 \qquad\qquad (23)$$

Separating the unobserved from the observed parameters we have:

$$\begin{bmatrix} x & 1 \end{bmatrix} \begin{bmatrix} dm \\ dc \end{bmatrix} + \begin{bmatrix} m & -1 \end{bmatrix} \begin{bmatrix} dx \\ dy \end{bmatrix} = 0$$

These equations can be written

$$\mathbf{A\,x + C\,s = 0}$$

The dimensions of the matrices and vectors for the line-fitting example are

$$\mathbf{A}\ \text{matrix} = 10 \times 2, \quad \mathbf{C}\ \text{matrix} = 10 \times 20$$
$$\mathbf{x}\ \text{vector} = 2 \times 1, \quad \mathbf{s}\ \text{vector} = 20 \times 1$$

Forming ten equations for every one of the ten observed points gives

$$\begin{bmatrix} 1 & 1 \\ 2 & 1 \\ 3 & 1 \\ 4 & 1 \\ 5 & 1 \\ 6 & 1 \\ 7 & 1 \\ 8 & 1 \\ 9 & 1 \\ 10 & 1 \\ (r \times n) \end{bmatrix} \begin{bmatrix} dm \\ dc \\ (n \times 1) \end{bmatrix} + \begin{bmatrix} 0.5 & -1 & 0 & 0 & \text{etc} \\ 0 & 0 & 0.5 & -1 \\ \text{etc} \\ & \text{Matrix C} \\ & \text{dimensions} \\ & 10 \times 20 \\ & & & 0.5 & -1 \\ (r \times m) \end{bmatrix} \begin{bmatrix} dx_1 \\ dy_1 \\ dx_2 \\ dy_2 \\ \\ 20 \times 1 \\ \\ dx_{10} \\ dy_{10} \end{bmatrix} = 0$$

In general there are r equations in m observed parameters to estimate n derived parameters.

Notice the change in the notation from the ordinary algebra of the problem, in this case m, c, x and y, to that of the matrix notation in which we always write the derived vector as **x** and the observed vector **s**.

As usual the small changes to the observed parameters are split into two parts: the known part **l**, and the unknown part **v**.
Thus for example

$$dx_1 = l_{x_1} + v_{x_1}$$

The final equations are of the form

$$\mathbf{A}\mathbf{x} + \mathbf{C}\mathbf{v} = -\mathbf{C}\mathbf{l}$$

$$\mathbf{A}\mathbf{x} + \mathbf{C}\mathbf{v} = \mathbf{b}$$

The reader should now refer to Appendix 5 where it is shown that by minimizing the sum of squares of weighted residuals, we obtain the normal equations

$$\mathbf{N}_1 \mathbf{x} = \mathbf{b}_1$$

where

$$\mathbf{N}_1 = \mathbf{A}^T \mathbf{N}^{-1} \mathbf{A} \qquad \text{and} \qquad \mathbf{b}_1 = \mathbf{A}^T \mathbf{N}^{-1} \mathbf{b}$$

$$\mathbf{N} = \mathbf{C} \mathbf{W}^{-1} \mathbf{C}^T \qquad \text{and} \qquad \mathbf{b} = \mathbf{C}\mathbf{l}$$

The weight matrix is **W**.

It is not practicable to give all the working of the above example here. Instead we give the key information only.

Approximate solution with weights

We assign weights of 1 to x and 2 to y giving a (20 x 20) weight matrix with these values alternating down the diagonal.

In this example, because the numbers have been chosen to simplify the arithmetic, the matrix **N** (10 x 10) is diagonal with one value (0.75) for all of its diagonal terms. Its inverse is therefore the simple:

$$\mathbf{N}^{-1} = (4/3)\,\mathbf{I}$$

where **I** is the unit matrix.

All values of the l vector are zero each of the x's because the provisional values were chosen to be the same as the observed values. It must be remembered that the provisional values have to be consistent and

$$F(\overset{*}{m}, \overset{*}{c} : x, \overset{*}{y}) = 0$$

The l values for the y's are the same as before.

The problem finally reduces to a solution of the normal equations

$$\mathbf{N}_1 \mathbf{x} = \mathbf{b}_1$$

which are

$$\begin{bmatrix} 513.33 & 73.33 \\ 73.33 & 13.33 \end{bmatrix} \begin{bmatrix} dm \\ dc \end{bmatrix} = \begin{bmatrix} -77.138 \\ -9.086 \end{bmatrix}$$

The solution is

$$dm = -0.246\,86$$

$$dc = 0.676$$

As would be expected, this solution is not very different from the previous method. In this case the residuals are perpendicular distances from the observed points to the line.

Case 3: solution with additional constraint
Sometimes new measurements have to be constrained to fit previous work upon which maps will have been plotted, or building construction has begun.

Suppose that the line just fitted to data has to pass through a fixed point (x', y').This means that there is one equation which must be satisfied exactly: i.e.

$$y' - mx' - c = 0$$

This is a constraint equation which the finally estimated parameters must also satisfy exactly, i.e. we have

$$y' - \hat{m} x' - \hat{c} = 0$$

One practical way to treat the problem is to assign a very high weight to the fixed point coordinates and treat them as observations in the usual way. If we hold the tenth point nearly fixed by assigning it a weight of 100, we obtain the solution:

$$dm = -0.290\ 1362$$
$$dc = 0.8348$$

Although this procedure is often acceptable in practice, it is incorrect theoretically because an infinite weight cannot be handled computationally.

However, the solution is seen to be acceptable when compared with the exact solution which follows.

Additional constraints
Since theoretically correct treatment is not difficult, it should certainly be employed in scientific studies. Two theoretically correct methods are available.

A conceptually simple method is to eliminate one parameter from the problem by expressing it in terms of the other. Although this would be the simplest way to treat the particular problem of our example, which has only two parameters, it is not convenient in complex cases. For this reason we present a general alternative method of treatment.

Constraints by Lagrangian multipliers
As explained in Appendix 5, the use of Lagrange's method yields equations of the form:

$$N_1 x + E^T k = b_1 \tag{24}$$

$$E x - d = 0 \tag{25}$$

These can be combined into one hyper-matrix as follows

$$\begin{bmatrix} N_1 & E^T \\ E & 0 \end{bmatrix} \begin{bmatrix} x \\ k \end{bmatrix} = \begin{bmatrix} b_1 \\ d \end{bmatrix}$$

If we hold point 10 fixed, its observation equation becomes the constraint equation E, and the hyper-matrix is:

$$\begin{bmatrix} 513.33 & 73.33 & 1 \\ 73.33 & 13.33 & 10 \\ 1 & 10 & 0 \end{bmatrix} \begin{bmatrix} dm \\ dc \\ k \end{bmatrix} = \begin{bmatrix} -77.138 \\ -9.086 \\ -2.080 \end{bmatrix}$$

the solution of which is

$$dm = -0.2922$$
$$dc = 0.8426$$

This compares very well with the previous weighted solution.

Comment
Readers are referred to Appendices 5 and 6 for more theory. A treatment of a level network by the traditional method of condition equations is given in Appendix 5. Worked examples may be found in many chapters of the book; especially in Chapter 2 section 2.8, Chapter 7 section 7.7, Chapter 8 sections 8.2 and 8.8, and Chapter 10 section 10.11.

6 Stellar and satellite systems

Chapter Summary

This chapter is concerned with the use of stars and satellites to provide the positions of points, and the orientation of lines, on the surface of the Earth. The treatment is to identify those principles which are not system dependent, and which are general enough to be used as the basis for the further study.

Introduction

Traditionally the stars, especially the Sun, planets and the Earth's moon, were observed by optical instruments at known times. With the advent of artificial Earth satellites different systems were adopted, using photography, as in stellar triangulation; ranging by electromagnetic distance measurement, as in the SECOR and GPS systems; range difference measurements, as with the Transit Doppler system; or interferometric systems, as with Minitrack or VLBI.

Heights are also obtained by radar altimetry from satellites, and from the stereoscopic analysis of remote sensing imagery and aerial photography.

Technology changes so quickly that it seems unwise to give detailed treatments of current systems. Instead, we use a simple spherical model to explain those mechanisms which underlie past, present and future systems. A brief study of traditional field astronomy assists with an understanding of basic concepts, if only for the reason that the Sun and stars are readily visible, whereas most artificial satellites are not. The reader is referred to any recent copy of the annual publication, *Star Almanac for Land Surveyors* (SALS), for a comprehensive summary of most aspects of field astronomy and relevant data.

Almanacs give the parameters (the ephemerides) needed to locate the positions of various bodies. This information is usually transmitted by satellite systems in real-time (*broadcast ephemeris*), or can be obtained later as a *precise ephemeris*. Much use is made of almanac data to program star observations and select favourable geometry for satellite and star fixes. Software packages are readily available for this purpose.

The almanac data is obtained from ground based stations specially located for the purpose: as for example the original Royal Observatory at Greenwich, England, or the GPS monitor sites. Techniques used at such stations can be very much more sophisticated than with portable equipment.

The astronomical model

The most important factor in modelling the apparent movements of the stars and Sun is that directions alone are significant. The stars are so distant that we can neglect distances almost completely. A simple model treating stars at infinity, with the Earth placed at the centre of a celestial sphere, is sufficient for much basic work. The simple model will be modified where necessary to deal with artificial satellites.

6.1 Time and longitude

The ability of an observer to determine longitude on the surface of the Earth is directly dependent on the accuracy with which *time* can be kept relative to another place. *Longitude* is the difference of time between the transits of two meridians by the same star. *Latitudes* are much simpler to determine without any need for time keeping.

Early astronomers and navigators could measure intervals of time at one place only by pendulum clocks, timing glasses and the human pulse, but could not measure instants relative to another place. Harrison's chronometers were the first to solve this problem in the 18th century: now radio signals and atomic clocks are capable of accuracies of one in 10^{11}.

Concepts of the day and the year are based on the daily spin of the Earth about its axis and its annual trip around the Sun. Today, the Earth is considered a poor time-keeper in relation to atomic repetitions. The whole subject is one of some complexity which is clearly summarized in SALS. However, we need some elementary ideas to understand the principles of astronomy and satellite position fixing. These may be summarized as follows:

(a) The Earth spins on its axis relative to the Sun in one *solar day* and relative to a distant star in one *sidereal day*. That these are different can be understood when one remembers that the Earth goes round the Sun but not round the other stars.

(b) The real Sun appears to move round its orbit (*the ecliptic*) at an irregular speed, so a harmonized fictitious Sun moving round the Equator *at constant speed* is used to define Greenwich Mean Time or *Universal Time* (UT). The true hour angle of the Sun is given by the expression

$$\text{GHA (Sun)} = \text{UT} + \text{E} \tag{1}$$

The quantity E is about 12 hours, to allow for the fact that stars are visible at night and the Sun by day. The exact values are tabulated in the SALS. Also tabulated are coefficients for polynomial interpolation.

(c) One annual trip around the Sun defines a *year*. There are 365.2422 *solar* days or 366.2422 *sidereal* days in the year, depending on the reference body used for the count.The Earth's trip around the Sun accounts for the one–day difference which cannot be counted from the Earth. The sidereal day is about four minutes shorter than the solar. The two are in step on about the 22nd September (the *autumnal equinox*) at the Greenwich meridian. The relationship between the two time systems is expressed by the equation

$$GST = UT + R \tag{2}$$

At the autumnal equinox, $R = 0$ gradually increasing by about 4 minutes per day throughout the year. The exact values are tabulated in the SALS which also gives polynomial coefficients for computer use.

(d) Civil time zones and the calendar are an administrative convenience. Observation times need to be corrected to UT for scientific use. The necessary corrections are obtained from time-signals transmitted at various locations: see SALS for details.

(e) Orbit time is based on the period of one trip of a satellite round its orbit. It is usually reckoned from the orbital point closest to the Earth (*perigee*), which in turn is known relative to the *ascending node* of the orbit (*perigee argument*). For simplified circular orbits the node is used directly as a marker.

6.2 Orbits

It was established by Kepler and others that an object orbiting round a central mass behaves approximately as follows;

(a) The orbit lies in a plane relatively fixed *in direction* in space. It is usually defined by the inclination i of its orbital plane to the plane of the equator.

(b) The shape of the orbit is an ellipse with the central mass at one focus.

(c) The period T of the rotating object is directly linked to its distance r from the central attracting mass and a constant C by the expression

$$T^2 = C r^3$$

(d) The radius vector r, joining the focus of the ellipse to the satellite, sweeps out areas at a constant rate. The *areal* velocity is constant.

Practical departures from these Keplerian rules arise from the action of other forces on satellites and the fact that the Earth's gravity field is irregular. Because most satellites used for geodetic purposes lie in nearly circular orbits, whose treatment is simple and illuminating, we shall adopt a very simple model for most of our discussions here. In this way the surveyor can gain sufficient insight into the basic geometry of satellite systems to understand software and plan practical missions.

Motion in a circular orbit

Following Kepler's law of areal velocity, if the orbit is a circle with constant radius r, then the satellite of mass m moves round the central mass M, the Earth, with a constant angular speed of ω radians per second. Thus the force F away from M due to the motion is

$$F = m \, \omega^2 r$$

But $\omega = 2\pi/T$, where T is the period of the rotation, so

$$F = \frac{m r 4 \pi^2}{T^2} \tag{3}$$

By Newton's law of gravitational attraction, F is also given by

$$F = \frac{k M m}{r^2} \tag{4}$$

When r = R, the radius of the Earth, F = mg, and $kM = g R^2$. Thus we have the useful relationship

$$T^2 = \frac{4 \pi^2 r^3}{g R^2} \tag{5}$$

Substituting values for g = 9.8 and π and converting time to minutes we obtain the useful working formulae

$$T = 84.6 \, (r/R)^{3/2} \tag{6}$$

$$r = R \, e^{(2/3) \ln \left(\frac{T}{84.6}\right)} \tag{7}$$

From equation (6), putting R = 6.4 x 10³ km, we calculate the distance r for a GPS satellite with a 12 hour orbit, as 26 678 km, giving the height h = r – R = 20 000 km. The Earth's moon with a 28 day period is about 384 000 km distant.

Likewise we find from equation (7) that a Doppler satellite with 108 min period is at a height of about 1000 km.

Geocentric reduction

Before applying the formulae of spherical trigonometry to near Earth satellites it is convenient to reduce actual or topocentric altitudes to geocentric values. Refer to Figure 6.1 which shows a satellite S at an observed altitude H on a spherical Earth of radius R. Neglect refraction effects.

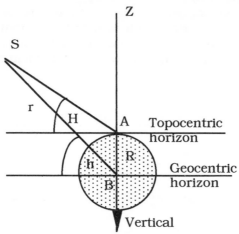

Figure 6.1

To find the geocentric altitude h from the centre of the Earth B we use:

$$\sin(h - H) = \frac{R \cos H}{r} \tag{8}$$

For example, if H = 18°, R = 6.4 and r = 26.67 (GPS satellite) we have:

$$h = 31°$$

This is the altitude of the satellite which would have been observed from the centre of the Earth. The converse calculation uses

$$\tan H = \tan h - \frac{R}{r \cos h} \tag{9}$$

6.2 Astronomy theory

We now give a short summary of the theory of field astronomy because it is needed by the surveyor, who may still require to determine azimuth or position from the stars, and because it is a convenient way to understand the simple theory of satellite geodesy.

Due to the axial rotation of the Earth, the period of which defines the day, the stars appear to rotate round the Earth. The Earth also revolves around the Sun in an elliptical orbit in a period of time – the year. Since the stars are so distant from the Earth, both the Earth and its orbit can be considered as a point in space, and the stars can be considered as fixed to the inside surface of a huge sphere – the celestial sphere – at whose centre is the Earth and its orbit.

The use of geocentric directions to satellites enables the simple model of the celestial sphere to be applied also to near Earth satellites. Only directions on the Earth have significance.

The *three fundamental directions* in field astronomy are

(a) The *axis* of the Earth.
(b) The direction of the plumb line : the *vertical.*
(c) The *line of nodes*: the intersection of an orbital plane with the plane of the equator.

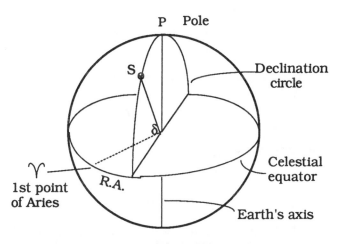

Figure 6.2

Figures 6.2 and 6.3 show the celestial sphere as seen from a point outside it and above the North Pole.

System defined by the spin axis of the Earth

Figure 6.2 shows the following information:

(a) The North and South celestial *poles* about which the stars appear to revolve.

(b) The plane of the *celestial equator* which is perpendicular to the axis and passes through the centre of the Earth.

(c) The *declination* δ of a star (Sun) or satellite - the angle between the equator and the star, measured in the great circle passing through the star and the pole - the *declination circle*. A South declination is negative.

(d) The *right ascension* (RA) of the star or satellite; the angle measured anticlockwise along the celestial equator from a reference point γ – the *first point of Aries* – to the declination circle through the star. The directions clockwise and anticlockwise are as viewed from outside the celestial sphere and above the North Pole. The first point of Aries is also known as the *Equinoctial point*.

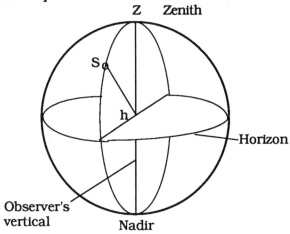

Figure 6.3

The quantities declination δ and right ascension (RA) are tabulated in almanacs from data obtained by direct observations at observatories. The almanac specially designed for land surveyors for all but geodetic work is *The Star Almanac for Land Surveyors*, published annually. Both stellar parameters, declination and right ascension, change very slowly throughout the year. The Sun changes more rapidly than the other stars. Satellite parameters alter very quickly.

The argument used in the SALS for various quantities is generally Universal Time (UT). The UT is obtained from radio time signals or telephone time. In precise work the UT has to be corrected to Atomic time defined with reference to the mean position of the Pole and the non uniform rotation of the Earth. Details are given in SALS.

The system defined by the observer's vertical
Figure 6.3 shows the following information defined by the observer's vertical:

(a) The *zenith* and *nadir* - the points vertically above and below the observer respectively.

(b) The *topocentric horizon* - the plane through the observer perpendicular to the vertical. Sometimes a distinction needs to be made between this and the parallel plane through the Earth's centre - the geocentric horizon.

(c) The *altitude* (h) of a star - the angle between the horizon and the star measured in the great circle through the star and the zenith - i.e. in the *vertical circle* through the star.

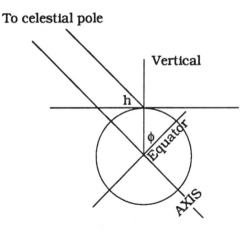

Figure 6.4

Combined systems
The two systems defined by the vertical and the Earth's spin axis are combined together via the following:

(a) The great circle containing the zenith and the poles – the *meridian*.

(b) The angle between the Earth's axis and the vertical – the *co-latitude* c. The latitude, $\phi = 90 - c$, is the angle between the vertical and the equator, measured in the meridian (see Figure 6.4). It will be seen that the altitude of the elevated pole equals the observer's latitude.

Figure 6.5 shows the two systems combined together, with the plane of the meridian in the plane of the paper, and the pole to the right of the zenith. If P is the North Pole, the North and South cardinal points are defined by the intersection of the meridian and the horizon, and points east and west are at right angles to this north-south line.

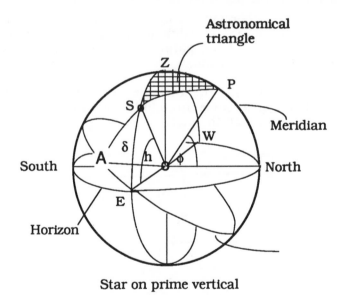

Star on prime vertical

Figure 6.5

When a star or satellite crosses the meridian it is said to *transit*. Since a star does so twice in one revolution of the Earth, we distinguish *upper transit* when the star transits on the same side of the pole as the zenith, and *lower transit* when the transit is on the opposite side of the pole from the zenith.

6.3 The astronomical triangle ZPS

The spherical triangle formed by the zenith Z, the pole P, and the star or satellite S is the astronomical triangle ZPS (see Figure 6.6). The angle Z is called the *azimuth angle*, the angle t the *hour angle*, and the angle S the *parallactic angle*. When the azimuth angle Z is 90° the star is said to be on the *prime vertical* (PV) and when the parallactic angle S = 90°, the star is at *elongation*. A star cannot elongate unless PS < PZ i.e.   90 – δ < 90 – φ   or   δ > φ.

The *azimuth* α of a star is the angle between the meridian plane and the plane of the vertical circle through the star, reckoned clockwise from North, even in the southern hemisphere. Hence for an eastern star, Z = α and for a western star, Z = 360 – α. The azimuth of a reference object (RO) is found by observing the horizontal angle between the star and the RO at the time of observation.

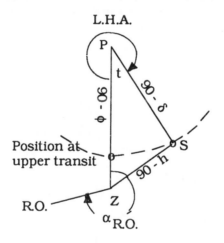

Figure 6.6

The Local Hour Angle (LHA) of a star is the angle between the plane of the meridian and the plane of the declination circle through the star, measured clockwise from the star's position at upper transit. Hence, for an eastern star, t = 24 hours – LHA and for a western star t = LHA. Hour angles are reckoned in units of time which may have to be converted to units of arc to obtain the values of their trigonometrical functions. Computer algorithms will use arguments in radians.

The *sides* of the astronomical triangle are

$$PZ = 90 - \phi \qquad PS = 90 - \delta \qquad ZS = 90 - h$$

The sides and angles are related by the formulae of spherical trigonometry: especially the cosine and cotangent formulae which avoid the sign ambiguities which arise from the sine formulae. Napier's rules for right-angled triangles are also useful. See Appendix 2.

The plane of the orbit and line of nodes
The orbit of the Sun or moon or artificial Earth satellite is inclined to the plane of the equator by an angle i : the *inclination* of the orbit. In the case of the Sun this angle is 23.5', for Doppler satellites it is about 90' and GPS satellites 55'. The line of intersection of the orbital and equatorial planes is the *line of nodes*. The ascending node is where the satellite crosses the equator from south to north. Its position is given by its Right Ascension.

In the case of the Sun the first point of aries γ, the zero point for Right Ascension, lies on the line of the Sun's nodes, called the *spring* and *autumn equinoxes*, since day and night are of equal length then. The plane of the Sun's orbit is called the *ecliptic*.

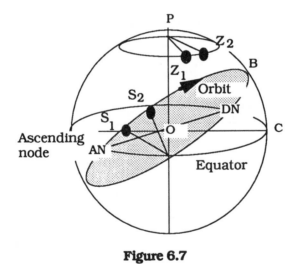

Figure 6.7

Figure 6.7 shows a GPS satellite at two successive positions S_1 and S_2 with the observer's zenith at Z_1 and Z_2 at corresponding instants.

6.4 The determination of azimuth from the Sun and stars

Although the azimuth between two points can be determined from two satellite positions, it is still a viable, and cheaper, alternative to do so from theodolite observations of the Sun or stars. In certain circumstances it may be the only way. An isolated oil platform, for instance, is unsuited to north–seeking gyropscopic methods because of instability, and another platform may not be visible from it.

There is also some merit in using Black's method of computation to obtain a geodetic azimuth. This method also yields the components of the deviation of the vertical as by-products.

The process of levelling a theodolite sets the primary axis parallel to the direction of the gravity field. Therefore *astronomical* positions and azimuths result. If it were possible to tilt the theodolite primary axis parallel to the ellipsoidal normal, then *geodetic* parameters would result.

For many practical purposes, such as in cadastral surveying, the distinction between astronomical and geodetic azimuths is academic and can be ignored.

We now outline the simple methods available to the surveyor.

Azimuth determination
The simplest method is to observe horizontal and vertical angles by theodolite, noting the approximate time only to interpolate data from SALS. This method is not capable of the highest accuracy, but can easily give results to about 5" arc.

The best method is to observe time and horizontal angle, noting the approximate vertical angles for possible i tanh corrections.

It is now practicable to observe all three quantities, horizontal angles, vertical angles and time, with electronic theodolites on–line to a field computer. The author has obtained first order results by this method from Sun observations.

Azimuth by simultaneous observation of horizontal and vertical angles
Applying the cosine formula of spherical trigonometry to the astronomical triangle ZPS gives:

$$\cos Z = \frac{\sin \delta - \sin \phi \sin h}{\cos \phi \cos h} \qquad (10)$$

from which Z and therefore α is derived. The angle θ to the RO is also known, and thence the azimuth of the RO is found.

Observation of altitude h
In star or Sun observations the telescope is pointed in such a manner that the star will appear to cross the cross-hairs at the required point without the observer requiring to move the slow motion screws at the time of the observation. When both horizontal and vertical angles are required, this procedure is difficult, and at least one slow motion screw has to be moved to bring about simultaneous bisection of both cross-hairs.

Care must be exercised in observing the Sun since the direct light passing through the telescope can cause serious damage to the eye if viewed even momentarily without the dark glass in position. If such a glass is not available, the image of both cross-hairs and the Sun can be focused on a piece of white card and the observation carried out viewing the card. Many observers, including the author, prefer this method.

When picking up the Sun in the field of view of the telescope, the best way is to view the shadows of the open sights of the telescope: when they are coincident the Sun is in the field of view. Picking up a star is facilitated if the open sights are illuminated by holding a torch high up behind one's head to illuminate the sights.

In all astro work involving altitudes, the vertical index error of the theodolite should be determined by observing a fixed object on both faces. If necessary the results can be computed for the separate observations on each face. Great care must be taken to level the theodolite properly using the vertical bubble, and check that the vertical circle bubble is level when taking vertical angles. Instruments with compensators are a great advantage in astronomy.

Corrections to an observed altitude $\overset{o}{h}$

If the star is observed on one face, the index error of the theodolite has to be applied. If both faces are used (as is normal in azimuth work) the mean of both faces is acceptable, provided the two observations are carried out within not more than five minutes of time. A correction for the curvature of the stellar path may have to be applied in precise work.

Due to the refracting or bending of the light path passing from the star to the observer, the theodolite vertical angle is too high and a correction amounting to about

$$58'' \cot \overset{o}{h}$$

has to be subtracted from the observed altitude. A better value of this refraction correction is obtained from tables in SALS which allow for the effects of atmospheric pressure P and the temperature t.

Example

$$\overset{o}{h} = 26° 27' 32'', \quad P = 29.9 \text{ inches of mercury, and } t = 75 \text{ F}°.$$

SALS gives the refraction correction $r = r_0 f$, where f is obtained from a table entering with P and t, and in this case $f = 0.95$. The angle r_0 is obtained from a critical table whose argument is the observed altitude.

The necessary critical entries in this case are

$$\overset{o}{h} = 26°26' \qquad r_o = 116'' \qquad \overset{o}{h} = 26°38'$$

Thus

$$r = 0.95 \times 116'' = 110'' = 01'50''$$

The final altitude corrected for refraction is then

$$h = 26°27'32'' - 01'50'' = 26°25'42''$$

Since the refraction correction increases rapidly as the observed altitude decreases, the altitude should not be less than 15° if possible, though for Sun observations at high latitudes, altitudes lower than this may have to be observed in winter.

Parallax correction
A parallax correction of $+ 8''.9 \cos h°$ is applied to all Sun altitudes if considered significant. In this example the correction is $+ 8''$.

Observation
Because the Sun appears so large in the field of the telescope, accurate bisection is not possible without special attachments. Normally, observations are made with the Sun tangential to the cross-hairs as shown in Figure 6.8.

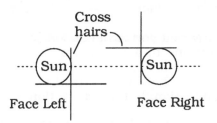

Cross hairs

Sun · · · · · · · · · · Sun

Face Left   Face Right

Omitting movement of Sun

Figure 6.8

Observing on face-left, the limbs of the Sun are placed tangentially to the cross-hairs in the top left hand quadrant, and on face-right in the quadrant diametrically opposite. The means of the horizontal and of the vertical angles give the position of the centre of the Sun.

If an observation is made in this manner on one face only, a correction for the Sun's semi-diameter (s.d.) is applied. The correction to a horizontal pointing at altitude h is:

$$\frac{s.d.}{2} \sec h$$

The s.d. which is about 16 minutes of arc is tabulated for each month in SALS. Notice that there is a potential danger of confusion with the stadia hairs which are 34' apart.

An attachment which enables the centre of the Sun to be bisected is the *Roelof's solar prism*. This prism splits the Sun into four images when viewed in the telescope, the overlapping positions of which form a bright star shape which is bisected to make the observation (Figure 6.9).

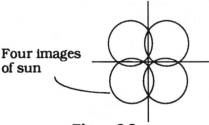

Four images
of sun

Figure 6.9

This prism attachment gives more accurate work in Sun observations than the method of limbs. Corrections have to be applied if observations are made on one face only using the prism. Some observers using the white card projection method claim equal accuracy as for the solar prism.

Corrections to an observed horizontal direction
The following corrections to a horizontal circle reading (direction) may be required:

(a) Horizontal collimation: if the observation is made on one face.

(b) If the observation is on one face to the limb of the Sun the correction for the semi-diameter is 16' sec h which becomes excessively large as h > 90°. Observation on both faces reduces errors from this source.

(c) If the primary axis of the theodolite is tilted relative to the vertical, the correction to be applied to the horizontal reading to the elevated star is + i tan h. This effect is not eliminated by changing face. To determine i, the plate bubble is read when making a pointing to the star. Both ends of the bubble are read (i.e. left L and right R as viewed from the observer's position). See Chapter 3, Figure 3.4.

Example
For example the bubble readings are as follows

	Face left			Face right	
	Bubble L	Bubble R		Bubble L	Bubble R
	+2.5	+2.5		+3.0	+2.0

Then i is given by

$$i = \frac{(\Sigma L - \Sigma R)\; D}{4}$$

both in magnitude and sign, where D is the value of one division of the particular bubble in question.
In the example

$$D = 20",$$
$$h = 26° \; 27'$$

The complete correction C is:

$$C = + \; \frac{(\Sigma L - \Sigma R)}{4} \; D \tan h$$
$$= + \; (+\, 2.5 \; + 3.0 - 2.5 - 2.0)\; 20" \; \times 0.497/4$$
$$= + 2"$$

which is negligible in this Sun observation.

Table 6.1

Field Book
Sun azimuth by altitudes

Theodolite No 26   Observer ALA   Booker DJM   Date 24 th September

At stn 1   RO   stn 2   Value of plate bubble divn = 20 "

Temp 24° C Pressure 1012 mb

Obs No	Notes		Hor Read ° ' "			Bubble L	R	Vert Read ° ' "			Approx UT
1	RO	FR	271	40	04						
2	S	FR	58	11	40	3.0	2.0	62	56	47	14h 39 m
3	S	FL	238	02	47	2.5	2.5	297	22	20	
4	RO	FL	91	39	16						
5	RO	FL	00	10	28						
6	S	FL	147	43	47	2.5	2.5	296	49	16	14h 48m
7	S	FR	328	43	40	3.0	2.0	63	54	12	
8	RO	FR	180	11	18						

The effect of axis tilt is small if h is kept low. This conflicts with the requirement that h has to be kept above 15° to avoid large refraction corrections to the vertical angle. Thus if vertical angles are observed, the value of h is kept between 15° and 60° as a compromise.

Note, the method of determining the tilt by vertical circle readings is not always feasible in astronomical observations when the azimuth of the star is changing rapidly. It is practicable however with Polaris observations. In any case, the theodolite should be levelled as accurately as possible to avoid the axis tilt error.

Complete example of Sun azimuth
In Table 6.1 a complete example of a Sun observation is shown, together with the calculation of the final azimuth of the RO in Table 6.2. Normally at least six observations are made, with the resultant azimuths agreeing to within a minute of arc at the outside.

For best results a set of balanced observations should be made at similar times before and after transit. This procedure almost eliminates systematic errors which arise from using an inaccurate latitude and from errors in the refraction model.

In this case the value of the latitude was scaled from a map, and the declination obtained from SALS. The tabular entries in the Table for the Sun are

24th September

UT = 12 h   Dec = South 00° 34'.8
UT = 18 h   Dec = South 00° 40'.6

Since the approximate time of the observation was 14 h 46 m U.T. the interpolated declination is South 00° 37'.3 which is treated as a negative quantity in the computation, i.e.

$$\delta = -00° \, 37' \, 18''$$

Azimuth by time and horizontal angles
This method of determining azimuth is potentially more accurate than by altitudes, since the observation involves the bisection of only the vertical cross hair (horizontal angle); vertical refraction does not affect the method; and the stars or Sun may be observed low down close to the horizon thus reducing the errors due to axis tilt.

Theory
From the observed time the hour angle t is calculated: the declination δ is obtained from SALS and the latitude and longitude of the observer are scaled from a map or obtained from the survey

coordinates. Applying the cotangent formula to the astronomical triangle gives

$$\cot Z = \frac{\cos \phi \tan \delta - \cos t \sin \phi}{\sin t} \tag{11}$$

The azimuth is obtained from Z and the horizontal angle to the RO.

Table 6.2

Computation of Sun Azimuth by altitudes

Formula

$$\cos z = \frac{\sin \delta - \sin \phi \sin h}{\cos \phi \cos h}$$

$\phi = 51° 35' 30"$ Sun West RO East

Part One Bubble correction tan h = 0.51

ΣL– ΣR	1.0			1.0		
i tanh			2.5			2. 5
	°	'	"	°	'	"
Mean Hor on Sun	58	07	13.5	148	13	43.5
Corrected	58	07	16	148	13	46
Mean Hor on RO	271	39	40	00	10	53
θ angle to Sun	146	27	36	148	02	53

Part Two Azimuth of RO

Mean Obs h	27	12	46	26	27	32
Refn corrn		– 01	46		– 01	50
Parallax			+ 08			
						+08
h	27	11	08	26	25	50
Approx UT		14h	39m		14h	46m
Dec	–00	37	10	–00	37	18
Z Sun West	131	51	56	130	16	31
θ	146	27	36	148	02	53
	278	19	32	278	19	24
Azimuth of RO	81	40	28	81	40	36
Mean	81	40	32			

Derivation of the hour angle t

An accurate chronometer is required to give times to 0.1 sec. The chronometer correction to UT is found by comparison with a radio time–signal or telephone time. A stop watch enables an accurate determination of the chronometer error to be made. Consider a determination using the BBC time-signal. The sixth pip of a series is the exact time of the signal. For example if the signal is transmitted at 09 h 00 m 00.0 s UT the first pip is emitted at 08 h 59 m 55.0 s. As the pips are heard, the observer beats time with them keeping the stop watch ready. On the sixth pip he starts the watch, which now theoretically continues the exact UT of the signal. The watch is then stopped at some convenient exact second on the chronometer, thus the UT of this chronometer second is the UT of the signal plus the watch reading, and the chronometer correction to UT is thus known. Note it is safer to refer to Fast and Slow than plus or minus which can be ambiguous.

Example

UT of the signal	$09^h\ 00^m\ 00.0^s$
stop watch reading	23.7^s
sum	$09^h\ 00^m\ 23.7^s$
chronometer reading	$09^h\ 10^m\ 10.0^s$
chronometer FAST on UT by	$09^m\ 46.3^s$

The chronometer time of an observation is obtained in a similar manner using the stop watch. When the star crosses the vertical hair the observer starts the watch, and later stops it on an exact second of the chronometer. The chronometer time of the observation is therefore this chronometer time minus the reading on the watch.

Example

stop watch reading	11.8^s
chronometer reading	$11^h\ 23^m\ 20.0^s$
chron time of observation	$11^h\ 23^m\ 08.2^s$
correction to UT	$-\ 09^m\ 46.3^s$
UT of the observation	$11^h\ 13^m\ 21.9^s$

After the UT of the observation has been calculated, the procedure to obtain the hour angle differs for the Sun and the stars.

Hour angle of the Sun

The Greenwich hour angle of the Sun (GHA) is first obtained from

$$GHA = UT + E$$

where E is obtained from SALS. For example if the UT is $14^h\ 22^m\ 43.8^s$ the requisite portion of the table for the Sun is

24th September

$$12^h\ UT \quad E = 12^h\ 08^m\ 02.3^s$$
$$18^h\ UT \quad E = 12^h\ 08^m\ 07.5^s$$

and the interpolated value of E for the observed time $14^h\ 22^m\ 43.8^s$ is

$$E = 12^h\ 08^m\ 04.3^s$$

Thus the GHA is

$$14^h\ 22^m\ 43.8^s + 12^h\ 08^m\ 04.3^s$$
$$= 26^h\ 30^m\ 48.1^s = 02^h\ 30^m\ 48.1^s$$

The local hour angle (LHA) is given by

$$LHA = GHA + \lambda$$

where λ is the longitude of the observer with respect to Greenwich, reckoned positive eastwards and negative westwards.
   In the example

$$\lambda = 2.5^s\ West = -2.5^s$$

whence the LHA (Sun) is

$$02^h\ 30^m\ 45.6^s$$

Since the Sun is in the West

$$t = LHA = 02^h\ 30^m\ 45.6^s$$

If the Sun had been in the East then

$$t = 24h - LHA$$

Hour angle of a star
To find the LHA of a star the Greenwich Sidereal Time (GST) is first found from

$$GST = UT + R$$

where R is tabulated in SALS.

Example
For the UT $21^h\ 45^m\ 28.7^s$ on the 24th September the relevant tabular values of SALS are from the table for the Sun:

$$
\begin{array}{ll}
\text{U T} & \text{R} \\
18 \text{ h} & 00^{\text{h}} \ 14^{\text{m}} \ 22.3^{\text{s}} \\
24 \text{ h} & 00^{\text{h}} \ 15^{\text{m}} \ 21.5^{\text{s}}
\end{array}
$$

from which the interpolated value of R is $00^{\text{h}} \ 14^{\text{m}} \ 59.3^{\text{s}}$ and the

$$\text{GST} = 21^{\text{h}} \ 45^{\text{m}} \ 28.7^{\text{s}} + 00^{\text{h}} \ 14^{\text{m}} \ 59.3^{\text{s}} = 22^{\text{h}} \ 00^{\text{m}} \ 28.0^{\text{s}}$$

Table 6.3

Field Book
Sun azimuth by hour angle

Theodolite No 26   Observer ALA   Booker DJM   Date 24 th September

At stn 1   RO   stn 2   Value of plate bubble divn = 20 "

Temp 24° C Pressure 1012 mb

No	Notes		Hor Read ° ' "	Bubble L R	Chron h m s	SW s	Obs h m s
1	RO	FR	271 40 07				
2	S	FR	55 45 08	3.0 2.0	14 13 30	08.0	14 13 22.0
3	S	FL	235 30 50	3.0 2.0	14 15 10	11.5	14 14 58.5
4	RO	FL	91 39 19				
5	RO	FL	91 39 18				
6	S	FL	236 23 59	2.5 2.5	14 18 40	11.6	14 18 28.4
7	S	FR	57 26 15	3.0 2.0	14 20 10	06.9	14 20 03.1
8	RO	FR	271 40 02				

Chronometer data	h	m	s
UT signal	14	11	10.0
Stop watch			42.0
	14	11	52.0
Chron	13	58	00.0
Slow on UT		13	52.0

The GHA of the star is finally obtained from the GST, and the right ascension (RA) of the star also obtained from SALS, as follows:

$$
\begin{array}{ll}
\text{RA} & 11^{\text{h}} \ 20^{\text{m}} \ 11.0^{\text{s}} \\
\text{GST} & 22^{\text{h}} \ 00^{\text{m}} \ 28.0^{\text{s}}
\end{array}
$$

$$\text{GHA} \quad 10^h \, 40^m \, 17.0^s$$

The LHA is found from the GHA as for the Sun, i.e. from

$$\text{LHA} = \text{GHA} + \lambda$$

SALS contains tables to assist in the interpolation of the various quantities E, R, δ, and RA. It also contains coefficients of polynomials to interpolate for R and E within a computer program.

Practical example of azimuth by timed horizontal angles of the Sun
Table 6.3 gives the observed angles, time and chronometer data for the computation of the azimuth of the RO in Table 6.4. The error of the chronometer was determined by comparison with telephone time which is accurate to 1/20 second.

Error analysis
The following equations are of importance in the error analysis of azimuth determination from stars or the Sun:

$$dz/dh = \cot S \sec h \tag{12}$$

$$dz/dt = - \cos S \cos \delta \sec h \tag{13}$$

If the parallactic angle at the star is 90°, the star is at elongation and the best results are possible. Stars however cannot elongate at latitudes greater than their declinations; which is particularly unfortunate in the case of the Sun. A position close to the prime vertical (at azimuths of 90° or 270°) is therefore selected.

In the altitude method, the star cannot be close to the meridian where cotS becomes very large. Reasonable results can be obtained from the hour–angle method close to the meridian since cosS is only a maximum of unity.

In the altitude method, h cannot be too low because of refraction. A minimum of 15° is usually adopted. By hour–angle, there is no such restriction.

In both methods the altitude should not be too great, because of the sech term and a possible i tanh tilt correction. A maximum restriction to 60° is usually taken.

To minimise the effect of an error in the latitude, observations are balanced East and West of the meridian. In the case of the Sun this means waiting to take morning and afternoon observations.

Only in the case of elongation is the effect of an error in longitude eliminated from the hour angle method. Good results are obtainable by hour angle from close circumpolar stars, such as Polaris in the northern hemisphere or Sigma Octantis in the southern (because δ tends to 90°). These stars appear to move very slowly in azimuth and,

Table 6.4

Computation of Sun Azimuth by hour angle

Formula

$$\cot z = \frac{\cos \phi \tan \delta - \sin \phi \cos t}{\sin t}$$

$\phi = 51°\ 35'\ 30''$　　Sun West : RO East

$\lambda = 02.5^s$ West

Part One : Bubble correction　tan h = 0.51

	2.0	3.0
$\Sigma L - \Sigma R$		
i tanh	5'	8'
Mean Hor on Sun	55° 37' 59"	236° 55' 07"
Corrected	55° 38' 04"	236° 55' 15"
Mean Hor on RO	271° 39' 43"	91° 39' 40"
θ angle to Sun	143° 58' 21"	145° 15' 35"

Part Two: Azimuth of RO

	h　m　s	h　m　s
Chron time	14　14　10.2	14　19　15.8
Chron corrn	+ 13　52.0	+ 13　52.0
UT obs	14　28　02.2	14　33　07.8
E at 12 h	12　08　02.3	12　08　02.3
Δ for E	+ 02.1	+ 02.1
GHA Sun	26　36　06.6	26　41　12.3
λ	−2.5	−2.5
LHA Sun = t	02　36　04.1	02　41　09.8
t arc	39° 01' 01	40° 17' 27
δ	−00　37　12	−00　37　12
Z	134　21　21	133　04　00
θ	143　58　21	145　15　35
	278　19　42	178　19　35
Azimuth of RO	81　40　18	81　40　25
Mean	81　40　22	

because of their proximity to the poles, special simplified computation is possible using specially designed tables, such as the Pole Star Tables of SALS, although computer processing has made the use of tables obsolescent.

6.4 Position fixation by vertical angles and time

Timed vertical angles to stars enable the observer's position to be determined to about 15 m without much difficulty. This is consistent with an accuracy of 0.5" in determining the direction of the vertical. If the method is used in conjunction with a satellite position fixation, which is not based on the direction of the vertical, the local slope of the geoid relative to the ellipsoid is also known. This technique forms the basis of the astro-geodetic determination of the geoid.

The calculations involved are quite similar to those for satellite prediction and the selection of good geometry for position fixing.

Example
Vertical angles, corrected for refraction, are made to the five stars listed in Table 6.5, at known times. These observation times are reduced to UT via clock corrections, as for the azimuth determination. Also recorded are the approximate azimuths of the stars.

The method follows the standard procedures for least squares estimation, by assuming initial coordinates (latitude and longitude) for the position of the observer, which are corrected to fit the observations.

Table 6.5

star	Z˙	h˙	dec (rad)	t (rad)
18	76	37	0.625 1370	1.271 8270
19	93	61	0.712 4478	0.694 1822
24	-133	39	0.141 8776	0.611 5502
16	-66	53	0.877 4818	1.035 7851
6	180	25	-0.261 7994	0.000 0000

In this type of observation it is usual to observe on one face only, making at least six rapid observations to each star. These observations are tested for blunders, either by plotting manually, or from a check on residuals and sample tests. The vertical index E of the theodolite should be determined separately from observations on both faces to a fixed object. However it can be modelled in the analysis. We shall do this here, having first applied an index of 25" to all vertical angles.

The example data of Table 6.5 are taken from the standard data set of Chapter 2.

In actual star work, the declinations of the stars are obtained by interpolation from SALS. In the example, they have been calculated from the azimuths and vertical angles using the formula:

$$\sin \delta = \sin \phi \sinh + \cos \phi \cos h \cos Z \qquad (14)$$

In practice, the hour-angle t, of a star, is obtained from

$$LHA = UT + R - RA + \lambda$$

$$\text{where} \quad t\,(\text{West}) = LHA \quad \text{or} \quad t\,(\text{East}) = 24 - LHA$$

In this simulated case each value of t is calculated from:

$$\cos t = \sin h \sec \phi \sec \delta - \tan \phi \tan \delta \qquad (15)$$

[Note:These equations (14) and (15) are also used in practical astronomy to check the identification of stars; the right ascension being the unknown part of t in equation (15).]

Table 6. 6

star	t'	h' deg	O–C "	O–C + 25"
18	1.271 8470	36.999 963	0.1343	25.1343
19	0.694 2022	60.999 118	3.1758	28.1758
24	0.611 5302	38.998 629	4.9355	29.9355
16	1.035 7651	53.001 812	–6.5219	18.4780
6	0.000 0200	24.997 200	10.0800	35.0800

In the example, the known position of the observer is 50˚ North latitude, 1˚ West longitude. Assuming a *provisional position* of

50˚ 00' 10" latitude, and – 00˚ 59' 55.86" longitude

the simulated vertical angles which would have been obtained to the stars from this position at the observation times are calculated from

$$\sin h = \sin \phi \sin \delta + \cos \phi \cos \delta \cos t' \qquad (16)$$

The hour angles t' are obtained from t with a longitude change of 0.000 02 rad (4.14"). The results are listed in Table 6.6 which shows the O–C values to which 25" have been added for the vertical index E.

Differentiating equation (16), assuming the declination is error-free, gives

$$dh = \cos Z \, d\phi + \cos \phi \sin Z \, dt \qquad (17)$$

The small differential part dh is separated into the index parameter E and the the usual parts, $\overset{o}{h} - \overset{\bullet}{h} = 1$ (observed – provisional) and the residual v. The term dt is equated to the change in longitude $d\lambda$, although timing errors could be modelled separately. Thus we obtain the observation equation of the simple kind in the form

$$\mathbf{A x = l + v}$$

with a typical equation

$$\cos Z \, d\phi + \cos \phi \sin Z \, d\lambda - E = l + v \qquad (18)$$

$$P d\phi + Q d\lambda - E = l + v$$

The five observation equations in the three parameters are given in Table 6.7. The normal equations, their inverse and solution are given in Table 6. 8 in the standard forms

$$\mathbf{N x = b}$$
$$\mathbf{x = N^{-1} b}$$

Table 6.7

P	Q		1
0.241 922	0.623 658	−1	25.1343
−0.052 336	0.641 869	−1	28.1758
−0.681 998	−0.470 080	−1	29.9355
0.406 737	−0.587 182	−1	18.4780
−1.000 000	0.000 000	−1	35.0800

Because the data of the example are perfect, the residuals are all zero and the standard error factor cannot be estimated. If we assume this to be unity, the inverse of the normals is the dispersion matrix of the parameters, and the sum of the first two diagonal elements (1.5) is called the *Positional Dilution Of Precision* (PDOP) of the solution.

[Note: Although we have used the simpler least squares model to save space, it is sometimes worthwhile to adopt the general approach, because there are really two observations (altitude and time) in equation (17)]

Table 6.8

	Normals		b
1.691 8217	0.199 0476	1.085 6758	−43.374 37
0.199 0476	1.366 7004	−0.208 2678	8.838 53
1.085 6758	−0.208 2680	5.000 0000	−136.803 82
	Inverse		Soln
0.708 1420	−0.127 3745	−0.159 0681	−10.08
−0.127 3745	0.759 2744	0.059 2840	4.12
−0.159 0681	0.059 2840	0.237 0087	−25.00

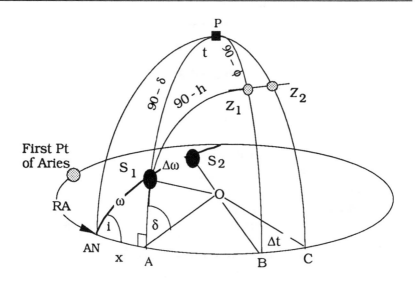

Figure 6.10

6.5 Example of satellite prediction

Although Earth satellites are no longer observed by angular methods, their positions relative to the measurement station are required to calculate good geometry for range or range-difference measurements. See Chapter 8 section 8.4.

However, the simple prediction of satellite positions follows closely along similar lines to the position-line calculations of the above section. The declinations and right ascensions of satellites vary more quickly than those of stars or the Sun. For example, the period of a GPS satellite is 12 hours, and its angle of inclination is 55°.

The key information about a satellite is the right ascension of its ascending node AN. This relates the orbit to the meridian of Greenwich via the right ascension of Greenwich, better known as the GST.

Figure 6.10 shows the basic elements of a simple model to predict GPS satellite positions for further use in an error prediction model. If a satellite's geocentric altitude h and its azimuth z are known at a position whose zenith is at Z_1 (latitude ϕ), its declination and hour angle can be calculated as for position-line stars from equations (14) and (15). From the UT the GST is found. The only missing link is the position of the ascending node AN. This is found from a knowledge of the inclination of the orbit i in the right-angled spherical triangle formed by AN, S_1, and A. The parts labelled x and ω are solved from equations

$$\sin x = \tan \delta \cot i \qquad (19)$$
$$\sin \omega = \sin \delta \operatorname{cosec} i \qquad (20)$$

Thus the right ascension of the ascending node is the RA of the satellite minus x. The angle ω is the distance along the orbit which the satellite has gone since at the node. After a known time Δt it will travel a further angular distance $\Delta \omega$ given by

$$\Delta \omega = \frac{\Delta t}{T}$$

where T is the period of the satellite (approx 12 hours ST for a GPS satellite). Thus the new value of its declination and RA can be calculated from a second right-angled triangle using formulae (20) and (19) in that order.

Meantime the observing position has moved to Z_2 with t increased by Δt. Finally the new altitude and azimuth of the satellite are calculated from equations (11) and (16). The topocentric altitude is computed from equation (8). The range to the satellite is computed from the radius of the Earth R and the known orbital distance r. This range is then compared with the measured range to give the absolute term of a variation of coordinates estimation process. As the whole process is quite lengthy, an example in two dimensions is given in Chapter 8 to illustrate the basic principles of position determination by range difference such as is used by GPS.

The data of Table 6.9 is for successive positions of the GPS satellite vehicle number 19 on 30th March 1992.

Table 6.9

UT	$H°$	$z°$	$h°$	$\delta°$	$t°$	$x°$	$\omega°$
07.30	60	116	67	36	26	31	46
08.00	61	93	68	44	31	42	58
09.00	48	60	57	55	55	84	87
10.00	18	58	31	43	91	138	123

Sky charts
The azimuth and altitude data for a satellite or star are often used to compile a sky chart depicting the geometry of the observation period. Figure 6.11 shows the above data for satellite 19 together with four others used in a GPS fixation. The altitudes and azimuths for satellites 18, 25, 24, 16, and 6 are given in Table 6.10.

The good cuts of the position lines at various azimuths give good planimetric strength, and the different altitudes indicate good height fixes. The PDOP = 3 at about 08.00 hrs indicates a good position. A figure greater than 10 would be unacceptable.

Table 6.10

sat 18					sat 24		
UT	07.30	08.00	08.45	09.00	10.00	07.30	8.00	09.00
$H°$	47	37	15	21	44	47	39	15
$Z°$	70	76	145	143	123	239	227	210

sat 16				sat 6			
UT	07.30	08.00	09.00	10.00	07.45	08.00	09.00	10.00
$H°$	43	53	74	56	18	25	55	84
$Z°$	296	294	248	185	179	180	182	121

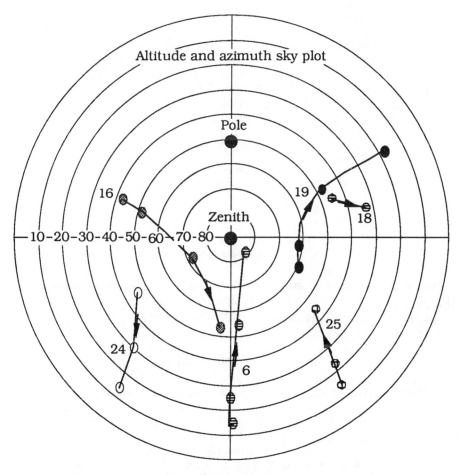

Figure 6.11

6.6 Geodetic azimuth by Black's method

The calculation of azimuth by the method due to Black (reference 4), solves for the misalignment of the ellipsoidal normal relative to the observer's vertical. It uses the I tan h effect on a horizontal direction to determine the unknown tilt I and its azimuth A. See Chapters 3 and 4. First we create a typical set of "observation" data resulting from the tilt of a theodolite axis, then use it to derived the tilt components as in a real case, thus explaining principles and verifying their effect.

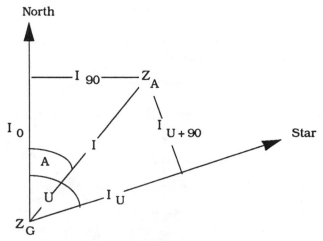

Figure 6.12

Figure 6.12 shows the angular relationship I, at azimuth A, between the astronomical and geodetic zeniths Z_A and Z_G. The meridian component is I_0 and the prime vertical component is I_{90}.

At any azimuth U we have

$$I_{U+90} = I_0 \sin U - I_{90} \cos U$$

The effect of this cross tilt on the horizontal pointing is

$$- I_{U+90} \tan h_U$$

This pointing error will enter directly into the azimuth of the RO, assuming its altitude is close to zero. If not, tilt effects on it also have to be allowed for in the same way. At least four stars at altitudes of about 50° are selected at azimuths varying by about 90° round the horizon. The star azimuths are found by the hour angle method using the *geodetic* latitude and *geodetic* longitude of the station. These are values computed on the ellipsoid using any data available.

When linking the star direction to the RO, the tilt effect differs according to the values of –sin U tan h and cos U tan h. Thus the different values obtained for the azimuth of the RO can be used in reverse to find the tilt components as well as the geodetic azimuth of the RO. An example should make the method clear.

Example of Black's azimuth

We assume that the standard data set are geodetic values. Assume the astronomical zenith is north and east of the geodetic zenith by 100" at an azimuth of 30°. The tilts at the various azimuths U will be :

$$I_U = 100'' \cos (U-30°)$$

and perpendicular to these directions:

$$I_{U+90} = 100'' \sin (U-30°)$$

The former alters altitudes directly from h_G to h_A, and the latter alters directions by:

$$dU = 100'' \sin (U-30°) \tan h$$

Table 6.11 gives all the values calculated from these expressions.

Table 6.11

star	U°	U–30°	I_U ''	I_{U+90}	h_G°	h_A ° ' ''	dU ''
18	76	46	69	71.93	37	37 01 09	54.20
19	93	63	45	89.10	61	61 00 45	160.74
24	227	197	–96	–29.24	39	38 58 24	–23.68
16	294	264	–10	–99.45	53	52 59 50	–131.97
6	180	150	–87	50.00	25	24 58 33	23.32
						Mean	16.52

If the geodetic azimuth of the RO is 0°, the mean astronomical azimuth will be 16.52''. This is the result which would have been obtained from a theodolite tilted by 100'' relative to the ellipsoidal normal. Notice that this is from a perfectly levelled theodolite.
In practice, the problem is the reverse of the model just set up. The observation equations will be of the form:

$$- I_0 \sin U \tan h + I_{90} \cos U \tan h + K = l + v$$

The extra parameter K is the correction to the astronomical azimuth to give the geodetic azimuth. The observation equation coefficients are listed in Table 6.12. The provisional value of the azimuth is taken as 16.52'' giving the C–O terms shown. Not surprisingly the solution is perfect since the model had no random errors introduced.
  The expected corrections are

$$I_0 = -100 \cos 30° = -86.60 \qquad I_{90} = -100 \sin 30 = -50.00 \qquad K = -16.52$$

Black's method is extremely practical since it merely involves the simple process of observing good quality azimuth from several timed stars. The procedure is similar to that for the Sun by the hour angle method. A scheme of three rapid face left and right observations to each star is acceptable, with times to 0.1s. It is usual to include observations to four low altitude stars in the set of eight. These give good values of the azimuth itself but contribute little to the determination of the components of the deviation of the vertical.

Black's method has the very considerable advantage over the Laplace equation, see Chapter 4, in that it does not require astronomical longitude to be observed.

Table 6.12

Observation equations			C–O = l
–0.7312	0.1823	1	37.68
–1.8016	–0.0944	1	144.22
0.5922	–0.5523	1	–40.20
1.2123	0.5398	1	–148.49
0	–0.4663	1	6.80
Normals			b
5.6008	0.3641	–0.7283	–491.19
0.3641	0.8560	–0.3909	–67.868
–0.7283	–0.3909	5	0
Inverse			Soln
0.1859	–0.0692	0.0217	–86.60
–0.0692	1.2372	0.0867	–50.00
0.0217	0.0867	0.2099	–16.52

Modern methods of field astronomy
At the time of writing, several developments are being reported on the use of CCD television cameras attached to theodolite or other motor-driven telescopes to produce very rapid observations of top quality for use in geoid determinations in conjunction with satellite positioning systems.

7 Electromagnetic distance measurement

Chapter Summary

This chapter is concerned with the use of electromagnetic waves to measure distances. The technique is described as EDM for short, meaning *electromagnetic distance measurement*. Systems operating on *microwave* radio carriers are called MDM, and those using *optical* carriers are called EODM.

The treatment is to identify those general principles which give some understanding of systems currently used by the surveyor and geodesist, but which should have some degree of permanence.

Also treated are the reduction of distances to datum, and the problem of EDM calibration.

The treatment of distances in coordinate computation will be found in Chapter 8.

7.1 Introduction to EDM

The most accurate measurement of distance using the electromagnetic spectrum (EM) is by *interferometry*. In this system the carrier waves themselves are used directly in an incremental counting system. The technique is employed in the laboratory and factory at *optical* wavelengths, and with *radio* frequencies at sea, from ground to satellite in differential Global Position fixing by Satellites (GPS), and from ground to stars in Very Long Base Interferometry (VLBI).

Most other practical instruments use *modulated waves* to measure distance by timing or phase difference measurement. In the modulation process a high frequency carrier wave is varied to create a more manageable, lower frequency.

The *sonar systems* of hydrographic and seismic surveying do not use the EM spectrum. They depend on *shock waves* which need a physical medium such as water or rock to transmit energy, which travels at a speed depending on this medium: 300 m/s in air and 1500 m/s in water etc. By contrast, EM waves travel through a vacuum at about 0.3×10^{-9} m/s.

Since the introduction of Bergstrand's visible-light Geodimeter in 1946, Wadley's micro-wave radio Tellurometer in 1956, and Holscher's near infra-red MA 100 Tellurometer in 1962, the use of electromagnetic waves for distance and range difference measurement has expanded into all branches of surveying and geodesy, especially into satellite systems.

Far from being distance measurers, EDM instruments are really *time measurers*, making use of the basic equation relating distance (<s>), speed of waves (c) and transit time (tim),

$$< s > = c \ tim \qquad (1)$$

To achieve useful results, time difference has to be measured to an accuracy of better than 10^{-9} sec (a nanosecond).Time differences are measured either

(a) directly, by time of flight measurement, or
(b) indirectly, by phase comparison.

The speed of EM waves, 299 792.458 m/s, is accurate to 10^{-9}, but the refractive index, about 1.0003 for air, is difficult to obtain to 10^{-6}, a figure of 0.5×10^{-6} being more practicable. This seriously limits the accuracy of EDM work.

In satellite to ground propagation, the limits are even worse. It does not make much sense to talk about absolute accuracies of 0.1 mm over a distance greater than 100 m unless very special equipment is available to measure the density and humidity of the air. Improvements to range differences can be achieved by the method of *length ratios*, which is of some value in monitoring *changes* in length between different epochs.

One way to improve matters is to use two different carrier waves for the measurement, because refractive index is wavelength dependent. Only one electro-optical distance measurer, the two-colour laser Terrameter, operates in this way. Because this instrument is exceedingly cumbersome and delicate to operate, its practical use is only for base-line measurement.

However, the dual frequency technique is quite standard for radio instruments, especially in satellite systems, where the additional problems of the ionospheric refraction are also tackled.

Direct measurement
This system has two synchronized clocks, one at each end of the line to be measured. The time difference between the transmitted and received signal, suitably marked for identification, is recorded as the difference of the two clock readings, tim. Such a simple concept depends on very accurate time keeping and clock stability, such as is found in the atomic clocks used by GPS satellites.

If there is a clock error E = d tim, the distances will be in error by

$$ds = c \ d \ tim = c \ E$$

and the proportion error is given by

$$\frac{ds}{s} = \frac{d\,tim}{tim}$$

Apart from a clock error, the precision of timing can be to 10^{-10}, which means that, for a GPS satellite some 26×10^6 m away from the Earth, a relative precision of 0.0026 m is possible. However the actual speed of the signal is known only to about 10^{-6}, because of refraction uncertainty, thus debasing the result to about 26 m. Ranging by this direct method is called *pseudo-ranging* because it still incorporates clock errors. Differential phase techniques can achieve the maximum accuracy of 2 mm, under certain conditions, thus yielding line vectors on the Earth to this accuracy.

Indirect measurement
In this method the carrier wave is modulated to a longer wavelength on which a phase comparison is made. Typically the modulation produces an effective wavelength of 10 m. The modulation is produced by one of three methods.

(a) *Amplitude modulation* (AM), in which the *intensity* of the signal is varied sinusoidally, is adopted by most infra-red instruments.

(b) *Frequency modulation* (FM), in which the *frequency* is varied sinusoidally, is used by radio instruments such as the MRA 7 Tellurometer and GPS satellites.

(c) *Elliptical polarization*, in which an optical signal is polarized for phase comparison, is used by the two Froome-derived instruments, Mekometer and Geomensor.

In modern electro-optical instruments, the infra-red or laser light sources are modulated directly at frequencies, usually in the region of 15 MHz, giving an effective 10 m wavelength. GPS uses two carriers at the 1575.42 and 1227.6 MHz frequencies.

When distance is varying with time, as in satellite or hydrographic applications, the signal is sampled over a selected period, and allowance made for the Doppler effect. Ground based instruments can also operate in a *tracking* mode provided signal lock is maintained. This feature is useful in setting out works.

7.2 Instrument design

Microwave radio instruments, with special omnidirectional aerials, generally require active receiver-transmitters at each line terminal.

Radar or stellar receivers are so large that they cannot be considered as surveying instruments. We concentrate attention on the electro-optical instruments used by most surveyors.

Electro-optical instruments are varied in design. Whatever the arrangement, all require a *pointing telescope*, a *transmitting optic* and a *receiving optic*.

Generally, all distances are obtained by subtracting a measurement to an internal reference point from the distance measured to the external reflector. This subtraction is to allow for the time drift between the effective electrical centre and the physical mounting of the instrument, and gives rise to the so-called *constant* index E.

Corner cubes are most commonly used as reflectors, although cat's eyes and plane mirrors have their applications. Some laser interferometers use expensive omnidirectional hemispherical reflectors.

Microwave radio instruments require two active units, one at each end of the line, and usually incorporate duplex telephones, or other telemetric links.

When an EDM instrument is mounted on, or incorporated in, a theodolite, the instrument is called a *total station*.

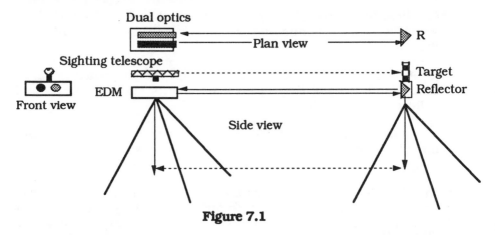

Figure 7.1

Dual optic instruments

The earliest instrument, the Aga Geodimeter, used a visible light carrier modulated externally by a Kerr cell. It had widely spaced dual optics and a separate sighting telescope, as shown in Figure 7.1. The transmitted light travelled to a corner cube reflector (see Chapter 3) to be returned to a receiving optic for phase comparison and subsequent calculation of distance off-line. The main problem was to deviate the returning light sufficiently to enter the receiving optic.

The returning light is deviated into the receiver by making the optics small in relation to the reflector or, in the case of one instrument, by the use of very large, and therefore very expensive, corner cubes. Instruments with dual optics do not suffer from electronic cross-talk within the instrument and are potentially the most accurate. At close range, they can be made to operate with the addition of a special adaptor, which effectively turns them into a coaxial instrument.

Figure 7.2

Coaxial instruments
Instruments with coaxial optics have the advantage that pointing is easier, close range operation is more accurate, and reflectors can be smaller. It is also an advantage if the sighting telescope is incorporated into the coaxial design. This feature is particularly useful in engineering work, because alignment can be carried out simultaneously with distance measurement.

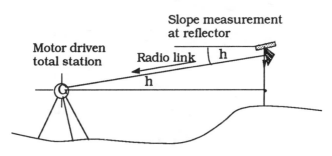

Figure 7.3

Total stations
The integrated instrument, capable of measuring angles and distances, has reached a high degree of sophistication. It may incorporate a computer processor to convert the basic quantities into derived results, such as coordinates, differences in height, horizontal distances; and instrumental effects, such as axis cross tilts, can be noted. Some total stations with motor driven axes, are able to respond to input commands, to point in a particular direction, or to seek out a corner cube reflector, assisted by a slope angle telemetered from the reflector by a radio link. It is also common practice to carry

out diagnostic blunder detection within software, and to off-load data to a memory device, for further processing.

Simpler, and cheaper, manually operated instruments consist of an EDM instrument mounted on a theodolite telescope. They extend the useful life of optical instruments, and have the singular advantage that angles can be observed if batteries are flat.

7.3 Principles of phase measuring EDM

An electro-optical distance measurer, illustrated in Figure 7.4, which is not to scale, shows the instrument reference point at G close to a physical mark A. A modulated signal (light) is transmitted along some path s until it is reflected back to G via a reflector whose reference point R is close to another physical mark B. The surveyor wishes to known the length of the line AB between the physical marks, which can be related to the ground.

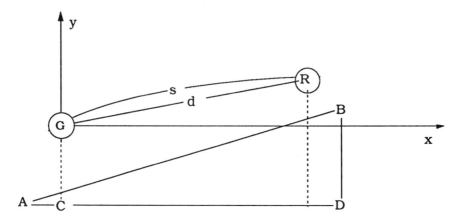

Figure 7.4

The EODM instrument measures the time interval tim which elapses between the transmission of the signal from G and its return along the double path via R. This time interval can be expressed in linear units as a distance <s> called the *optical path length* defined by

$$<s> = \tfrac{1}{2} c \text{ tim}$$

where c is the speed of light in vacuo, in which the refractive index n = 1 by definition. If the average refractive index for the air mass through which the signal passes is n_A, and if the recorded distance, called the *geometrical path length*, is s_A, Fermatt's principle of stationary time means that

$$n_A \, s_A \;=\; \langle s \rangle \;=\; \tfrac{1}{2}\, c \; \text{tim} \;=\; \text{constant} \;=\; K, \text{ say}$$

Thus

$$s_A \;=\; \langle s \rangle \, / \, n_A \;=\; K \, / \, n_A$$

The refractive index is calculated from readings of air pressure, temperature and humidity. To find the chord length d and ultimately the distance AB requires further small corrections.

Phase comparison

Suppose a transverse harmonic disturbance E moves with constant speed along an axis x. At any point on this path at time t let

$$E = f(x - v t)$$

Then

$$\frac{\partial E}{\partial x} = f'(x - v t) \qquad\qquad \frac{\partial E}{\partial t} = -v f'(x - v t)$$

$$\frac{\partial^2 E}{\partial x^2} = f''(x - v t) \qquad\qquad \frac{\partial^2 E}{\partial t^2} = v^2 f''(x - v t)$$

therefore

$$\frac{\partial^2 E}{\partial x^2} = \frac{1}{v^2}\, \frac{\partial^2 E}{\partial t^2}$$

A solution is

$$E = a \, \sin B (v t - x)$$

If f is the frequency in Hz, ω is the freqency in radians per second, and λ is the associated wavelength, all referring to the modulation, we have

$$E = a \sin (\omega t - 2 \pi M - 2 \pi \, \Delta\lambda / \lambda)$$

We can express x as

$$x = M \lambda + \Delta\lambda \tag{2}$$

If the phase of the receiver is compared with that of the transmitter, the system yields $\Delta\lambda / \lambda$ but not M. The kernel of the instrument is a system to modulate the carrier, its transmission and detection, and finally a phase measurement.

Since a phase angle can be measured to about 1 in 6000 by digital techniques, an EDM instrument with an effective wavelength of 1.5 m can resolve 0.25 mm, and the more usual 10 m can resolve about

2 mm. These precisions are achieved by current technology with infra-red amplitude modulation.

The elliptically polarised light instruments can do better to 0.1 mm. Improvements on this accuracy are possible.

Cyclic errors
The measurement of phase can be subject to small cyclic errors due to cross-talk between the two signals and other factors. Although this effect is now very small (2 mm), if the very best is required of an instrument it can be detected and allowed for in measurements.

Frequency standard
The unit of length stems from the fundamental modulation frequency used by the instrument. This must be calibrated to guard against scale error. A laboratory check against a frequency counter controlled by a fundamental radio standard is quite easy to carry out. Some manufacturers provide equipment for this purpose. Experience shows that frequencies are now very stable to at least 10^{-6} over periods of years. In precise work, direct calibration should be carried out at least against a known distance, if the frequency is not tested.

Range
Although the range of microwave radio systems (MDM) is virtually limitless, provided there is radio line of sight, electro-optical (EODM) systems suffer from *attenuation* or absorption of the signal, which limits their operational distance. The reduction dI in signal of intensity I due to absorption can be expressed by Lambert's Law

$$dI / I = -K\,dx$$

where K is an absorption coefficient and x is the range. K varies with weather conditions and the carrier wavelength which is selected to utilize windows in the absorption spectrum. A wavelength of 0.93 μm is much used by infra-red instruments. From empirical data, a typical infra-red instrument gave the following equation relating range x in km achievable from a number N of standard Aga circular prisms

$$N = \exp(1.22(x - 0.7))$$

If N = 1, x = 1.5, if N = 3, x = 1.6, which shows the severe practical limitations on range. A realistic maximum number of prisms is six.

Range can be extended by increasing the energy transmitted and by narrowing the beam width, for example by pulsed laser instruments, which can reach 15 km in clear weather.

Propagation and refraction
The speed of light in a medium varies with the carrier wavelength. The phenomenon is known as *dispersion*. For normal dispersion, ie if $dn/d\lambda$ is negative, the refractive index n can be expressed by Cauchy's equation

$$n = A + \frac{B}{\lambda^2} + \frac{C}{\lambda^4} + \qquad (3)$$

For air under the conditions 0°C, 760 mm pressure and 0.3% carbon dioxide content, the refractive index for optical wavelengths is given by

$$N = (n-1) \; 10^{-6} = 287.604 + \frac{1.6288}{\lambda^2} + \frac{0.0136}{\lambda^4}$$

For a typical visible-light instrument,

$$\lambda = 0.56 \qquad \text{giving} \qquad N = 292.9$$

and for a typical infra-red instrument,

$$\lambda = 0.92 \qquad \text{giving} \qquad N = 289.5$$

The dispersion effect is used to great advantage in two-colour instruments to measure the refraction effect with high precision.

Two-colour measurement
Suppose the distances measured by a synchronous phase instrument on two colours red and blue are s_R and s_B and the refractive indices are n_R and n_B, we have

$$\frac{<s>}{1} = \frac{s_R}{n_R} = \frac{s_B}{n_B}$$

$$\frac{s_R - s_B}{n_R - n_B} = \frac{<s> - s_B}{n_B}$$

which on rearranging gives

$$<s> = s_B + \frac{1 - n_B}{n_R - n_B}(s_R - s_B)$$

$$<s> = s_B + A(s_R - s_B)$$

Using the figures for the visible and infra-red wavelengths mentioned above, we have A = 86. For an actual instrument, the Terrameter which uses blue and red light, A = 55. This means that the difference ($s_R - s_B$) has to be measured to very high precision (0.1 mm) to give a required length < s > of 5 km to a precision of 10^{-6}.

A similar dual frequency principle is involved in improving the micro-wave radio measurements used in satellite systems.

Group velocity
Because the distance is achieved on a modulation pattern, the speed of this modulation is required, not that of the carrier. This speed is obtained from the group refractive index n_G given by differentiating equation (3).

$$n_G = n - \lambda \frac{dn}{d\lambda}$$

where

$$\frac{dn}{d\lambda} = -\frac{2B}{\lambda^3} - \frac{4C}{\lambda^5}$$

The correction is of the order of 10^{-5}.

Correction for refractive index
Built into every EDM system is a standard refractive index n_o, which enables it to give results s_o in length units, from the equation

$$s_o = \frac{c}{n_o} \text{ tim}$$

We need the actual distance s_A, using the actual refractive index n_A, from the equation

$$s_A = \frac{c}{n_A} \text{ tim}$$

Thus we have

$$s_A = \frac{n_o s_o}{n_A} \tag{4}$$

Optical wavelengths
The refractive index can be expressed in terms of meteorological parameters T K (t'C + 273), atmospheric pressure P and water vapour pressure e (both in mm of Mercury), and the instrumental standard refractive index n_s by the equation due to Barrel and Sears

$$n_A - 1 = (n_s - 1) \frac{273}{T} \frac{P}{760} + \frac{15.02\, e}{T} \times 10^{-6} \tag{5}$$

The standard refractive index, which depends on the the carrier wavelength, is provided by the manufacturer.

Since n = 1.0003 approx, the quantity $N = n \times 10^{-6} = 300$ is often used for simplicity. The following results obtained by differentiation of (4) are significant

$$dN / dt = -1 \qquad dN / dP = 0.4 \qquad dN / de = -0.05 \qquad (6)$$

From these equations, it can be seen that the humidity effect can be neglected in all but the most accurate work, and only temperature and pressure need be measured, to 1° C and 3 mm Mercury (4 mb) respectively. Thermometers and barometers need to be calibrated for reliable results.

Microwave radio corrections
The important difference between radio and optical waves is that the former are seriously affected by humidity. Therefore the relative humidity must be measured by wet and dry bulb thermometers. The corresponding refractive index formula, due to Essen and Froome, is

$$N = 103.49 \frac{P}{T} - 17.23 \frac{e}{T} + 49.59 \frac{e}{T^2} \times 10^4 \qquad (7)$$

The partial pressure of water vapour e is obtained from

$$e = e' - C (t - t')/ 755$$

where C = 0.5 for a water-covered thermometer bulb and 0.43 for an ice-covered bulb. The saturation vapour pressure is obtained from Marvin's or Smithsonian tables, or estimated by the formula

$$e' \approx \exp (1.526 + 0.072\ 43\ t' - 0.2945\ t'^2 \times 10^3)$$

Example

If P = 725 mm, t = 9°C and t' = 8°C, we find

$$e' = 8.211 \qquad e = 8.210 \qquad N = 317$$

Differentiation of equation (7) gives the equations

$$dN/ dT = -1 \quad dN / dP = 0.4 \qquad dN / de = 6 \qquad (8)$$

The need to obtain e to 0.2 mm causes problems with psychrometers using wet and dry bulb thermometers, especially in hot dry, or in very cold, climates.

Meteorological readings
It is customary to make pressure readings with a suitable barometer which must be calibrated against a mercury standard. If a surveyor's altimeter is used, it must be remembered that an index has been applied to the scale, to avoid negative height readings. The conversion from altimetric heights to equivalent pressure is made with the aid of the Smithsonian Table no 51. See Appendix 1.

Temperature readings are made with traditional glass or solid state thermometers which should also be calibrated. Taking wet bulb readings with a hand whirling psychrometer needs great care if systematically biased readings, and therefore lengths, are to be avoided. The water should be pure, distilled, and the sleeve covered. Rapid readings are essential after stopping the rotary motion, which should be vigorous, because the whole theory depends on a fast minimum speed being exceeded.

Care should be taken to ensure that thermometers are recording genuine air temperatures and have not been lying in sunlight.

At about 0°C the wet bulb, coated with supercooled water, will give erroneous results. The water should be triggered to form ice by touching, before making readings.

Satellite ranging corrections
The troposphere and ionosphere both seriously degrade ranging to satellites, the former by as much as 10 m depending on the angle of elevation, and the latter by as much as 150 m depending on sunspot activity. Refraction models for dual freqency systems and tropospheric effects are complex. However, as most systems are able to be employed in differential modes, many of these effects are removed when measuring relatively short (up to 20 km) ground distances. In fact there is evidence to support the view that attempts to apply topospheric corrections for lines in excess of 20 km debases, rather than enhances, results.

Integer counts
The integer M is derived by most systems by introducing different frequencies (or wavelengths) to yield different values for the fractional part of the distance $(\Delta\lambda)$ on each frequency. These are subtracted and scaled to give M. The principle is like a vernier scale used in reverse.

Refer to Figure 7.5 which shows two scales A and B divided into ten and nine parts respectively. The B scale is divided into units which are 10/9 of scale A. The separation between the two scales increases, as we move away from the zero point, by 1/9 of a B unit or 1/10 of an A unit for every unit of the distance. Thus we would obtain a measurement of 4.4 units on the A scale and 4.0 on the B scale for an intermediate distance.The difference of these readings (4/10 of scale A units) therefore tells us the we are 4 A units from the zero mark. For lengths greater than 10 A units, a third scale C in which the units are

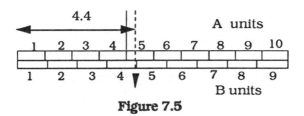

Figure 7.5

100/99 of the A units would give, by subtraction, the hundreds of A units away from zero. Usually four different units (frequencies) are sufficient to resolve the integer count M. It can of course be obtained from an estimated distance from another source, such as a directly ranged distance, or from other data. A wavelength miscount on the GPS system is called a *cycle slip*.

Example
Suppose a distance of 70 418 m is measured by an instrument using successive wavelengths A, B, C, and D of 10, 9/10, 99/100, and 999/1000 m respectively. The results would be respectively 70 418, 63 376.2, 69 713.82, and 70 347.582. Subtracting these in turn gives

$$
\begin{aligned}
A &= 70418 \\
A - B &= 7041.8 \\
A - C &= 704.18 \\
A - D &= 70.418
\end{aligned}
$$

But since the system records only the fractional part of a wavelength (Δλ) the records will be

$$
\begin{aligned}
A &= 18 \\
A - B &= 42 \\
C - C &= 04 \\
A - D &= 70
\end{aligned}
$$

Thus the complete distance can be obtained from these differences, adopting each pair of figures from top to bottom, and left to right.

Microwave instruments
Although microwaves have their widest use in satellite ranging systems there are still applications for ground to ground instruments, where long lines in excess of 15 km are to be measured.

They also have the immense advantage of all-weather operation, including thick fog and heavy rain. Errors produced by multi-path problems have been reduced by polarization techniques, so that accuracies of 5 mm plus refraction uncertainties are possible. Excellent results have been obtained by length ratio techniques.

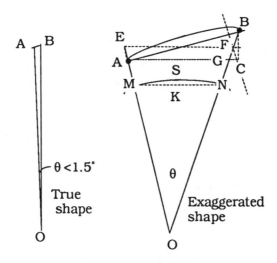

Figure 7.6

7.4 Reduction of slant ranges to the ellipsoid

In contrast with GPS and industrial measurements, it is unusual to calculate topographical coordinates in a three-dimensional system. The two-plus-one system of plan and height is usual. Therefore lines measured by EDM have to be reduced to a datum for incorporation into a survey network. The most elaborate case is the reduction of long lines to the ellipsoid which will be dealt with first.

Corrections have to be applied for the following factors.

(a) Curvature of signal path.
(b) Second refraction correction.
(c) Slope, C_1.
(d) Height, C_2.
(e) Chord to arc.

It is usual to combine the three small corrections (a), (b) and (e) into one C_3.

(a) Curvature of signal path
The refractive index formulae give values only for the line terminals where the meteorological (MET) data are recorded. Due to refraction, the path taken by the signal is curved. The curve is usually taken to

a circular arc of radius σ related to the radius of the Earth R by the expression:

$$k = \frac{R}{2\sigma}$$

where k is called the *coefficient of refraction*. (Note some authorities define K as R/σ). If the angle subtended by the arc AB length L at the centre of curvature of the ellipsoid is θ, see Figure 7.6, and the refraction angle at A (the angle between the curve and the chord at A) is r, we have

$$r = k\theta = \frac{kL}{R}$$

also

$$L = 2r\sigma$$

Average values for k_L and k_R are 0.065 (1/16) and 0.125 (1/8) respectively, giving approximate values of the radii of curvature for light, $\sigma_L = 8R$ and radio, $\sigma_R = 4R$.

The chord length AB = D is given by

$$D = 2\sigma\sin r$$
$$= 2\sigma(r - \frac{r^3}{6} + \text{etc})$$

Which gives the correction for path curvature

$$D - L = -\frac{L^3 k^2}{6 R^2} \text{ etc}$$

(b) Second refraction correction
Under ideal conditions, the index of refraction n may be considered constant over a spherical surface at a height h whose radius is R + h.

Conversely, if the refractive index derived for the terminals is taken to be typical of the line as a whole, it is assumed that the curvature of the signal path is constant and equals the radius of the Earth, if we neglect the height h which is small in relation to R.

A small correction has to be applied to allow for the actual curvature σ and the consequent change in n for the line as a whole. Figure 7.7 shows a small element dL of the signal path of radius σ passing from one spherical surface to another, whose refractive indices differ by dn and whose linear separation is dh. The signal path makes an angle i with the normal at P.

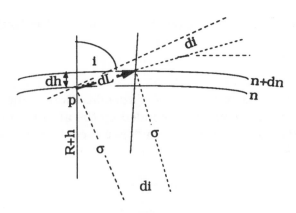

Figure 7.7

By the law of refraction

$$n \sin i = \text{constant}$$

therefore

$$\frac{dn}{n} = -\cot i \, di \approx -\cos i \, di \qquad (\text{since } i \approx 90^\circ)$$

$$\approx -dh / \sigma$$

The maximum separation between two circular arcs of radii R and σ of length L is

$$-\frac{L^2}{8}\left(\frac{1}{R} - \frac{1}{\sigma}\right) = -\frac{L^2}{8R}(1 - 2k)$$

and the mean separation (by Simpson's rule) is two thirds of this maximum ie

$$-\frac{L^2}{12R}(1 - 2k)$$

Now

$$\frac{dL}{L} \approx \frac{dD}{D} = -\frac{dn}{n}$$

thus the correction dD to be applied to D is

$$dD = D\frac{dh}{\sigma} = -\frac{D^3 k}{6R^2}(1 - 2k)$$

(c) Slope correction C_1

In Figure 7.6 the chord AB of length D is to be reduced to the ellipsoidal arc S. The heights of A and B are respectively H_A and H_B and their difference ΔH. Since the arc subtended at the centre of the Earth is never more than 2' it is quite permissible to put $\Delta H = BG = BC$. Then

$$AC = D_1 = \sqrt{(D^2 - \Delta H^2)} = D - (\beta + \frac{\beta^2}{2D} \text{ etc })$$

where

$$\beta = \frac{\Delta H^2}{2D}$$

therefore

$$D_1 = D - C_1$$

(d) Height correction C_2

We are considering the ellipsoidal heights of A and B. In Figure 7.6 since $AC = EF$, the chord length K is obtained by reducing D_1 from the mean height

$$H_M = \frac{1}{2}(H_A + H_B)$$

thus

$$K = \frac{R D_1}{R + H_M} = D_1 - \frac{D_1 H_M}{R}$$

with ample accuracy. Strictly R is the radius of curvature of the ellipsoid at latitude ϕ and azimuth α of the line AB given by Euler's theorem (see Chapter 4),

$$\frac{1}{R} = \frac{\cos^2 \alpha}{\rho} + \frac{\sin^2 \alpha}{v}$$

Typically the quantity $1/2R \approx 7.85 \times 10^{-8}$ at latitude 30° and azimuth 45°, gives the correction

$$C_1 \approx 7.85 D_1 (H_A + H_B) 10^{-8}$$

(e) Chord to arc correction

The arc length S is obtained, with sufficient accuracy, from the expression

$$S \approx K + \frac{K^3}{24 R^2}$$

Combining this correction with the first two (a) and (b) we obtain a small correction

$$C_3 = \frac{D^3}{24 R^2} (1 - 2k)^2$$

which for radio instruments putting $k = \frac{1}{8}$ gives

$$C_3 = \frac{D^3}{43\,R^2}$$

This is the working formula for the small correction, which amounts to 0.072 m for a line of 50 km, and is just within the precision of microwave instruments.

Summary
The final ellipsoidal arc S is given by

$$S = D - C_1 - C_2 + C_3$$

Note there may still be a scale factor to apply if coordinates are treated on a projection. Note also that if the coordinates are being estimated by least squares in a three dimensional system, the EDM line may only need reduction to the slant chord distance D, but with the inclusion of the first two small corrections only.

Example
A microwave radio MDM instrument with a standard refractive index of 1.000 275 gives the line AB to be 41 998.26 m. The actual refractive index calculated from meteorological data is 1.000 3182. The heights of A and B are respectively 90 m and 46 m above the ellipsoid, at a mean latitude of 51° and an azimuth of 66°. We calculate the ellipsoidal length S.

$$D = \frac{1.000\ 275 \times 41\ 998.26}{1.000\ 3182} = 41\ 996.45$$

$$C_1 = 0.02 \qquad C_2 = 0.45 \qquad C_3 = 0.04$$

$$S = D - C_1 - C_2 + C_3 = 41\ 996.02$$

Alternative formulae
When reducing very long lines, the following two alternative formulae should be considered, provided precision of computation is sufficient.

(a) The arc S can be obtained from the angle θ subtended at the centre curvature of the ellipsoid from

$$S = R\theta \qquad \text{where} \quad \sin\frac{1}{2}\theta = \sqrt{\frac{(s-a)(s-b)}{ab}}$$

and

$$a = R + H_A \qquad b = R + H_B \qquad 2s = a + b + D$$

(b) The chord K can be computed from

$$K^2 = \frac{(D - \Delta H)(D + \Delta H)}{(1 + \dfrac{H_A}{R})(1 + \dfrac{H_B}{R})}$$

In both formulae the correct ellipsoidal value for R has to be used.

7.5 Reduction of short lines

In the reduction of short distances of less than 1 km the small corrections can usually be neglected. Attention is drawn however to the need for care with steep slope reduction, and the accurate measurement of slopes or height differences. If a precision of 10^{-5} is sought, the accuracy with which slope angles have to be measured is given by

$$d\,\theta'' = 2.062\ 65\ \text{cosec}\ \theta$$

If θ takes values of 10°, 20° and 30° the accuracies needed are 12", 6" and 4" respectively.

Slope correction for eccentricty

As in all distance reduction, it greatly simplifies slope corrections if the instrument and reflector heights are set equally above the tripod, as in forced centering systems.

However an EDM instrument may be mounted above the theodolite, either on a separate bracket with a separate tilting axis, or on the telescope itself with which it tilts. The corner cube may have a tilt axis, for pointing at the EDM, so that the maximum signal can be obtained. Older reflectors, with no such axis, depend on the fact that the corner cube operates within ± 10° of the cube normal, and are set vertically on the tripod.

When the EDM and the theodolite have separate axes, and point to different target heights, the theoretical solution to the problem is not so simple, but gives rise to a quadratic equation, or iterative solution. When, as is generally the case, slopes are less than 5°, approximate solutions may be adequate.

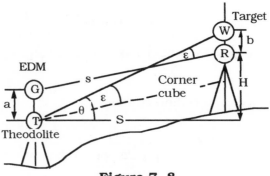

Figure 7. 8

General case

Figure 7.8 shows the EDM instrument G mounted on the standards of the theodolite T so that it can rotate independently to measure the line s = GR. The slope angle θ is observed to a target W placed above the reflector. The two instruments G and T, and target and reflector, may have been set up quite separately, and even on separate occasions. Offsets a and b, and s and θ are known. We require to find S and H.

$$\sin \varepsilon = \frac{(a + b) \cos \theta}{s}$$

$$S = s \cos (\theta - \varepsilon)$$

$$H = S \tan \theta - b$$

Typical figures are

s = 100.678 m,   a = 0.2 m,   b = 0.3 m, $\theta = 5°$

giving the solutions

$$\varepsilon = 0° \ 17' \ 0.48''$$

$$S = 100.337$$

and H = 8.478

Telescope-mounted EDM

When the EDM is mounted directly on the telescope and offset by an amount d, the optical axes of both instruments are usually made parallel for ease in pointing. The EDM is aligned on the reflector to measure s by sighting an approximate distance d below the reflector.

Since the vertical angle is taken by sighting the centre of the reflector, in theory, a small correction, $d^2/2s$, needs to be applied to the EDM distance s, to give the theodolite-to-reflector distance.

If $s = 40$ m and $d = 0.2$ m, this correction is a mere 0.5 mm which is usually ignored.

Steep sights
Industrial and engineering surveying surveys often involve very steep sights, with the result that the reflector is difficult to see, when tilted down, unless a special mount is made or the tripod tilted over. A simple arrangement is to mount the reflector on a theodolite telecope lens cap, as shown in Figure 7.9.

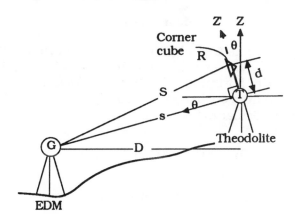

Figure 7.9

The theodolite is centered accurately, the EDM at G is sighted and the telescope pointed perpendicular to this slope using the vertical circle. The reflector is placed on the cap and distance GR = S measured. The reflector offset d is also measured by tape. The distance GT = s is given by

$$s = \sqrt{(S^2 - d^2)}$$

and the horizontal distance $D = s \cos\theta$, the slope angle θ being measured at T.

If $S = 10$ m and $d = 0.2$ m, $s = 9.998$ m

If much work of this kind is undertaken, it is worthwhile making a special lens cap with a 5/8 inch male screw for attaching the reflector, and which can be clamped to the telescope.

Differential measurement

To maximize the accuracy of line measurement, particularly steeply sloping lines, the differential method should be used. Figure 7.10 shows the two reflectors R_1 and R_2 and the EDM G. The required distance is $R_1 R_2 = S_2$.

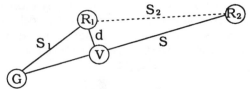

Figure 7.10

From G, the distances $GR_1 = S_1$ and $GR_2 = S$ are measured, together with the offset $d = R_1 V$. This offset is in the plane of the three points G $R_1 R_2$ and is not the horizontal distance. Provided d is kept small, a tape can be viewed from G when sighting at R_2.

Since d is perpendicular to S we have

$$GV = \sqrt{(S_1^2 - d^2)} \approx S_1 - \frac{d^2}{2S_1}$$

$$VR_2 = S - GV \approx S - S_1 + \frac{d^2}{2S_1}$$

$$S_2 \approx VR_2 + \frac{d^2}{2VR_2}$$

If $S_1 = 10$ m, $S = 60$ m and $d = 0.2$ m, we have $S_2 = 50.0033$ m

The accuracy with which d must be measured depends on the overall geometry of the figure. Normally d can be less than 0.5 m and need be accurate to about 5 mm. The slope line S_2 has to be further reduced in the normal way. If the reflectors are mounted on precise lens-caps, the whole operation can be achieved to sub-mm accuracy.

Double measurement

To double the accuracy of most EDMs over short lines (< 50 m) the double measurement technique is applied. This needs a precise front-silvered mirror attachment to a theodolite telescope. This is easily made from an old stereoscope mirror. The lens cap mirror attachment can be clamped in position to great accuracy. Because the offset distance MT = p enters directly into the line measurement, it has to be determined accurately.

The arrangement is shown in Figure 7.11. The line GMR = 2a is measured by the EDM, with R is placed centrally below the EDM with the aid of the plummet. The separation d is taped. The slope angle to O, the mid point of GR, is h_M measured by the theodolite T. In practice the mean angle to G and R is sufficient.

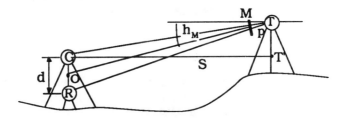

Figure 7.11

The horizontal distance S = G T is given exactly by

$$S = p \cos h_M + 0.5 \,(2a) / \sqrt{(1 + \tan^2 h_M - 0.25 \,(d/a)^2)}$$

In practice this can be simplified to the obvious

$$S \approx (a + p) \cos h_M$$

Suppose GMR = 2a = 120 m, d = 0.5 m, h_M = 5° and p = 0.1 m, the exact formula gives S = 59.8708, and the simple formula S = 59.8713.

In practice the mirror has to be adjusted slightly after pointing the telescope to O, so as to give the maximum EDM signal. The slow motion screws of the theodolite make this task easy.

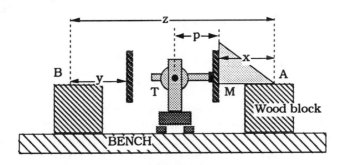

Figure 7.12

If the standard error of the double distance is σ, the standard error of the calculated single distance by this method is $\sigma/\sqrt{2}$. Note however that the method does nothing to improve instrument scale error.

The mirror offset p can be measured to submillimetre accuracy by the method illustrated in Figure 7.12.

The theodolite T with mirror M is set up on a bench together with two wood blocks or boxes. Two fine marks, A and B, are made on paper fixed to the blocks. Assisted by a set-square, the distance MA = x is measured by steel rule. The telescope is transitted and length y measured. After the theodolite has been removed, AB = z is measured, giving finally

$$2p = z - x - y$$

Folded lines

Measuring very steep lines by a combined differential and mirror folded technique is also useful and accurate, standard errors of 2 mm being quite normal with most EDMs. Although the explanation may sound complicated, the practical procedure is quite simple.

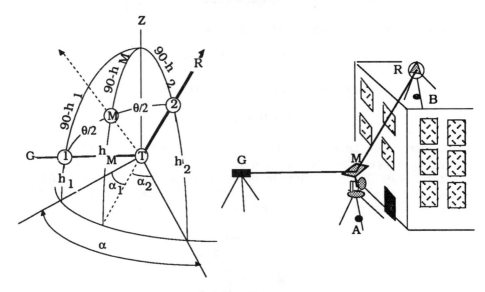

Figure 7.13

The principle is to fold a line into two parts with the aid of a front-silvered mirror. If the mirror is mounted on two axes lying in the plane of reflection, the actual distance measurement process is very straight forward. With a telescope-mounted mirror, some thought has to be given to the effect of the mirror eccentricity. As will be seen, this turns out to be uncomplicated in practice, although the exact theory is complex. The advantages which stem from mounting the mirror on the telescope are:

(a) exact pointing of the mirror is very easy, and

(b) the subsequent incorporation of the slope distance into the survey network is both simple and very accurate.

Figure 7.13 shows, on the right diagram, the arrangement of EDM G, mirror M on the theodolite T measuring the folded line GMR, to reflector R. From GMR we subtract the direct distance GT, measured as described above, to yield TR, which is finally reduced to horizontal and vertical components between the ground marks A and B, through the vertical angle at T. Because the theodolite at T remains on station during the process its centering can be checked and the EDM distance is directly related to angles without any fear of disturbance.

In the left diagram of Figure 7.13 the geometry of the measurement is modelled by the spherical triangles, Z1M and ZM2, based on the vertical through T. The horizontal and vertical angles to G and R, and when the mirror is aligned, to M, are read directly. Because the reflected rays at the mirror are coplanar with its normal, the slant plane GMR contains all the lines required, as illustrated in Figure 7.14.

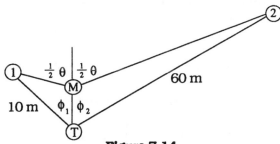

Figure 7.14

The mirror is aligned by a simple field procedure or from calculation of the look-angles at T, and the distance GMR measured. GT is measured either before or after the folded line.

The first stage of the calculation is to derive θ, the angle in the tilted plane, from the cosine formula.

$$\cos \theta = \sin h_1 \sin h_2 + \cos h_1 \cos h_2 \cos \alpha$$

If $h_1 = 10°$, $h_2 = 50°$, $\alpha = 75°$,   then $\theta = 72° 43' 51.13''$

Solving for angle 1 by the sine formula,

$$\sin \text{angle } 1 = \frac{\cos h_2 \sin \alpha}{\sin \theta}$$

gives

$$\text{angle } 1 = 40° 33' 22.81''$$

The mirror vertical angle h_M can be calculated from

$$\sin h_M = \sin h_1 \cos \tfrac{1}{2}\theta + \cos h_1 \sin \tfrac{1}{2}\theta \cos \text{angle } 1$$

In this case we have

$$h_M = 35° \, 41' \, 43.85''$$

But if the mirror was aligned in the field, this angle is read directly on the theodolite. The only angle that need be calculated is θ.

It now remains to correct the distances for the mirror offset. Suppose we measure GMR = 69.8387 m and GT = 10.0000 m and the mirror offset d = 0.1 m. The practical formula is

$$GM = GT - d \cos \tfrac{1}{2}\theta = 9.9195$$

The precise formula via ϕ_1 gives   GM = 9.9193.
The slant range

$$TR = GMR - GM + d \cos \tfrac{1}{2}\theta = 59.9999$$

It will be seen that the working formula is simply

$$TR = GMR - GT + 2 d \cos \tfrac{1}{2}\theta$$

Since $h_2 = 50°$, the horizontal and vertical components of the line are respectively 38.5673 and 45.9627 m.

7.6 The calibration and testing of EODM

The following sections deal with the calibration and testing of electro-optical distance measurers. Similar procedures, dealing with the MDM used in ground and in satellite systems, are less likely to be met with by the ordinary surveyor. Objectives, and methods of achieving them, are described, such as cyclic error testing by least squares, and laboratory and baseline methods of determining scale. Also considered are the statistical methods used to assess results.

Calibration and testing form part of the much wider topic of *quality assessment*, in which the characteristics of an instrument are established and verified. General characteristics can include reliability, robustness, range and accuracy of an instrument.

In this chapter we concern ourselves only with the accuracy of the instrument and its testing and calibration.

In *calibration*, instrument parameters are assessed with a view to improving performance by the application of corrections. For example, a prism-instrument index will be quantified and applied to all distances measured with the instrument.

In *testing* we simply seek to find if the instrument is performing to a specification, usually claimed by the manufacturer. If it does not meet the criteria, then it may be possible to calibrate it so that it does. This latter process is usually the task of the manufacturer.

Calibration objectives

Calibration is the process of establishing the accuracy performance of an instrument within stated and limited criteria. It is usually performed by the manufacturer or by an accredited calibration laboratory. The procedures adopted must conform to an acceptable standard, and within statistically stated rules, otherwise the results are unclear.

It is customary to employ cyclic error tests with least squares procedures, and to give detailed statistical results to assess reliability and accuracy. The results of these procedures are presented in a Calibration Certificate which might be produced in court by the instrument user to show his care in carrying out the work. Needless to say, the court will also need to assured that the instrument has also been used in a proper manner, and in the correct place.

Testing objectives

Testing, on the other hand, is a simpler process used to find out if the instrument is performing to the manufacturer's specification. It too needs to make use of some stated statistical basis for judgment as to whether the instrument meets the specification or not. In the event of a failure to meet the specification, the instrument may be in need of further calibration; otherwise it can be used with confidence.

General points

It is always assumed that correct operational procedures, such as instrument pointing and alignment of reflectors, are adopted during calibration and testing work.

It is comparatively easy to test the precision or repeatability of an instrument, but not to establish its accuracy.

Precision is expressed in terms of the dispersion of a series of observations about a best estimate. It is usually stated in terms of the standard error, whose reliability in turn can be expressed at an agreed confidence level: usually 95%. See Chapter 5 for details.

Accuracy of a result, on the other hand, is nearness to the truth, and since the true value may only be obtained by accident and without our knowledge, our ideas of the true value need to be expressed clearly.

Usually we mean a value which is much better than one obtained by the instrument under test. In the context of EODM calibration, this requires that we have access to a better system of measurement, such as a high accuracy instrument or a laser interferometer, with which to compare a conventional instrument. In the last analysis, comparison is made against the very best possible standards set up by a laser interferometer or the two-colour Terrameter.

Alternatively, the quality of the instrument can be judged in quite a different way by some *laboratory test* of its intrinsic functional mechanisms; that is of

(a) its frequency standards, and
(b) its phase interpolation performance.

It has already been shown that the functional principles of typical EODMs can be reduced to two simple mathematical models relating the actual measured distance s_A to the distance recorded by the instrument s_o, assuming a standard refractive index n_o, as follows:

$$s_o = M\lambda_o + \Delta\lambda_o \tag{9}$$

$$s_A = s_o \frac{n_o}{n_A} \tag{10}$$

where n_A is the actual refractive index for the line.

Finally the distance S between the survey marks, that is the distance between the mechanical reference points on the instrument and its reflector, is expressed by

$$S = s_A + E \tag{11}$$

where E is the instrument-prism index. Note that E is a combined instrument and reflector index. Changing either may produce a different result. Normally prisms of the same type have similar indices. Measurement with each one in turn will reveal any differences, which may need to be corrected.

Calibration

The calibration process is required:

(a) To establish the length of the standard wavelength λ_o ; that is, to check the scale of the instrument.

(b) To ascertain if the interpolated fractional part of a wavelength $\Delta\lambda_o$ shows a cyclic trend ; in other words, to see if the interpolated readings of the instrument are exactly linear over the full wavelength.

(c) To measure the instrument-prism index E.

(d) To check that any ancillary data directly applied within the instrument processor, such as the refractive index, instrument index and slope correction, are also correct.

(e) To see if the calibration values thus obtained are stable with time.

The calibration process can seek to isolate one particular effect at a time or, as with the multiple stationed base line, can seek to detect all three effects, scale, linearity and index, from one series of observations.

In practice, the index error is the most troublesome, because instruments can be used with reflectors of several kinds, and with different mounts, each of which may have different calibration values. Also, the algorithms hard-wired into the instrument may permit these indices to be changed, at the same time exposing the measurement process to the possibility of a mistake.

The various algorithms can easily be verified by checking results against separate values calculated from the same basic data by some external calculation. One instrument manufacturer has been known to apply a correctly calculated refractive index with the wrong sign.

Whilst pillars are desirable for a permanent calibration facility, it is feasible to use tempory stands, provided they remain stable during the measurements. Heavy metrology stands, or wall mounted brackets, are ideal.

Calibration does take some time and may also be costly. It is therefore practicable to establish a hierarchy of calibration and testing procedures, using the most exact only when necessary. It does mean that the performance of an instrument is ultimately traceable to an international standard.

There is also the problem of guaranteeing a standard after an instrument has been transported some distance to the field, or into a hazardous environment. For this situation, simple field testing methods are useful.

Scale checks
The scale of an instrument can be wrong for two reasons:

(a) the refractive index n_A is incorrect, or has been incorrectly applied, and
(b) the fundamental frequency of the oscillator, from which the effective wavelength λ_0 is derived , is out of specification.

Bearing in mind that n ≈ 1.0003, a specification of 10^{-6} implies that, for an accuracy of 0.1 mm, a distance of 100 m lies at the limit of our competence. Since modern surveying instruments do have this high performance, this matter is not entirely theoretical. At the accuracy level of 5 mm the problems of refraction are solvable, provided due care is taken in measuring the meteorological parameters.

There are two ways to check the frequency:

(a) Directly by measuring the fundamental frequency of the generator producing the wavelength.

(b) Indirectly by comparing a measured distance against a better known value, usually obtained from a laser interferometer, or a more accurate EDM, or by taping.

Where possible it is prudent to use both methods to give an independent check on the work, and to ensure that no effect is left untested. The second method has the advantage that the complete performance of the system is being calibrated, but the disadvantage that other errors may swamp the scale effect.

Cyclic errors

We assume the simplest mathematical model for the cyclic error to be expressed by

$$y_i = a \, \sin (\, \theta_i + \varepsilon) \tag{12}$$

This expresses the fact that the interpolation of fractional parts of a wavelength is incorrect, and that the error is systematic and predictable. If the two unknowns, the amplitude a, and the initial phase error ε, are found, any reading on the instrument can be corrected. Most instruments do not exhibit this effect to a serious degree. A small amplitude of 2 mm, will not matter if a 5 mm accuracy is wanted, and this is the manufacturer's claim. Thus the cyclic error is usually the subject of a test, but not of subsequent calibration, unless the instrument is faulty. The practical surveyor does not want to apply a cyclic correction to every reading; he wants an instrument that keeps the error below specification. An instrument with a serious cyclic error needs to be returned to the manufacturer for repair.

However, to squeeze the very most from an EDM instrument, a cyclic correction can be applied if the instrument performs in a consistent manner.

Frequency calibration

Suppose the modulation frequency used for the accurate distance measurement is 100 MHz, and the instrument is provided with a port for comparison with a frequency counter, which in turn is regulated by an "off air " radio standard. Typical comparison figures might show accuracies of 1 in 10^{10} for a new instrument. Ageing of crystals does cause frequency drift over many years, so annual checks are advised. If the frequency is greater than standard the instrument measures too long, and results have to be scaled down accordingly.

7.7 Field calibration and testing

In field testing or calibration we make measurements of a number of lines, whose lengths are selected to sample the effective pattern wavelength of the modulation at stepped intervals. There are advantages and disadvantages in making these steps equal. It is assumed that the EDM is always properly operated, slope corrections applied, and that refractive indices are set to the standard used by the manufacturer. Meteorological readings are taken to allow the actual refractive index to be applied later. In this way the work can be checked. Stability of instrument stands is essential, pillars being ideal provided they are properly constructed and monitored for movement. Other marks are suitable, such as ground or wall marks on well-established concrete or stone. With such marks the transfer of centres has to be done with great care. Reflectors mounted on theodolite telescopes looking at the zenith are ideal for this purpose.

Testing methods are of two types:

(a) those which do not need a known length,
(b) those which depend on the length of a baseline being known to a high accuracy.

It is impossible to say anything about scale error unless a known length is available, or alternatively unless the instrument frequency is monitored. However it is possible to determine the index error E and to *detect* the cyclic error in this case. If no scale error is present, the cyclic effect can also be *measured*.

Since the more general case involves lines of known lengths, it will be described in full. Three methods of increasing complexity are usually identified. The names are attributed to the authors who first published the methods in the literature. See references (5) and (6).

(a) Schwendener method, SCH, 28 lines measured, Figure 7.15.
(b) Sprent method SPR, 10 lines measured, Figure 7.16.
(c) Simple method, SIM, 4 lines measured, Figure 7.17.

EDM calibration model
Before describing these methods, the mathematical model used to analyse results need to be explained. Although more complex models can be used, the following is adequate for most work.

$$\hat{s} = \overset{o}{s} + v = (1+k)s + E + a\sin\frac{(x-x_0)2\pi}{\lambda} \tag{13}$$

where
  s is the measured length,
  $\hat{s}$ is the least squares estimate of the corrected length,
  $\overset{o}{s}$ is the measured length s corrected by the model,
  v is the residual,

  $x = s - \lambda$ int $\dfrac{s}{\lambda}$, the fractional part of the wavelength,

  x_0 is an initial phase error converted to length.

To avoid upsetting any manufacturer, and to give figures which are easily worked, we have created a theoretical instrument BADEM, (effective λ = 10 m), which has the following exceptionally poor and unlikely error characteristics;

  (a) scale error $1 + k$ =1.0006,
  (b) index error E = 0.03 m,
  (c) the amplitude of the cyclic error a = 0.1
  (d) the initial phase error converted to length x_0 = – 0.2 m.

This instrument is calibrated by the three methods to illustrate their advantages and disadvantages. The numerical version of the error equation is

$$S = (1.0006)s + 0.03 + 0.1\sin\frac{(x+0.2)\,2\pi}{10}$$

Since x is a function of s, we can create the standard value S. For example if s = 75.0900,

$$x = 75.0900 - 10 \text{ int}(75.0900/10) = 5.0900$$

$$S = 1.0006 \times 75.0900 + 0.03 + 0.1\sin(5.29 \times 36°) = 75.1469$$

However, the reverse calculation needs iteration, because x is a function of s not of S. The calculation is to iterate from

$$(1.0006)s = S - 0.03 - 0.1\sin(x+0.2)\,2\pi/10$$

In the above example, the best first approximation to s is

$$S - 0.03 / 1.0006 = 75.0719, \qquad \text{giving } x = 5.0719,$$

and the cyclic correction + 0.001 699 8, giving the new value of

s = 75.088 86, and so on to converge on s = 75.0900.

We use this error equation to create fictitious measurements for selected lengths. To give more reality, we have selected examples from the simple 150 m base line at University College London, and the eight-pillar base of Thames Water Utilities, which has nationally accredited NAMAS recognition, and has been measured by the two-colour Terrameter. Measurements have also been carried out with real instruments. The fictitious instrument BADEM is useful to test computer software on any base.

Least squares model
The least squares model is obtained by differentation of equation (12) to give equations of the form

$$dy_i - \overset{*}{a} \sin(\theta_i + \overset{*}{\varepsilon})\,da - \overset{*}{a}\cos(\theta_i + \overset{*}{\varepsilon})\,d\varepsilon - \overset{*}{E}dE - (1 + \overset{*}{k})\,dk = 0 \quad (14)$$

where

$$dy_i = 1 + v \qquad 1 = S - \hat{s}$$

S is the accepted standard base figure assumed errorless

$\overset{*}{\varepsilon}$ is the provisional value of ε etc.

Taking provisional values for a and E of 1, and zero for ε and k, we obtain m observation equations of the form

$$\mathbf{A}\mathbf{x} = 1 + \mathbf{v}$$

The assumption is that all m measurements are of equal weight, and there are n = 4 parameters to be estimated.

The best estimates of the parameters $\hat{\mathbf{x}}$ and residuals $\mathbf{v}$ are given by

$$\mathbf{A}^T\mathbf{A}\hat{\mathbf{x}} = \mathbf{A}^T 1$$

or

$$\mathbf{N}\hat{\mathbf{x}} = \mathbf{b}$$
$$\mathbf{v} = \mathbf{A}\hat{\mathbf{x}} - 1$$

The variance factor σ_0^2 is given by

$$\sigma_0^2 = \frac{\mathbf{v}^T\mathbf{v}}{m - n}$$

The estimated dispersion matrices are :

of parameters $D_x = \sigma_o^2 \; N^{-1}$

of observed parameters $D_s = A D_x A^T$

of residuals $D_v = D_o - D_s = \sigma_o^2 \; I - D_s$

A statistical Tau test for the rejection of a measurement is a standardised residual (v/σ_o) divided by its standard error σ_v obtained from D_v. A 5% confidence level is customary.

When the testing model is used, a Fisher F test on the estimated variance σ_o^2 against the manufacturer's quoted variance σ_m^2 may be used to verify the latter's claim, again at the 5% confidence level.

A graph indicating the cyclic error can be obtained, provided the scale factor and index error are first applied to the observed values before obtaining a new set of "residuals" for plotting.

Unless a series of random errors is applied to the fictitious readings on BADEM, all residuals are zero, and these statistical checks yield nothing. In using BADEM to test software this is a useful first stage, afterwards testing again with errors applied.

The Schwendener method

Figure 7.15 shows the layout and approximate lengths of the Thames Base which has eight pillars, instead of the seven of Schwendener's original. All 28 lines have been "measured ". In practice this can be achieved in one day.

**Schwendener type baseline
eight pillars**

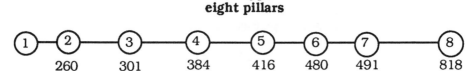

All 28 combinations are measured

Figure 7.15

Table 7.1 shows the base lengths and those observations created for BADEM.

Table 7.1

Observed	Standard		Observed	Standard
260.6170	260.8528		114.7076	114.8124
301.6181	301.9207		178.3541	178.4116
383.7810	384.1016		189.8146	189.9593
416.5427	416.7331		516.7647	517.0097
480.0017	480.3323		32.4819	32.6315
491.4679	491.88		96.2214	96.2307
818.4836	818.9304		107.7800	107.7784
40.9469	41.0679		434.5204	434.8288
123.0553	123.2488		63.4557	63.5992
155.8168	155.8803		75.0900	75.1469
219.3463	219.4795		401.8298	402.1973
230.7996	231.0272		11.4250	11.5477
557.8086	558.0776		338.4410	338.5981
82.00269	82.1809		326.9222	327.0504

This data can be processed by the method outlined above. The expected results can be obtained from any four data points since they are perfect.

The Sprent method
In Sprent's method, the EDM instrument is located at only two pillars close to each other, and 10 readings taken.

Sprent type baseline
Seven pillars

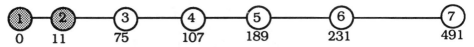

| 0 | 11 | 75 | 107 | 189 | 231 | 491 |

Only ten lines measured
1-3, 1-4, 1-5, 1-6, 1-7
2-3, 2-4, 2-5, 2-6, 2-7

Figure 7.16

Table 7.2

Obs	Standard	l	v	Tau
75.1472	75.1469	−0.0003	2.8915E−05	0.36
107.7792	107.7784	−0.0008	−0.00023170	2.72
189.9608	189.9593	−0.0015	−0.00029440	3.51
231.0293	231.0272	−0.0021	0.00017359	2.13
491.8833	491.8800	−0.0033	−0.00014330	2.24
63.5995	63.5992	−0.0003	3.6429E−06	0.06
96.2313	96.2307	−0.0006	5.9233E−05	0.74
178.4134	178.4116	−0.0018	0.00015173	1.73
219.4817	219.4795	−0.0022	0.00017287	1.93
480.3360	480.3323	−0.0037	7.9482E−05	1.05

Table 7.3

s		$\cos \theta$	$\sin \theta$	l
75.1472	1	−0.0923547	−0.99572620	−0.0003
107.7792	1	−0.9846548	0.17451336	−0.0008

Figure 7.16 shows the Sprent type base selected from the Thames base using pillars 7 and 6 as Sprent's 1 and 2. Table 7.2 shows the measures made by a run-of-the-mill instrument, ROTMI. This is a typical infra-red EDM with a 5 mm specification. The readings in the column marked Obs are the means of 10. This would lead us to expect a standard error (se) of about 2 mm in the means of ten.

The standard base lengths have been measured by the best EODM instruments and the two colour Terrameter. ROTMI shows up very well indeed. The first two observation equations are given in Table 7.3 for the reader to check.

The full normals matrix and its inverse are given in Table 7.4, and the solution in Table 7.5. From these values the residuals of Table 7.2 are calculated.

Table 7.4

Normal equations

672703.8	2133.861	339.2742	1169.232	–5.16522
2133.861	10	–0.451390	2.469980	–0.01650
339.2742	–0.45139	4.132839	0.180902	–0.00191
1169.232	2.469981	0.180902	5.867160	–0.01033

Table 7.5

Inverse of Normal matrix

1.0011E–05	–0.00189180	–0.00097730	–0.00116860
–0.0018918	0.46992334	0.19905117	0.17303160
–0.0009773	0.19905117	0.33953688	0.10049928
–0.0011686	0.17303160	0.10049928	0.32737503

Table 7.6

	x	se m
Scale	–6.4280E–06	3.2591E–07
Index	–0.00018230	7.0611E–05
a	6.3026E–05	6.0021E–05
e	–0.00040500	5.8936E–05

$\mathbf{v}^T\mathbf{v}$	σ_o^2
2.546E–07	1.061E–08

The leading submatrix of the residuals dispersion matrix is

$$
\begin{array}{ll}
0.6048 & 0.8884 \\
0.8884 & 0.6851
\end{array}
$$

Thus we calculate the first Tau ratio for the first observation from

$$\text{Tau} = v / \sqrt{(0.6048 \; \sigma_0^2)} = 2.8915E{-}05 / 8.01E{-}05 = 0.361$$

At the 95% acceptance level the critical value of Tau is about 3, so this measurement is very acceptable. Only the third observation gives some indication of a possible unacceptable value with a Tau value greater than 3.

**Simple baseline
No pillars**

Minimum of four lines measured
AB, AD, CB, CD

EDM at A and C, prisms at B and D

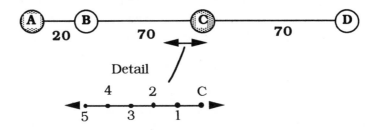

Figure 7.17

Simple base
A simple base, 150 m long, is marked by brass plates set in well-established concrete. The four measured lines give only scale and E. However, if a series of measures is taken at the mid station C over five points approximately one metre appart (see detail), a full analysis can be obtained.

The graph of Figure 7.18 shows the results obtained for BADEM without scale error applied.

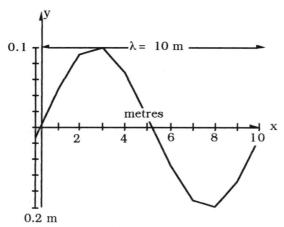

Figure 7.18

Simple base method to determine the index E

To determine the index error E a suitable line, longer than 20 m, is measured differentially and in two parts. Subtraction of these two measures of the line yields twice the error E. The arrangement of instrument stations A and C and the reflector points B and D is shown in Figure 7.17. The distance BD is approximately an integral number of wavelengths λ_o.

This procedure using four stations rather than three is much to be preferred over the alternative method of measuring the distance BD directly for comparison with the sum of BC+ CD. It is by far the easier to carry out and gives better results.

Great care has to be taken to ensure that the instrument and reflectors are stable during this procedure, and especially that the reference point of the EODM instrument when at C does not deviate from a constant position when the instrument is turned round to face in the opposite direction. It is difficult to maintain stability with ordinary survey tripods.

In addition, it can be more convenient to measure a line differentially, without the need to centre the instrument, nor is it necessary to have all stations exactly collinear. Small alignment corrections can be applied. The recommended procedure is as follows.

The reflectors are placed at positions B and D and the instrument first at A and next at C. It is good practice to compare the two reflector indices by direct comparison over some line. If they are not identical, their differential indices have to be applied.

Table 7.7

Meas No	BD = AD – AB		BD = CB + CD + 2 E	
	AB	AD	CA	CD
1	20.0525	170.06	75.0873	75.0655
2	0.0524	0.0599	0.0873	0.0656
3	0.0524	0.0597	0.0874	0.0654
4	0.0524	0.0601	0.0875	0.0655
5	0.0524	0.0605	0.0873	0.0655
6	0.0522	0.0596	0.0870	0.0655
7	0.0524	0.0601	0.0877	0.0655
8	0.0524	0.0596	0.0870	0.0654
9	0.0524	0.0598	0.0869	0.0656
10	0.0526	0.0607	0.0871	0.0657
Mean	20.0524	170.0600	75.0872	75.0655
Sigma	0.0001	0.0004	0.0002	0.0001
2E= (CB + CD)–(AD – AB)		0.1451		
E		0.072		
Scale factor k	1.000 000 85	0.0725	Compare	0.0724

With short lines up to 100 m, there are good reasons for mounting the reflectors on wall brackets if the procedure is to be repeated often. It is then easy to move one reflector on a slide to ensure that the distance BD is nearly an integral number of units of λ_0.

Example
A high accuracy (sd 0.5 mm) EDM instrument was calibrated by the simple method on the UCL base giving the results shown in Table 7.7. It was also calibrated on the Thames base. The indices by both calibrations compare very well. The UCL base length is found to be

$$BD = 170.0600 - 20.0524 = 150.0076$$

The instrument ROTHMI was also applied to the simple UCL base giving the results shown in Table 7.8. From the intermediate positions 5...1, C distances were measured to the reflectors at B and D in turn. The results do not exhibit any cyclic error.

Table 7.8

Cyclic Test of ROTMI

	Means	SUM	Means	
5B	70.039	150.063	80.024	5D
4B	71.034	150.063	79.029	4D
3B	72.036	150.063	78.027	3D
2B	73.033	150.064	77.031	2D
1B	74.031	150.064	76.033	1D
CB	75.037	150.064	75.027	CD
	Mean	150.0635		
	No cyclic error at 1mm level			
	Base length	150.0076		
	2E	0.0559		
	E	0.02795	Agrees with UCL value of 0.028	
	Index applied in Thames measures was		0.03	

General comments on these three methods

The mathematical models described here have been kept within practical limits. It is possible to identify more parameters and to take measurements for time variation of these and other factors. Readers should refer to the work of Reuger (reference (7)) for details of this work.

With the instigation of a system of quality assurance and its technical aspect, quality control, a three tier calibration hierarchy is reasonable using

(a) simple methods to detect errors of scale and index,
(b) more sophisticated methods to isolate cyclic effects, and
(c) the fundamental systems to accredit the other two systems.

After all the EDMs have to be working for their keep, not spending all their time being tested and calibrated.

Laboratory folded baselines

Figure 7.19 shows how a long line may be simulated indoors with the aid of front silvered mirrors. Within limits and depending on mirror quality, good results can be obtained from intensity modulated instruments up to three reflections. The elliptically polarized light instruments appear to work better and can tolerate many more reflections. Much depends on the quality of the mirrors, the stability of their mounts and atmospheric conditions.

The collimator (a level telescope set to infinity) is used to align each mirror in turn, until the standard EDM can be seen. The folded length is measured to the corner cube mounted on the collimator.

Another mirror is introduced to deviate the line to the reference instrument which is moved until the new folded line is nearly the same as the standard. The distance that each EDM is from the new mirror is taped carefully, and any small difference applied. The corner cube is then moved sequentially through the effective modulation wavelength (usually 10 m), with each EDM used in turn. Thus the test instrument is compared directly against the reference instrument.

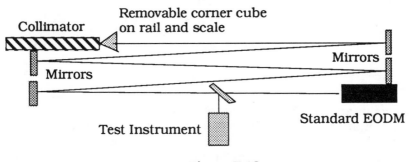

Figure 7.19

Mirrors mounted on theodolite telescopes make this method very portable and adaptable, for use on a distant location; the standard scale being created by a calibrated tape, levelling staff, or subtense bar.

8 Survey control

Chapter Summary

This chapter is concerned with survey control points from which the detailed mapping or setting out is performed. The emphasis between points fixed by ground methods and those fixed from satellite systems is continually changing. However small schemes, such as for building surveys etc, will continue to make little use of satellites.

Much was, and still is, possible without the use of satellites. The computation of control networks is now almost exclusively processed by variation of coordinates and least squares because of convenience and the desire to use all data efficiently.

Control points are usually marked physically on the ground for further use during a project, or by a national survey organisation for continual mapping or land boundary demarkation.

8.1 General

Control systems which cover the Earth as a whole, *geodetic* control, have become practicable through the use of artificial satellites.These in turn are linked, through various degrees of accuracy, to national frameworks to be used as the basis for topographical, engineering and cadastral maps.

Local control systems are adopted for special surveys connected with projects such as dams, roads, railways and pipe lines, large or small construction sites etc.

The purpose of a control system is to prevent the accumulation of errors, by connecting detail work to a consistent geometrical system of points which are accurate enough for the task. Thus for example, the overall squareness and verticality of a tall building will be controlled by specially surveyed points to which reference is made at frequent intervals. Great care is taken to ensure that this control is sufficiently accurate.

There was a time when geodetic control points consisted of triangulation networks marked by observation pillars. Because of the ease with which positions can be established by satellite systems, there is less need for so many such points, although some are required. Gradually, as the scope of the survey becomes smaller, we see the retention of non-satellite systems to provide control. Ultimately the matter is resolved on the basis of cost. Control underground, in urban streets and inside buildings is predominantly carried out by terrestrial methods.

Apart from their use with further ground survey operations, control points are also required to augment photogrammetric and remote sensing methods of mapping. Plan coordinates and heights of points identifiable on imagery are needed by all but the most sophisticated systems.

Control points are computed on two or three dimensional grids, or on a reference surface, such as an ellipsoid, sphere or inclined plane.

Photopoints
In ground surveys, detail points are tied to the control network by various methods such as radiation, minor traversing, and intersection. When the map is made from photography, identifiable coordinated points are required as control. This means that the ground control has to be linked to the photography via photopoints which can be clearly seen both on the ground and on the imagery.

Great care is required in this matter since the accuracy of the whole map will depend on the strength of the link between control and imagery.

The most common procedure is to select these points once the photography has been taken. Points required for plan control need to be well defined at photo scale: not too small and not too large. For example, if the centre of a symmetrical feature, such as a rectangular drain cover, is chosen, the photogrammetric operator must be able to place the floating mark central to it, without obliterating the mark.

Clear unambiguous station descriptions are needed, especially in bush country, snow fields and sand dunes. Sometimes special large scale photos have to be made from a helicopter to link the ground to very small scale photos.

In theory a better strategy is to *pre-mark* points with targets, or large crosses made of cloth, or coal dust, or stones, so that they appear on the photographs. Industrial practice is always to use targets, before photography. In topographical mapping, pre-marks are liable to disappear from natural or human intervention before photography can be taken. Road markings should be avoided because they are often changed during the period between photography and survey.

Points required for height control are often separate from the plan control. They require to be located on locally flat areas so that the floating mark can be place at ground level with the greatest accuracy. A road intersection makes a good height control point.

Field organization
The organization of field work forms a very important part of the surveyor's work. It entails the preparation of an efficient technical programme, which is both cost effective and acceptable to the workforce. Because surveying practice is still weather dependent the work cannot be scheduled to normal working hours. It is also affected by the site environment. Surveying near an operating railway or busy motorway is particularly hazardous: work in

tropical forests is a danger to health; sub–zero temperatures call for special clothing; and transport demands the proper care and maintenance of vehicles or boats. The provision of food, accommodation and fuel has to be attended to, not to mention the recruitment and payment of temporary staff and care for their health. Technical operations can often be the easiest part of any task.

Technical procedures
The surveyor must be fully conversant with his instrument if the best results are to be obtained. The centering of instruments and targets over reference marks needs continual vigilance, and all tripods and plummets need to be secure and adjusted.

For the best results on control work, horizontal and vertical angles should be taken separately, the former on various arcs or zeros. Independent pointings are necessary to remove systematic bias. A regular field booking procedure should be adopted to avoid omission of vital dimensions, such as heights of instruments and targets.

In the case of EDM, proper pointing procedures and meteorological measurements are essential, as is the attention to batteries and their charging.

A technical programme must have sufficient flexibility to respond to weather and other unpredictable factors. Radio communication adds greatly to efficiency and general security.

Planning a control scheme
All control schemes require careful planning to suit the task in hand and any likely future demands. Figure 8.1 shows a diagram to illustrate some documentation to support a continuing scheme to control aerial photography for topographical mapping. The figure depicts the current state of operations at any time during its execution.

Various survey methods are being used: GPS fixes to give differential vectors between stations; an astronomical station to give information about the geoid and azimuth control: EDM lines forming part of a lateration scheme; and some radiations to fix photogrammetric control points in plan and in height.

Other methods might be used such as intersection, resection and traversing, although not on the same kind of task. In past times the surveyor was somewhat inhibited in his choice of method to fix controls because of the tedium of computation. This situation has all but disappeared in most countries. Of particular interest in this respect is resection, now much employed throughout surveying because of the ease with which computer processing has replaced the thirty-minute calculation by hand machine.

The principal methods used by the ground surveyor to fix controls are now outlined.

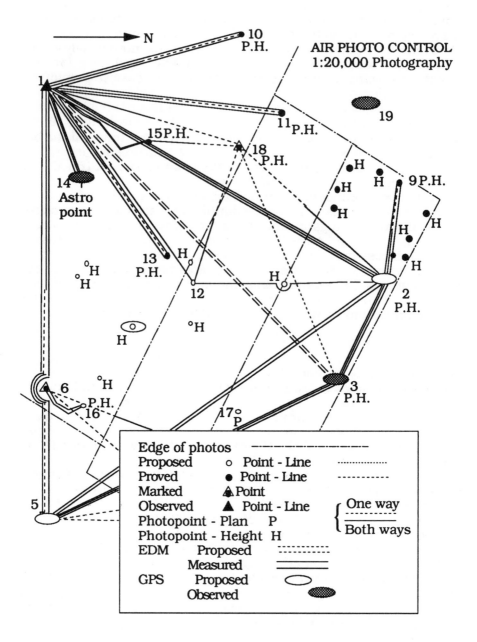

Figure 8.1

Point positioning
There is no really *absolute* point positioning method: even astronomical methods are based on stellar data derived from observatories located at positions determined astronomically. Thus all systems of control are datum dependent. Gradually various systems throughout history, some it very recent, have contributed to our knowledge of the Earth and its gravity field, so that a very consistent set of coordinates can be derived from satellite systems such as GPS. Ultimately positions can be represented in two dimensions by observation equations of the form

$$dE = l_E + v_E \qquad dN = l_N + v_N \qquad (1)$$

Usually these observations have associate dispersion matrices.

Examples of direct point positioning from observed vertical angles and from satellite ranging are given later in this chapter.

Network datums
To remove a rank defect in the least squares processing of data, some constraints have to be applied to the observations. Usually at least one point is held fixed, and one azimuth also held fixed. Other minimum combinations are possible. Thus we have at the datum point

$$dE = 0 \qquad dN = 0$$

For a line at a fixed bearing we have a limitation (constraint) on the four variables between the terminals, giving an equation of the form

$$P\,dE_i + Q\,dN_i - P\,dE_j - Q\,dN_j = 0$$

The coefficients P and Q are respectively the cosine and sine of the fixed bearing. Thus any one of the four variables can be expressed in terms of the other three, and may therefore be eliminated from the problem. Alternatively this constraint equation can be treated as described in Appendix 5.

However, a more practical way is to treat the equation as an observation with very high weight.

It is usual to establish the datum in this way, leaving the scale to be determined by the measurements themselves. To adopt two fixed points as datum for an independent scheme, constrains the scale unnecessarily.

8.2 Resection in two dimensions

A useful introduction to the processing of data in a control scheme is to consider the problem of resection.

If directions in the horizontal plane of a theodolite circle, are observed, at an unknown station P, to three known points A, B, and C, the coordinates of P in the horizontal plane can be calculated, provided all four points do not lie on a vertical cylinder. Considering only the horizontal plane through the theodolite, this condition means that all four points, P and the projections of A, B and C on this plane, must not lie on a circle.

This surveying technique, known as *resection*, is very convenient in the field when fixing photo control points, or positions at sea, or on engineering sites, where sudden decisions have to be taken to fix a point with reference to visible controls, but without visiting them. Coordinated points like church spires are excellent for this control.

The computation is usually effected by least squares and variation of coordinates, if software is available, and more than three control points are observed, as a check. Because resection illustrates principles which are applied more widely in other methods, it will be considered in some detail.

Variation of coordinates
The method of resection is explained with reference to the fixation of point A from control points 18, 19, 24, 16, and 6 of Table 8.1 and Figure 8.2.

Table 8.1

Plan coordinates of data points

Point	x (N)	y (E)
18	3431.47	3730.53
19	2946.97	3011.82
24	1824.27	739.19
16	3521.06	829.67
6	885.13	2000.00
A	3000.00	2000.00
B	2994.40	2003.55
C	2999.99	1996.41

Provisional coordinates are required to evaluate the coefficients and absolute terms of the observation equations. These coordinates can be obtained by prior computation, or graphically, by fitting a tracing paper drawing of the directions (U) to the plotted positions of the controls. The reader should see how this process can give a good approximation to point A fitted to the five controls (18,19,6,24,16). Suppose the provisional values for the Northings and Eastings of A are respectively

3001   2001

When fitting the tracing paper, the reader will see that the directions need to be *rotated* together, and *translated* to fit the control. This means that there are three variables in the problem: an orientation parameter Z and the two corrections to the provisional coordinates, dE and dN.

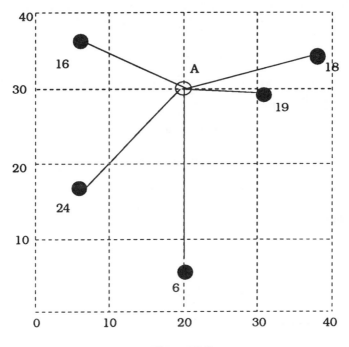

Figure 8.2

Although the observation equation for a direction U is derived easily by partial differentiation of the equation

$$\tan U = (E_2 - E_1) / (N_2 - N_1)$$

it is instructive, as will be seen later, to apply a semi-graphic approach. Consider the direction from A to 18 of Figure 8.2, an enlarged view of which is shown in Figure 8.3.

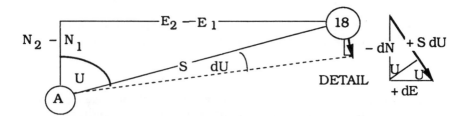

Figure 8.3

A change in the bearing U to U + dU moves the point 18 through a distance S dU, and the coordinates by – dN, + dE. These quantities are related by inspecting the detailed error diagram to give

$$S\,dU = -dN_{18}\,\sin U + dE_{18}\,\cos U \tag{2}$$

If the point 18 is held fixed and A is allowed to move, the equation of its movement is

$$S\,dU = dN_A\,\sin U - dE_A\,\cos U \tag{3}$$

If both ends of the line may move, we have the combined effect

$$S\,dU = (dN_A - dN_{18})\,\sin U - (dE_A - dE_{18})\,\cos U \tag{4}$$

In the resection problem, points such as 18 are kept fixed, leaving equation (3) as the appropriate case.

We may write this as
$$dU" = P\,dE_A + Q\,dN_A$$

where the coefficients P and Q are

$$P = -(N_2 - N_1)/S^2\sin 1" \quad \text{and} \quad Q = (E_2 - E_1)/S^2\sin 1"$$

To complete the equation, dU will contain a small residual v, and the difference l between the bearing $\overset{*}{U}$, computed from the assumed coordinates of A, and its observed value $\overset{o}{U}$.

Because the theodolite directions are only related to the setting of the circle zero, a bearing cannot be observed. Hence we need to say the the theodolite readings are in error by Z. With the usual notation we have

$$dU = l - Z + v$$

The sign of Z is chosen merely to give a positive coefficient in the observation equation for a direction

$$P\,dE\ +Q\,dN + Z\ =\ l + v \tag{5}$$

For convenience, the value of l for the reference direction, A to 18, is made zero, by assuming the initial observed bearing to be the same as the computed one. Other observed bearings are obtained by adding the observed angles to this initial bearing. Usually they differ from other computed bearings, except by chance.

Table 8.2

1	2	3	4	5	6
18	76.0234	0	0	0° 00' 00"	00.00
19	93.0597	17.0363	17° 02' 10.46"	17° 00' 32"	−98.46
6	180.0271	104.0037	104° 00' 13.23"	104° 00' 05"	−8.23
24	226.9982	150.9748	150° 58' 29.36"	151° 00 04"	94.64
16	293.9408	217.9174	217° 55' 02.60"	217° 59' 59"	296.40

The observed directions referred to a zero setting on station 18 are shown in column 5 of Table 8.2. Using the provisional coordinates of A we compute the bearings to the controls as shown in column 2 of the table. After conversion to sexagesimal degrees, we ultimately have the absolute vector (1) in column 6.

Table 8.3

1	2	3	4
−27.951	112.304	1	0
10.876	203.476	1	−98.46
97.485	−0.046	1	−8.24
81.535	−87.430	1	94.64
−65.310	−147.098	1	296.40
*43.216	*36.316	*2.236	*127.16

The full observation equation coefficients and the absolute vector are given in Table 8.3, computed from equation (3). Note that the number of figures for the coefficients can be misleading. They are simply computational. The equations are not linear to this precision.

In Table 8.4, the normal equations are given on the assumption that the directions are all of an equal weight still to be found. It will be noticed there is a sixth equation of Table 8.3, marked with asterisks. This Schreiber equation is explained later.

Table 8.4

1	2	3	4
21 316.22	1547.78	96.63	–13 514.98
1547.78	83 296.26	81.21	–71 909.17
96.63	81.21	5	284.34

Table 8.5 shows the inverse of the normal equations in columns 1 to 3, and the solution in column 4.

Table 8.5

1	2	3	4
5.1418E–05	1.359E–08	–0.000 993 96	–0.979
1.359E–08	1.2198E–05	–0.000 198 38	–0.934
–0.000 993 96	–0.000 198 38	0.222 431 99	90.95

The final coordinates are

$$2000.021 \qquad 3000.066$$

The residuals, in seconds of arc, are respectively

$$13.43, \; -11.12, \; 3.73, \; -1.88, \; \text{and} \; -4.77$$

giving the standard error 2.603".

Error ellipse and pedal curve
As explained in Appendix 4 the quality of the fixation may be represented graphically by an error ellipse and its pedal curve. First we find the direction T of the maximum variance from

$$2T = A \tan 2 (2 \sigma_{EN}, \sigma_N^2 - \sigma_E^2)$$

$$= A \tan 2 (2 \times 1.359E{-}08, 1.2198E{-}05 - 5.1418E{-}05) = 180.0°$$

Therefore

$$T = 90°$$

Now a and b, the semi–major and semi–minor axes of the error ellipse, are given by

$$2a^2 = \sigma_N^2 + \sigma_E^2 + \Delta \qquad \text{and} \quad 2b^2 = \sigma_N^2 + \sigma_E^2 - \Delta$$

where Δ is given by either of the equations

$$\Delta = 2 \sigma_{EN} \operatorname{cosec} 2T$$

$$\Delta = (\sigma_N^2 - \sigma_E^2) \sec 2T$$

From the second equation we find

$$\Delta = 3.922 E - 05$$

and a and b from

$$\sqrt{(1.2198E{-}05 + 5.1418E{-}05 \pm 3.922 E - 05)}/1.414$$

therefore

$$a = 0.072 \qquad \text{and} \qquad b = 0.035$$

The equation of the standard error σ_U at any direction U is given by

$$\sigma_U = \sqrt{((6.3616E{-}05) + (5.1418E{-}05) \cos 2(U - 90))}/ 1.414$$

It may be plotted by computer using this formula or by hand as described in Appendix 4.

Schreiber reduction
The orientation parameter Z is of very little interest to the surveyor. At every station where horizontal angles are observed, one such parameter has to be included, thus increasing the size of the matrix of

normal equations unnecessarily. The parameter, at each observation station, can be eliminated by Schreiber's method, as follows.

Compile an additional observation equation whose coefficients are the respective sums of all coefficents in each column, including the absolute term, divided by the square root of the number of directions. For example, the first coefficient of this Schreiber equation marked with an asterisk in Table 8.3 is

$$43.216 = (-27.951 + 10.876 + 97.485 + 81.535 - 65.310) / \sqrt{5}$$

The Schreiber normals are formed from the Schreiber observation equation in the usual way, giving the results of Table 8.6.

Table 8.6

1	2	3	4
1867.63	1569.45	96.63	5495.44
1569.45	1318.87	81.21	4618.04
96.63	81.21	5	284.34

If the Schreiber normal matrix is subtracted from the original normal matrix, the equation with the orientation correction z is eliminated, as shown in Table 8.7.

Table 8.7

1	2	3	4
19448.59	-21.67	0	-19010.42
-21.67	81977.39	0	-76527.21
0	0	0	0

Thus solving the smaller, 2 x 2, matrix of Table 8.7 missing out the zero terms, gives the same answer as before, in Table 8.8. The inverse of the 2 x 2 matrix is also given to show that the error ellipse parameters are unchanged.

Schreiber's method can be used in all cases where such a parameter is present, such as in distance measurement. It is useful where the capacity of the computer is being stretched to the limit.

Table 8.8

1	2	3
5.1418E–05	1.359E–08	–0.979
1.359E–08	1.2198E–05	–0.934

Direct resection solution

It is not always convenient nor sufficiently accurate to derive the provisional coordinates of the new point by the semi-graphic method. A direct solution is possible. The Snellius method has been found to give the most universal solution to the resection problem. Because the once-popular Tienstra solution fails in the common important case when the three control points are collinear, it is not advocated.

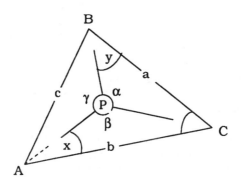

Figure 8. 4

The angles, α β γ, observed at P to three control points A, B, and C, and the side lengths a and b, are known. The problem is first to find an unknown orienting angle such as x or y, then to solve for the coordinates of P.

From Figure 8.4 we have

$$x + y = 360° - (\alpha + \beta + C) = S \text{ say, which is known.}$$

$$x = S - y$$

and

$$\frac{b \sin x}{\sin \beta} = PC = \frac{a \sin y}{\sin \alpha}$$

therefore

$$\frac{\sin x}{\sin y} = \frac{a \sin \beta}{b \sin \alpha} = K$$

say, which is also known, therefore

$$K = \frac{\sin(S-y)}{\sin x} = \sin S \cot y - \cos S$$

and finally

$$\cot y = \frac{K + \cos S}{\sin S} \qquad (6)$$

Note that although it is possible to write

$$\frac{\sin A}{\sin B} = \frac{a}{b}$$

this ratio, calculated from angles, is indeterminate when C lies on AB but not if the sides are used.

Once y, and therefore angle PCB, have been found, we calculate the coordinates of P by intersection from B and C.

It is most important to be systematic about the signs of the various angles, particularly if a computer algorithm is written. When the new point P lies outside the control triangle ABC, special care is needed. The following convention is acceptable.

(a) Letter the triangle ABC in a clockwise manner.
(b) Assign the angles α, β, and γ also clockwise. For example α is the clockwise angle between the directions PB and PC (i.e. those not involving A).

It is also prudent to anticipate what will happen to the arithmetic process when the point P lies on the *danger* circle through ABC. In this case x = y, K = 1, and S = 0, therefore sin S = 0 and cot y is indeterminate. No unique solution is possible.

8.3 Error analysis

The analysis of the errors in resection is best carried out by combining a graphic figural treatment to give general guidance, with a follow-up by a least squares analysis of the interesting cases.

Resection consists in general of the intersection of two circles defined by the angles subtended at two control points. In Figure 8.5 we have the two angles α and β subtended by the pairs of points A and B, and C and D, as shown.

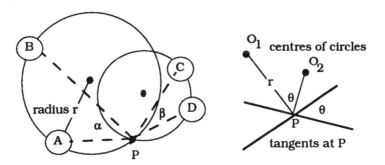

Figure 8.5

The angle α defines the circle, centre O_1 radius r, through the controls A and B. The other circle is defined by β. The point P lies at the intersection of these two circles. There are two solutions, one of which is correct. Usually this can be chosen from other information such as estimated distances. Of course if the angle BPC is also observed, there is no ambiguity. At sea, only two angles are usually observed by two sextants. The two position-circles aid the error analysis.

At P the solution consists of the intersection of the two tangents, which are perpendicular to their radii. Thus in the detailed diagram on the right of Figure 8.5 we see that P is fixed by two lines intersecting at an angle θ. This angle is that subtended by P at the two centres, O_1 and O_2. Thus the resection problem has been reduced to an equivalent distance intersection problem. If we decrease the angle α by δα the tangent at P moves away from the centre by dr given by

$$dr = \frac{ab}{p} d\alpha \sin 1''$$

where a, b and p are the lengths of the sides of the triangle ABP. In this way we can calculate the shifts to the radii, and estimate the strength of the positioning of P. If some close comparison is required of two values, the numerical least squares method is needed.

The figural approach is useful as a management tool in the field. It is also a useful way to locate a buried mark for which previous angles to the controls are known. Angles are taken at a likely point P close to the mark. The tangents can be marked out by strings because the angle BPO_1 = BAP – 90°. The shift is computed from the change in the angle between the old and new, and thus the position line for the mark layed out by an offset string. The same approach to the other angle β gives a second string line and the location of the mark. This procedure is much easier and more accurate than coordinating the new point and calculating the bearing and distance to the old mark. It will also be noted that the coordinates of the controls are not required, only the old angles to them.

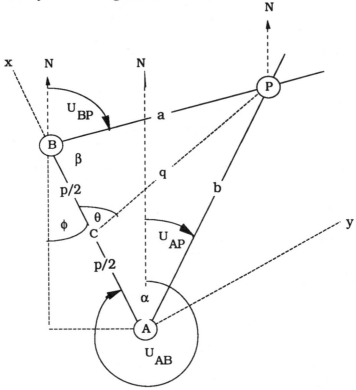

Figure 8.6

Intersection
The observation equation to be used for an intersection is equation (1) of the above section. Generally the treatment is identical, only that the orientation error should be very small, because a known bearing is used as the reference direction. Many surveyors set the theodolite circle to a bearing for this purpose and to ease the arithmetic. An example of this type of equation is given for the traverse computation later.

The provisional coordinates of an intersected point can be obtained graphically or by direct solution using two fixed points.

Consider Figure 8.6 in which the new point P is to be fixed by intersection from the known points A and B. The known quantities are angles α and β, and E_A, N_A, E_B and N_B.

The computation is carried out in the following stages.

(a) Obtain the bearing of AB from

$$U_{AB} = \text{ATAN2} (E_B - E_A , N_B - N_A)$$

(b) Obtain the bearings of AP and BP from

$$U_{AP} = U_{AB} + \alpha$$
$$U_{BP} = U_{AP} + 180° - \beta$$

When programming these formulae for a computer, strict adherence to sign convention is required. For example the angle β is best recorded as the clockwise angle between the bearings BA and BP. In this case β would be 360° - the β used in the above formula, giving

$$U_{BP} = U_{AB} + 180° + \beta$$

Trig functions of angles greater than 360° are correctly evaluated by most computers.

(c) Calculate the coordinates of P from the formulae

$$N_P = \frac{N_A \tan U_{AP} - N_B \tan U_{BP} - E_A + E_B}{\tan U_{AP} - \tan U_{BP}} \tag{7}$$

$$E_P = \frac{E_A \cot U_{AP} - E_B \cot U_{BP} - N_A + N_B}{\cot U_{AP} - \cot U_{BP}} \tag{8}$$

These formulae are proved simply by rearranging the expressions

$$\tan U_{AP} = \frac{E_P - E_A}{N_P - N_A} \qquad \tan U_{BP} = \frac{E_P - E_B}{N_P - N_B}$$

Alternative formula for angles

Alternative formulae for intersection in terms of the angles α and β are

$$N_P = \frac{N_A \cot \alpha + N_B \cot \beta + E_A - E_B}{\cot \alpha + \cot \beta} \tag{7a}$$

$$E_P = \frac{E_A \cot \alpha + E_B \cot \beta - N_A + N_B}{\cot \alpha + \cot \beta} \tag{8a}$$

Formulae (7a) and (8a) may be proved as follows.

Select a new set of axes (x,y) through an origin at A (0,0) making B (p,0). The new bearings are α and $180° - \beta$. Apply formulae (7) and (8) to give coordinates of P with respect to A and B. Then transform these (x,y) coordinates of P to the (E,N) system by a rotation of axes ϕ, and a shift of origin.

Error analysis

As mentioned already, guidance in the analysis of error can be derived from a simple graphic treatment, considering the angle of cut between the rays, and their likely lateral shifts. For example a simple graphic consideration gives the error information of Figure 8.7. A precise calculation can then be done, by the least squares method, to examine the error ellipses in the most interesting areas, such as close to the base line AB.

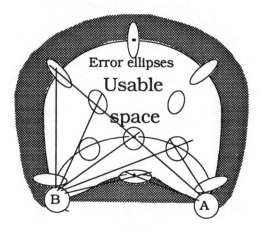

Figure 8.7

Figure 8.7 shows the usable space within which a satisfactory result may be obtained from the intersection of points from the two fixed stations A and B. The plotted criterion is usually the sum of the variances in E and N, called the PDOP, an acronym from *positional dilution of precision.*

$$PDOP = \sigma_E^2 + \sigma_N^2 = a^2 + b^2$$

Sometimes the maximum error, a, is used as an alternative criterion to illustrate the bounds of the area within which the required precision will be achievable.

8.4 Fixation by distances: or lateration

Fixing position by lengths alone is called *lateration*. It is more familiarly known a *trilateration* when three sides are measured to fix a triangle. The computational procedure is similar to other methods once the measured distances have been reduced to a common reference system, usually the ellipsoid or on a projection.

Provisional coordinates are obtained, graphically, in this case by circles drawn by compasses, or by prior exact computation.

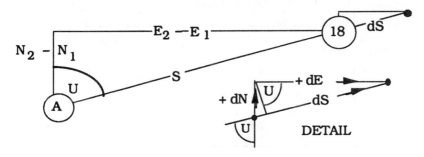

Figure 8.8

Variation of coordinates

Figure 8.8 shows how an observation equation for a measured distance S may be derived graphically with ease. By inspection we can write for the movement of point 18 relative to A

$$dS = dE_{18} \sin U + dN_{18} \cos U$$

and for A relative to 18

$$dS = - dE_A \sin U - dN_A \cos U \qquad (9)$$

If both terminals are allowed to vary we have

$$dS = (dE_{18} - dE_A) \sin U + (dN_{18} - dN_A) \cos U$$

Introducing an instrument index E and scale factor 1+k the general observation equation is

$$(dE_{18} - dE_A) \sin U + (dN_{18} - dN_A) \cos U + E + k S = 1 + v \qquad (10)$$

In most EDM work the index E and the scale factor 1+k, determined by calibration as described in Chapter 7, are applied to all distances at the outset. The terms in E and k are therefore left out of the equation, leaving the more usual working form

$$(dE_{18} - dE_A) \sin U + (dN_{18} - dN_A) \cos U = 1 + v \qquad (11)$$

However to give acceptable results in the GPS pseudo range mode, a clock error E at a receiver or satellite has to be modelled by equation (10). We take the simple model (11) for an example in two dimensions.

Table 8.9A

1	2	3	4
18	1783.48	0.061	0.055
19	1013.18	0.054	0.021
24	1723.79	0.061	0.075
16	1281.11	0.056	0.023
6	2114.78	0.065	−0.033

Table 8.9B

Observation equations

18	−0.9703	−0.2419	1.1820
19	−0.9986	0.0523	0.9139
24	0.7314	0.6820	−1.5663
16	0.9135	−0.4067	−0.4771
6	0.0000	1.0000	−1.0899

sigma zero = 0.979

Example of simple 2D model
Table 8.9A lists, in column 2, the horizontal distances from A to the control points 18 to 6. Their standard errors, in column 3, have been computed by the error model

$$\sigma_S = \sqrt{(0.05^2 + (20 \times S \, 10^{-6})^2)}$$

The standard errors of the instrument index and of the refractive index are taken here as 0.05 m and 20 parts per million (ppm) respectively. A more reasonable figure for a typical EDM instrument index would be 2.5 mm. The larger figure is taken merely to give clearer numbers in the example.

The observation equations are given in Table 8.9B with the residuals listed in column 4. The normal equations and the solution are given in Table 8.10.

Because the weights were estimated correctly, the value of sigma zero is very nearly one. The error ellipse parameters are also given in the Table 8.10.

Table 8.10

Normal equations

1001.8600	61.6551	– 1066.3954
61.6551	427.8760	– 541.3161

	Inverse	Solution
10.0708E–04	– 14.5116E–05	–0.995
– 14.5116E–05	23.5804E– 04	– 1.122

Error ellipse parameters

0.048	0.031	173.9°

Direct solution of lateration coordinates

Although provisional coordinates may be obtained from a solution of the triangle ABP, the following direct formulae are convenient.

$$E_P = \frac{1}{2}(E_A + E_B) + \frac{1}{2\,p^2}(a^2 - b^2)(E_A - E_B) - \frac{2\,\Delta}{p^2}(N_A - N_B) \quad (12)$$

$$N_P = \frac{1}{2}(N_A + N_B) + \frac{1}{2\,p^2}(a^2 - b^2)(N_A - N_B) + \frac{2\,\Delta}{p^2}(E_A - E_B) \quad (13)$$

where

$$\Delta = \sqrt{(S (S - a)(S - b)(S - p))}$$

$$2S = a + b + p$$

To outline a proof of these formulae refer to Figure 8.6 in which C is the mid point of AB, AB = p, PC = q. The easting of P and the area of the triangle ABP= Δ are given respectively by

$$E_P = E_C + q\cos(\theta - \phi) \qquad \Delta = \tfrac{1}{2}pa\sin\beta$$

It is easy to see that

$$\sin\theta = \frac{2\Delta}{pq} \qquad \cos\theta = -\frac{a^2 - b^2}{2pq}$$

$$\sin\phi = \frac{E_A - E_B}{p} \qquad \cos\phi = -\frac{N_A - N_B}{p}$$

Since C is the midpoint of AB we obtain the expression for the easting of P. The formula for northings is proved in the same way.

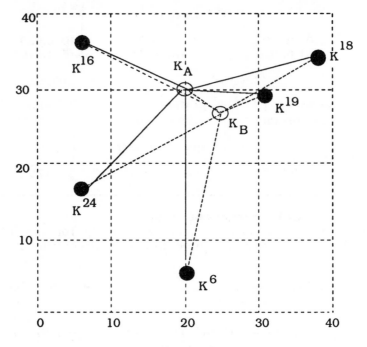

Figure 8.9

Pseudo and differential ranging

As with ground based EDM, range measurements, by direct and differential methods, have been used widely in hydrographic and satellite surveying, with improved results. The improvement is due largely to the elimination or reduction of index and refraction errors, and to the detection of blunders caused by miscounting the integer number of wavelengths contained in the line.

Ground to satellite systems are particularly prone to error because of the loss of signal lock between the receiver and transmitter, and the unpredictable effects of the ionosphere.

Instrument index error (clock bias) can be restricted to small limits using expensive clocks, and refraction modelled by dual frequencies, as explained in Chapter 7. However, large errors remain even after the most sophisticated modelling.

To improve matters, differencing is adopted as in all surveying. Figure 8.9 illustrates two new points A and B from which ranges S to satellites 18 to 6 are measured and reduced to a common epoch. The index errors due to clocks at A and B and residual tropospheric errors are denoted by K_A and K_B, and at the satellites by K^{18} etc, where clock and ionospheric errors are present. Since the ground distance AB (60 km) is very small compared with the Earth to satellite distances (26 000 km) the error model is reasonable. Thus we may express the range from A to satellite 18 in the form

$$S_A^{18} = r_A^{18} + K_A + K^{18} \tag{14}$$

The notation is: subscripts for ground stations and superscripts for high satellites. The range r_A^{18} is the crude distance measurement, often called the *pseudo-range*. This would be the same as measuring an EDM distance without applying index and refraction corrections.

The measurement and computation is usually carried out in three dimensions (x,y,z). Here we reduce the problem to two (E,N) to reduce the amount of calculation required by the reader when verifying examples, and to save space. The principles involved are valid in either two or three dimensions.

Table 8.11

Data points			Horizontal distances from	
N	E	pt	A	B
3431.47	3730.53	18	1783.5079	1781.4292
2946.97	3011.82	19	1013.2087	1009.3850
1824.27	739.19	24	1723.9440	1722.7334
3521.06	829.67	16	1281.0838	1286.6094
885.13	2000.00	6	2114.8700	2109.2740
3000.00	2000.00	A	0	6.6304
2994.40	2003.55	B	6.6304	0

Adopting the usual procedure for least squares estimation based on provisional coordinates etc we obtain linear equations of the form:

$$(dE_{18} - dE_A) \sin U_A^{18} + (dN_{18} - dN_A) \cos U_A^{18} - K_A - K^{18}$$
$$= 1_A^{18} + v_A^{18} \tag{15}$$

Usually, the satellite coordinates are considered fixed, so we have

$$- dE_A \sin U_A^{18} - dN_A \cos U_A^{18} - K_A - K^{18} = 1_A^{18} + v_A^{18} \tag{16}$$

In the example which follows, we have assigned respective values of:

$$K_A = 0.2 : K_B = -0.3$$

$$K^{18} = 10 : K^{19} = 11 : K^{24} = 12 : K^{16} = 13 : K^6 = 14$$

Table 8.11 gives the standard coordinates from which the other data such as bearings, are calculated. See also Table 8.2.

Table 8.11

	A +0.2	B –0.3	OBS A	OBS B
18 +10	1783.707 90	1781.129 23	1793.707 90	1791.129 23
19 +11	1013.408 70	1009.084 96	1024.408 71	1020.084 96
24 +12	1724.143 99	1722.433 42	1736.143 99	1734.433 42
16 +13	1281.283 85	1286.309 89	1294.283 85	1299.309 89
6 +14	2115.070 00	2108.972 99	2129.070 00	2122.972 99

Table 8.12 shows the calculation of errors applied to the horizontal distances. For example the range from A to 18 is changed by

$$K_A + K^{18} = 0.2 + 10$$

We will demonstrate that the effects of such errors can be eliminated from the solution by the methods of differencing.

Provisional coordinates adopted for A and B are:

A	3000.1	2000.1
B	2994.4	2003.55

No change has been made to point B simply to ease the arithmetic.

Table 8. 13

	Observation equations		From A	
	– cos U	– sinU	Obs S	1
18	–0.241 9881	–0.970 2792	1793.707 90	10.272 8332
19	0.052 2454	–0.998 6343	1024.408 71	11.305 0924
24	0.681 9402	0.731 4079	1736.143 99	12.195 0590
16	–0.406 7699	0.913 5307	1294.283 85	13.067 9709
6	1	4.7286E–05	2129.070 00	14.299 9976

The observation equations for the direct solution for A and B from pseudo ranges are given in Tables 8.13 and 8.15 and their solutions in Tables 8.14 and 8. 16.

Table 8.14

Normals		b	Soln
1.69179189	0.30984896	15.4053774	9.28735223
0.30984896	3.30820811	–0.3989376	–0.99044980

Table 8.15

Stn	Observation equations		From B	
	– cos U	– sinU	Obs S	1
18	–0.24534790	–0.9694351	1791.12923	9.7
19	0.04698901	–0.9988954	1020.08496	10.7
24	0.67922871	0.73392667	1734.43342	11.7
16	–0.40933930	0.91238223	1299.30989	12.7
6	0.99999858	0.00168304	2122.97299	13.7

Not surprisingly there are considerable changes to coordinates because of the huge errors in the lengths and the failure to model them. (We could have modelled these errors in a similar manner to the orientation equation using Schreiber's method.)

Table 8.16

Normals		b	Soln
1.691 311	0.317 625	14.571 2547	8.767 4296
0.317 625	3.308 690	0.105 5529	−0.809 7473

Single differences

To eliminate the effect of the large errors at each satellite, the equations to A and B are subtracted before solution giving a typical range-difference equation:

$$- dE_A \sin U_A^{18} - dN_A \cos U_A^{18} + dE_B \sin U_B^{18} + dN_B \cos U_B^{18} - K_A + K_B$$
$$= 1_A^{18} - 1_B^{18} + v_A^{18} - v_B^{18} = 1_{AB}^{18} + v_{AB}^{18} \qquad (17)$$

The observation equations of this type are given in Table 8.17 and their solution in Table 8.18.

Table 8.17

	Single difference A − B				1
18	−0.241 988	−0.970 279	0.245 347	0.969 435	0.5728
19	0.052 245	−0.998 634	−0.046 989	0.998 895	0.6051
24	0.681 940	0.731 407	−0.679 228	−0.733 926	0.4951
16	−0.406 769	0.913 530	0.409 339	−0.912 382	0.3679
6	1.000 000	4.72E−05	−0.999 998	−0.001 683	0.5999

The very large changes are due to the assigned errors at A and B being large in relation to the short line AB and no attempt being made to model them. However, the *differences* in coordinates between A and B are preserved well and little affected by the very large errors introduced. Such large errors occur in practice on account of the uncertainties in ionospheric and tropospheric refraction in satellite systems, or instability of reference marks in monitoring surveys.

Table 8.18

Normals				b
1.691 791	0.309 848	−1.691 525	−0.313 451	0.680 913
0.309 848	3.308 208	−0.314 027	−3.308 443	−0.461 803
−1.691 525	−0.314 027	1.691 311	0.317 625	−0.673 519
−0.313 451	−3.308 443	0.317 625	3.308 688	0.459 671

Inverse				soln
23 878.098	30 322.594	23 821.688	302 95.636	137.526
30 322.594	331 639.785	29 620.216	331 644.309	−7.478
23 821.688	29 620.216	23 767.371	29 593.180	137.161
30 295.636	331 644.309	29 593.180	331 649.176	−7.477

Double differences

If each equation for the single difference is now subtracted from one reference equation at a station, in this case the first, the remaining unmodelled station errors are eliminated. These *double difference* equations are given in Table 8.19 and their solution in Table 8.20.

Table 8.19

	Double differences				l
18	0	0	0	0	0
19	0.294 23	−0.028 35	−0.2923	0.029 46	0.032 25
24	0.923 92	1.701 68	−0.9245	−1.703 36	−0.077 77
16	−0.164 78	1.883 80	0.1639	−1.881 81	−0.204 86
6	1.241 98	0.970 32	−1.2453	−0.971 11	0.027 16

In this case, the solution is almost perfect, exactly as predicted, showing that the effects of the very large imposed errors have been eradicated.

In practice, it is important to make simultaneous measurements with two or more receivers to allow the differencing process to proceed, and to make sure that a sufficient number of satellites has been observed to secure the necessary redundancy of measurement.

Guidance as to potential accuracy is obtained from the pseudo-range pre-analysis using the orbital prediction methods outlined in Chapter 6. Practical systems include computer software for this prediction.

Table 8. 20

	Normals			b
2.509 904	2.458 610	−2.513 986	−2.461 143	0.005 129
2.458 610	7.386 816	−2.464 515	−7.386 712	−0.492 825
−2.513 986	−2.464 515	2.518 084	2.467 052	−0.004 947
−2.461 143	−7.386 712	2.467 052	7.386 616	0.492 561
	Inverse			Soln
120 999.3	−28 692.7	120 891.1	−28 753.7	0.100 865
−28 692.7	218 515.9	−28 988.1	218 640.4	−0.101 241
120 891.1	−28 988.1	120 784.1	−29 049.5	0.000 865
−28 753.7	218 640.4	−29 049.5	218 765.1	−0.001 241

Other differencing strategies could be adopted, such as swapping the receivers over and repeating measurements at a later epoch. Or differencing the double differences between two epochs: *triple differencing*. This latter helps to isolate an individual error in a line due to an integer miscount (due to cycle slips).

It is probable that many of these computation methods will become obsolete as the measurement systems improve and the error sources are eliminated.

Correlations
As was pointed out when treating directions, if the orientation parameter is eliminated by differencing to give angle equations, the new dispersion matrix of the reduced observations also has to be found. This is also true of range differencing techniques.

8.5 Resection by vertical angles

The simplest and cheapest method of finding the position of a point on the surface of the Earth is to observe vertical angles to some stars at known times. This method depends on the fact that the angle between the direction of the observer's vertical and a star varies from place to place on the surface of the Earth. As explained in Chapter 2, a vertical angle defines a cone whose axis in this case is a line from the Earth's centre to a star. This cone strikes the Earth in a circle upon which the observer lies. Two suitable circles will locate him. Neglecting the effect of the Earth's motion, this angle differs from place to place because of the curvature of the Earth. Position is therefore found by comparing observed vertical angles with those computed for a provisional assumed place. See Chapter 6.

The method, however, has greater generality. Position can be determined from vertical observations to any elevated points of known coordinates. We shall use the data set to give an example of this. Figure 8. 10 also depicts the geometry of the observation. Points A and its provisional position at A' are constrained to lie on a plane.

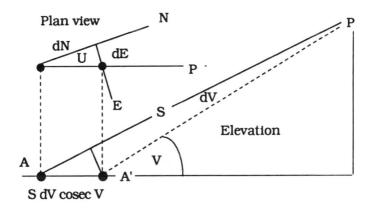

Figure 8. 10

From the assumed position of A (3001, 2001) we obtain vertical angles to the controls as before. Subtraction from the "observed" verticals at A gives the absolute vector l. By inspection of Figure 8.10 the observation equation is obtained as

$$\sin U \, dE + \cos U \, dN = S \cosec V \, dV + v$$

which is of the general form

$$\mathbf{A} \mathbf{x} = \mathbf{l} + \mathbf{v}$$

All data, observation equations and their solution are given in Table 8.21. The term $S \operatorname{cosec} V\, dV$ is called the *offset* in the table.

Table 8.21

AZ	Comp V	Obs V	dV	Slope S	Offset
76	37.0187	37	−0.000 3269	2233.19	−1.213
93	61.0226	61	−0.000 3943	2089.90	−0.942
227	38.9770	39	0.000 4016	2218.30	1.416
294	52.9892	53	0.000 1892	2128.71	0.504
180	24.9896	25	0.000 1811	2333.50	1.000

Obs Equns

sin U	cosU	l	Soln	Expect
0.9703	0.2419	1.213	0.999	1.000
0.9986	−0.0523	0.942	1.002	1.000
−0.7314	−0.6820	−1.416		
−0.9135	0.4067	−0.504		
0.0000	−1	−1.000		

In the astronomical version, because each equation contains two observed parameters, V and time, it is theoretically correct to apply the general least squares method.

Resection from two control points
If vertical angles and horizontal directions are observed to only two points, one of which must be elevated, a solution is possible. If both are elevated, a check on the work is available.

If the height of the unknown station is known, as could be the case on a factory floor, the solution is direct. If the height is not known, a three dimensional treatment is appropriate, or an iterative approach in two dimensions may be preferred. The approach is to use the direction and distance equations already derived. The pseudo-distance measurement is derived from the vertical angle V as follows:

$$S = \Delta z \cot V$$

where Δz is the height difference between A and the control point. If this is unknown, we assume a likely value by estimation. Differentiation gives

$$dS = -\Delta z \operatorname{cosec}^2 V\, dV + \cot V\, dz$$

Substituting for

$$dS = -dE_A \sin U - dN_A \cos U$$

we have

$$-\Delta z \operatorname{cosec}^2 V\, dV + \cot V\, dz = -dE_A \sin U - dN_A \cos U$$

which reduces to the form

$$P\, dE + Q\, dN + R\, dz = 1 + v$$

where

$$P = \frac{\sin U}{\Delta z \operatorname{cosec}^2 V} \qquad Q = \frac{\cos U}{\Delta z \operatorname{cosec}^2 V} \qquad R = -\frac{\cot V}{\Delta z \operatorname{cosec}^2 V}$$

If three dimensional software is not available, the problem can be treated iteratively to give a solution in two dimensions.

Table 8.22

Point	E	N	$V \pm 5''$	$U \pm 5''$
16	829.67	3521.06	$53° 00' 00''$	$293° 59' 57''$
6	2000.00	885.13	$25° 00' 00''$	$180° 00' 30''$
A	2001.00	3001.00		

	Approx height diff	Pseudo-distance
16	1701	1281.80
6	987	2116.63

	Final co-ords	
A	2000.42	2999.75

From an estimated height difference calculate an approximate pseudo-distance from

$$S = \Delta z \cot V$$

Thereafter the problem is treated in terms of direction and distance equations. We will demonstrate the latter with reference to point A being fixed from stations 16 and 6 of Figure 8.1. These stations have been selected to give good cuts from the position lines.

Table 8. 22 gives the vertical angles (± 5″) observed at A to 16 and 6. The equivalent pseudo-distance computed from these vertical angles and assumed height differences, (in error by a metre) are also listed, together with standard errors estimated using the fomula

$$dS = -\Delta z \operatorname{cosec}^2 V \, dV + \cot V \, dz$$

These distances, together with the previous direction equations, form the observation equations of Table 8.23 and the normals, inverse and solution of Table 8.24. If the solution is repeated with the coordinates, distances and their standard errors updated to the new values the changes to the new coordinates are only a few millimetres, showing that the system has converged to a solution.

Table 8.23

Point	Distance equations			v
16	0.9140	– 4058	0.2089	–0.23
6	0.0005	1.0000	0.7601	–2.01
Point	Direction equations			
16	97.4847	– 0.046	0.00	0.04
6	– 65.3099	–147.098	277.65	– 0.04

Table 8.24

Normal equations		
531.52	478.13	– 903.66
478.13	433.02	– 816.54

Inverse of Normals and solution

27.93	– 30.84	– 0.578
– 30.84	34.29	– 1.247

8.6 The inaccessible base

It sometimes happens that only two control points such as A and B of Figure 8.11 are visible from a point C whose coordinates are required. If there is no great difference in elevation between the points, the vertical angles are not of much use. The problem is solved by

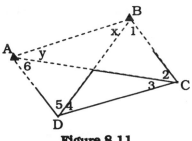

Figure 8.11

introducing an auxiliary fourth point D so that a well conditioned quadrilateral is formed. Angles 2, 3, 4, 5, at C and D are observed, from which angles 1 and 6 can be deduced. To find the orienting angles x and y we proceed as for the resection problem. In this case

$$\frac{\sin x}{\sin y} = \frac{\sin 1 \sin 3 \sin 5}{\sin 2 \sin 4 \sin 6} = K$$

The solution then follows identical lines to the three point resection. See equation(6). If the distance CD is also measured, it acts as an independent check on the work. If this length is not measured, no unique solution is possible if either ADC or BCD are collinear.

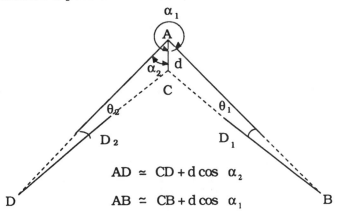

$$AD \simeq CD + d \cos \alpha_2$$

$$AB \simeq CB + d \cos \alpha_1$$

Figure 8.12

8.7 Reduction to centre

It often happens that the theodolite and target cannot be centred over the same mark: for example if a flagpole is used as a target. Either they are treated as two separate stations in the network or the directions to one are reduced to what they would have been at the other by applying an eccentric correction every direction. Consider Figure 8.12 in which the theodolite is at A. Directions are observed to B, C and D, and the short distance AC = d is measured. If C is the permanent station to which readings are to be reduced we have the correction to the direction to B given by

$$\sin \theta_1 = \frac{d \sin \alpha_1}{D_1} \tag{18}$$

The distance CB = D_1 is obtained from a provisional calculation of the points C and B or from EDM.

A little thought will show that the accuracy required in the measurement of d and of D_1 is not great if the correction is kept small by keeping d as short as possible.

The worst case is when $\alpha = 90°$ or $\sin \alpha = 1$. Then

$$\frac{\delta \theta}{\theta} = \frac{\delta d}{d} = -\frac{\delta D}{D}$$

If we want directions to be accurate to 1" the maximum error in the eccentric correction will be given by

$$\delta \theta = \frac{\delta d}{D} = \frac{1}{206265}$$

Thus if D = 1 km the distance d must be measured to about 5 mm.

A special case of reduction to centre, the Weisbach triangle, is often used in engineering work to transfer bearings from two close targets. See Chapter 13 for details.

8.8 Combined systems

Most survey networks consist of measurements combined together by least squares to yield the best estimates of coordinates. Observation equations are formed for each measured parameter, weights are assigned and a solution obtained from the normal equations. When using coordinates a datum has to be imposed. Usually this datum takes the form of one fixed point and a fixed bearing or azimuth. The coordinates of the fixed point may have been obtained from astronomical observations, or from satellite position fixing such as Doppler or GPS pseudo-ranging. The orientation could have been obtained from an astronomical or gyroscopic azimuth, or from the direction between two fixed points.

Quite often this datum is purely arbitrary and is not tied to any natural system such as north. The datum is assumed merely to establish the coordinate system. Perhaps the most typical example is the control traverse of an engineering scheme. (Cadastral traverses have to be controlled by azimuths.)

The most common combined system used to establish control is the traverse, which will now be used by way of more general illustration.

Traverse computation

The traverse is a series of points joined together by a chain of distances and directions. Figure 8.13 shows the layout of a typical traverse, the data for which are given in Table 8.25. Provisional coordinates are obtained either from a scale drawing (which has much to commend it if only to detect blunders), or by prior computation by bearing and distance in succession.

There are only two checks on the work.

(a) That the bearings taken round in a loop agree to within an expected limit.

(b) That the extra distance connecting the last two points agrees within expected limits.

As a guide in the field, bearings usually have to be within $3\,\sigma\,\sqrt{2n}$ where there are n observation stations, and a *direction* standard error is σ. This value of sigma is selected from previous experience, or from a computer simulation of the task in hand. For example if $n = 6$ and $\sigma = 5''$, bearings have to agree to $52''$. If they do not, a blunder is suspected, and search made for the likely source. If an error is made in a direction at one station, it can be found by computing the traverse clockwise and anticlockwise. The station with the error will be the only one with identical coordinates.

The distance check should meet the criterion of $3\,\sigma\sqrt{n}$, where σ is the standard error of a typical length. For example if $\sigma = 2.5$ mm and if $n = 6$, the limit of tolerance is 18 mm. If a blunder has been made in one length, and no other mistakes are present, the final compared distances will differ by a vector which has the same size and direction as the line in which the blunder lies.

The computation of a simple traverse needs an orderly approach to successive bearing and distance computations. If a least squares treatment is to be given, no alteration of the observation data should be made, because the hand computation is only to provide provisional coordinates. There is some justification for reconciling small discrepancies within the closure bounds, such as averaging the bearing error, recomputing and then averaging the final vector misclosure (Bowditch method). If this so-called "adjustment" makes a material difference to the answer, it should not be done, but the work repeated to a proper standard. The arithmetic of traverse computation is much simplified if the theodolite circle is set to an approximate bearing at each station as the work proceeds.

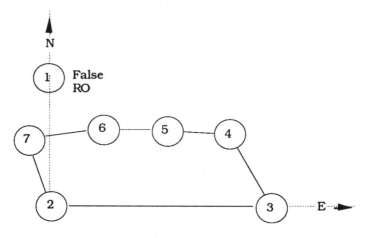

Figure 8.13

When approximate coordinates have been obtained, direction and distance equations are formed, and the solution effected. The example below uses the Schreiber method of eliminating the station orientation parameters, and the simple distance model.

Two fixed points are assumed to be at 2, the initial traverse point, and at 1, a fictitious point not actually observed. The reason for adopting this procedure is that the software was written on the assumption that the traverse would always be tied to a fixed control point to give it orientation, as in cadastral work.This computational device does not affect the relative results in any way. In this example the traverse is oriented by an arbitrary direction.

All directions have been assigned a cautious 5" standard error, being the means of three rounds taken on a *one–second* optical scale theodolite over steep slopes. All distances, measured by a precise EDM instrument, have been reduced for slope and index and have been given standard errors of ±1 mm ± 20 ppm.

Example of traverse computation

Figure 8.13 shows the layout of points of the traverse between points 1 to 7. Since this was an isolated engineering control traverse which was not related to any other coordinates, an arbitrary reference system was adopted. Point 2, the effective origin of the traverse, has been given non-zero values to avoid negative coordinates later on. Its arbitrary eastings differ from the northings to assist identification and to avoid swapping values by mistake. Plane coordinates are used without a projection scale factor or arc-to-chord corrections. All directions are means of six faces. Distances have been reduced to the horizontal after the application of index and refraction corrections.

Table 8.25

	U	t	S	E	N
1				2000.000	2000.000
1–2		180.0000		2000.000	1000.000
2–3	90.0000	90.0000	263.209	2263.209	1000.000
3–4	59.7200	329.7200	41.593	2242.237	1035.919
4–5	127.2489	276.9689	90.828	2152.080	1046.939
5–6	166.3600	263.3289	118.030	2034.849	1033.227
6–7	183.3817	266.7106	70.294	1964.671	1029.194
7–2	42.8619	129.5725	45.829	1999.997	999.998
2–7	309.5992	309.5725	closure	0.003	0.002
	Bg closure	0.0267	vector	0.004	
		96.12"			

The entries of column U are the clockwise *angles* (differences of direction) at each station, from which bearings t are derived in succession from the starting bearing, which was computed from the fixed points 1 and 2. For example

$$t_{3\text{-}4} = t_{2\text{-}3} + 180° - U_{3\text{-}2} + U_{3\text{-}4}$$

$$t_{3\text{-}4} = 90 + 180 + 59.7200 = 329.7200$$

The coordinates are obtained by adding the successive differences for each line, computed from the bearings t and distances S using the formulae

$$E_{i+1} = E_i + S \sin t \quad \text{and} \quad N_{i+1} = N_i + S \cos t$$

The positional misclosure is good.

Because there are only three redundant parameters, the reliablity of a simple closed traverse is poor. If redundant crossing rays, such as 7–3, are possible, they should be observed to increase reliability, even if they add little to the accuracy.

Table 8.26

	U	t	S	E	N
1				2000.000	2000.000
1–2		180.0000		2000.000	1000.000
2–3	90.000	90.0000	263.209	2263.209	1000.000
3–4	59.7200	329.7200	**42.593**	2241.732	1036.782
4–5	127.2489	276.9689	90.828	2151.576	1047.802
5–6	166.3600	263.3289	118.03	2034.345	1034.091
6–7	183.3817	266.7106	70.294	1964.166	1030.057
7–2	42.8619	129.5725	45.829	1999.492	1000.862
2–7	309.5992	309.5725	closure	0.508	–0.862
	Bg closure	0.0267	vector	1.002	
		96.12"			

Traverse blunders

If a mistake is made in measuring one leg, the effect can usually be detected and the line checked. Table 8.26 shows the effect of introducing a metre error into the line 3–4. The misclosure vector is one metre plus random error, and at the same bearing as the line 3–4. The process can therefore be used in reverse to locate the likely leg in which a mistake is made.

In a similar manner, the effect of a mistake of a degree at station 4 swings the rest of the traverse to create a 4 metre misclosure perpendicular to the line 4–2. See Table 8.27. Thus the perpendicular bisector of the vector misclosure passes through the point at which the mistake was made, and so the the blunder station can be found.

Another way to locate this is to compute the traverse in the reverse direction, from 2 to 7 and so on, clockwise back to 2. Only the point 4 will have the same coordinates in both calculations, thus proving it is the guilty point. Some surveyors compute every traverse in both directions to give some idea of its intermediate consistency.

Traversing is often the only way to fix control, particularly inside buildings, tunnels or large pipes, and in urban streets, and forests. It has therefore to be done with care and a system of quality checking the basic data. If it is inevitable that the traverse cannot be closed, another return route should be taken to close back on the starting point. Inside a tunnel this should be a mirror image of the inward route, to balance the effects of refraction on lines of sight, which can be as much as 10" per station in a systematic manner.

Table 8.27

	U	t	S	E	N
1				2000.000	2000.000
1–2		180.0000		2000.000	1000.000
2–3	90.0000	90.0000	263.209	2263.209	1000.000
3–4	59.7200	329.7200	41.593	2242.236	1035.919
4–5	**128.2489**	277.9689	90.828	2152.286	1048.511
5–6	166.3600	264.3289	118.03	2034.833	1036.847
6–7	183.3817	267.7106	70.294	1964.596	1034.039
7–2	42.8619	130.5725	45.829	1999.407	1004.231
	309.5992	310.5725	vector	0.593	–4.231
	Bg closure	–0.9733		4.273	
		–3503.88			

Least squares computation of a traverse
As an example of a combined network of directions and distances the above traverse is treated by least squares. Although the process is complicated, it poses no problems if computer software is available.

Such software will generally use the method variation of coordinates and observation equations, even though the method of condition equations has much to commend it, such as having only three normals equations to solve, irrespective of the number of stations in the traverse. The formation of condition equations can be troublesome, whilst observation equations are easy to program and have great flexibility.

Table 8.28

2 Z	3 E	N	Z	4 E	N	Z	5 E	N	Z	6 E	N	Z	7 E	N	Z
							Direction equations								
1													A	A	
1															
1	A	A													
	A	A	1												
	A	A	1	A	A										
	A	A		A	A	1									
				A	A	1	A	A							
				A	A		A	A	1						
							A	A	1	A	A				
							A	A		A	A	1			
										A	A	1	A	A	
										A	A		A	A	1
													A	A	1
							Schreiber equations								
√3	C	C											C	C	
	2B	2B	√2	B	B										
	B	B		2B	2B	√2	B	B							
				B	B		2B	2B	√2	B	B				
							B	B		2B	2B	√2	B	B	
										B	B		2B	2B	√2
							Distance equations								
	A	A													
	A	A		A	A										
				A	A		A	A							
							A	A		A	A				
										A	A		A	A	
													A	A	

Example

There is insufficient space to list the thirteen direction and six distance equations in the sixteen variables formed from the data of Table 8. 25. They are given in symbolic form in Table 8.28. An A indicates any coefficient other than zero. The symbols B and C of the Schreiber equations mean $A/\sqrt{2}$ and $A/\sqrt{3}$ respectively.

Only coefficients of the first four direction and one distance equations are given, since they are the most interesting. Table 8. 28 shows the structure of the direction equations matrix, and the Schreiber equations formed from them. The solution and statistical results are also listed.

Because the first point 2 is held fixed, its only variable is the orientation parameter, Z, hence there are no other coefficients than zero in the first line.

The second equation links 2 and 7 giving the coefficients and 1

| 1 | 2866.22 | 3465.53 | 22.22 |

The third equation links 2 and 3, giving coefficients and 1

| 1 | 0 | −783.68 | 0 |

The observed value is used as the provisional value.

The fourth equation, for direction 3 to 4 takes on the standard form for the remaining equations, with coefficients

| −4282.60, | 2500.29, | −4282.60, | 2500.29 |

and a zero absolute term again.

The fifth equation, for direction 4 to 3 has identical coefficients to the equation for the direction 3 to 4. This repetition of coefficients may seem strange until it is realised that

$$\sin(t + 180°) = -\sin t \quad \text{and} \quad \cos(t + 180°) = -\cos t$$

Eliminating the six orientation parameters leaves a 10 x 10 matrix of weighted normals equations, which begins with

$$\begin{matrix} +1.040E+6 & -0.102E+6 \\ -0.102E+6 & +1.002E+6 \end{matrix}$$

The *a-priori* weights were: 5" per direction, and ±1 mm ± 20 ppm per distance.The solution dE, dN for points 3 to 7 is

| 0.001 | −0.005 | | 0.045 | −0.072 | | 0.028 | −0.056 |
| 0.026 | −0.034 | | 0.022 | −0.013 | | | |

Sigma zero is 3.641 and the one sigma error ellipse parameters a, b and θ are

0.021	0.011	5.5°		0.020	0.011	174°
0.014	0.010	165°		0.007	0.006	139°
0.004	0.004	161°				

Error propagation in traverses

The precise way to determine the errors propagated through a traverse, or for that matter in any network, is by considering the error ellipses at points and between points. The former are dependent upon the choice of datum, usually the starting point of the traverse, whilst the latter give information between pairs of points.

As a rough guide in the field the simple analysis of a straight traverse can be of assistance in planning work. Consider a straight traverse of n legs each of length s running along the x axis. The linear standard error in x between the ends of the traverse is

$$\sigma_x = \sigma_s \sqrt{n}$$

and in y is

$$\sigma_y = s\,\sigma_\theta \sin 1'' \sqrt{\tfrac{1}{6}\, n\,(\,n+1\,)(\,2n+1\,)}$$

Suppose the standard error of a leg is 2.5 mm, the 16 legs are each 50 m long, and the standard error in an angle is 10", we have

$$\sigma_x = 10\,\text{mm}$$
$$\sigma_y = 9.4\,\text{mm}$$

Comment

The provision of control points can involve many survey methods and instruments. All measurements are carried out carefully and to a designed tolerance usually established by a network simulation of the proposed system. This simulation follows the lines of the least squares processing of data, but without actual measurements. Likely dispersion matrices are used in the process. Various designs can be tried, tested and costed before actual measurements begin.

Although it is desirable to complete the control computations before coordinates are used for mapping or setting out, sometimes this sequence has to be broken. The result is that maps are out of sympathy with the final control. Such is the case with much cadastral mapping throughout the world. The process of reconciling old to new mapping is one of constant need. In engineering tasks, datum differences can lead to disaster.

The needs of modern society, as manifest in geographical and engineering information systems, navigation by satellite, and the location of underground services, demand better homogeneous control points to be established. Computer treatment of data is able to achieve this goal.

9 Heights and levels

Chapter Summary

This chapter is concerned with the use of terrestrial methods to provide the heights of points on the surface of the Earth by ground survey methods. It deals only with levelling and trigonometric heighting.

9.1 Introduction

A knowledge of the heights and levels of points on the surface of the Earth is required for all or some of following purposes:

(a) To enable survey measurements, aerial photographs and satellite imagery to be reduced to a datum surface such as sea level, or a reference ellipsoid.

(b) To create digital terrain models (DTMs) to be used to solve such problems as intervisibility between points on the ground.

(c) To enable drainage and other water works to be surveyed so that water may flow in desired directions.

(d) To enable relief to be depicted on topographical and air maps.

(e) To enable roads and railways to be constructed in such a manner that steep hills are avoided.

(f) To provide information for scientists concerning the shape and structure of the Earth, and tectonic movements in its surface.

The vertical and levels
The vertical is the direction which a plumb line takes up when it hangs freely under the effect of the Earth's gravitational pull. A horizontal surface is a *level surface* at right angles to the vertical. The height of a point may be defined as the linear distance from the point up or down the vertical through the point, to a reference horizontal surface or datum. Later in this chapter height is defined purely in terms of gravitational force.

If the horizontal datum surface covers a small portion of the Earth it may be sufficient to consider it as a plane surface throughout its entirety, but when treating the surface of the Earth as a whole, the horizontal surface will be curved and be everywhere at right angles to the verticals, as in Figure 9.1.

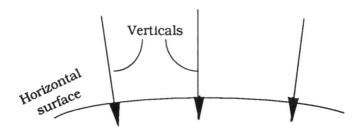

Figure 9.1

The horizontal datum surface for heights is one which closely approximates to mean sea level (MSL), determined from observations on tide gauges averaged over a period of 19 years. This period allows the various tide raising forces to repeat.

Since most vertical angles between ground stations are of the order of 1° or less, computational approximations are usually permissible. Most heighting processes deal with differences in height, observed over comparatively short lines, carried forward over long distances by a chain of many measurements. For this reason, small systematic errors may often accumulate in serious proportions and must be avoided where possible. Accuracies of 2 mm per km are possible by levelling.

Artificial Earth satellites can provide geometrical height differences to about 3 mm accuracy but need to be connected to a level datum, either by transfomations to fit levelled points, or to less accuracy, by a knowledge of the Earth's gravity field.

Methods of determining relative heights
The following four ground methods are used to determine the relative heights of survey points:

    (a) Hydrostatic levelling.
    (b) Spirit levelling.
    (c) Trigonometrical heighting.
    (d) Barometric heighting.

Depending on the flying height of the aircraft and other factors, topographic stereo-photogrammetry, the principal way to derive heights over large areas of land, can regularly achieve practical accuracies of about 0.1 m. Ground survey heights are usually needed to control this technique.

Hydrostatic levelling

This is probably the oldest form of levelling. A tube containing water will readily define two points A and B at the same height due to the balance set up by the water under the force of gravity (see Figure 9.2). The heights found by this method are 'dynamic' heights.

Figure 9.2

This method of levelling was much used in Holland where differences in height are very small and great accuracy is essential. Pipes of up to 10 km long have been used. For convenience, flexible pipes are laid out along the bed of a canal by a survey ship. The main problems associated with this method are to do with eliminating air bubbles from the pipe when filling it with water or dual liquids, and controlling the temperature of the fluids. Miniature versions of high accuracy hydrostatic levelling are employed in industrial surveying; an ordinary water hose is useful for work in forested areas, for example to mark out the future shore line of an artificial lake.

Spirit levelling

In this method a horizontal line of sight is established by the observer with the aid of a spirit bubble, or a plumb line or a freely suspended compensator system, which enables him to sight through a telescope in a horizontal direction. Although bubbles have largely been replace by compensators in all but the most precise work, the term *spirit levelling* is retained here.

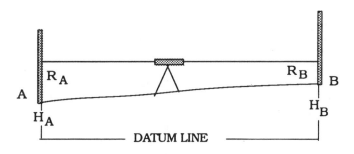

Figure 9.3

For ease of operation, spirit levelling is used in conjunction with one or two graduated rods or staves to determine the difference in height between two points. In Figure 9.3 the height of B with respect to that of A is found by taking readings R_A and R_B on the vertical staves at A and B. For

$$R_A + H_A = R_B + H_B$$

therefore

$$H_B = H_A + R_A - R_B$$

Since A is the backward point if we are moving from A to B, the reading on A, i.e. R_A, is called the 'back sight reading' or simply the *backsight*. The reading on B, i.e. R_B, is called the 'fore sight

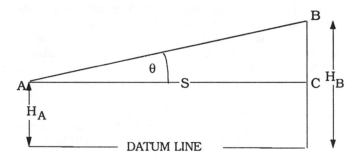

Figure 9.4

reading' or simply the *foresight*. It will be noted that the height difference between A and B is given both in *magnitude* and *sign*. We always consider the height difference in the sense

Difference in height = Backsight – Foresight

After making a reading on B, the staff at B is turned round to face the new instrument position and the staff at A is moved forward to the new instrument position C. Thus the new backsight and foresight readings are r_B and r_C. The foot of the staff at B should not change height in turning the staff round, hence it is necessary that some point whose height is well defined should be chosen for the *change point* B. Such a point would be the head of a rivet on a drain cover, a kerb stone, etc. In open fields a large screwdriver driven into the ground makes a good change point. Special plates are also available. If a point is heighted but not used as a change point, it is called an *intermediate point*, and the sight to it is an intermediate sight (IS).

Trigonometrical heighting
In Figure 9.4 the difference in height between the points A and B is given by

$$\Delta H = S \tan \theta$$

where S is the horizontal distance AC, and θ is the vertical angle recorded by a theodolite circle against a horizontal datum, determined by a bubble or compensator. Since $\tan \theta$ is negative when θ is negative, and positive when it is positive, for angles less than 90°, the equation gives both the *magnitude* and *sign* of ΔH. If either the height of A or the height of B is known, the height of the other station may be calculated. In precise work, vertical angles are observed at both ends of the line, i.e. at both A and B. The distances sighted in this method may be anything up to 50 km, though accurate results are limited to lines of 10 km or less. Recent work shows that if lines are restricted to 200 m accuracies of 2 mm can be maintained.

Barometric heighting
This method of measuring height differences depends on the variation of atmospheric pressure with height above sea level. Pressure P is related to the acceleration due to gravity g, the density of the air ρ and the height H by the functional expression

$$P = F(g, \rho, H)$$

If P is measured at two places, where g and ρ are sensibly constant the pressure height relationship for two points is

$$\Delta H = H_2 - H_1 = K \ln \frac{P_1}{P_2}$$

where K is a constant. If one height is known we may find the other.

The barometric method of heighting is especially convenient where points are not intervisible, such as in forested country, or where rapid, but relatively inaccurate results are required. Accuracies of about 1 m are attainable with care.

Accuracies of heighting methods
It is impossible to give figures for the accuracy attainable by a survey technique which will apply to all circumstances, since the accuracy of a method depends, to some extent, on the care with which the survey is carried out, the time spent on the operation, and various imponderable factors, such as the weather conditions pertaining at

the time of the measurements. However, as a guide, the following accuracies may be expected if normal technique is adopted.

(a) Hydrostatic and geodetic levelling : a standard error of 1.5 $\sqrt{K}$ mm, where K is the length of the line in km.

(b) Ordinary spirit levelling : a standard error of 6 $\sqrt{K}$ mm.

(c) Trigonometrical heighting: a standard error of 60 $\sqrt{K}$ mm, though accuracies as high as for geodetic spirit levelling can be obtained with special care.

(d) Barometric heighting: about 1.5 m, provided operations are limited to an area of about 10 km square and a difference of elevation of not more than 500 m.

Bench marks
A bench mark is a permanent survey mark whose height is known, and from which levelling and height are controlled. Wherever possible a line of levels or series of trig heights should be tied into bench marks; but it is essential that barometric heighting is controlled by bench marks. Great care is needed in the siting of these marks on ground that is thought to be stable and relatively free from vandalism.

9.2 Spirit levelling

The principle of levelling with the aid of a staff or staves has been explained above. The same simple technique is common to all types of spirit levelling irrespective of the accuracy required. The following section explains the basic field work involved, the manner in which the observations are booked in the field, and the various simple checks that are applied to the arithmetical operations required to produce the final heights of the stations. In many books, the subject of geodetic levelling is described after that of simple or engineering levelling. Since these have so much in common, we shall treat the most accurate work first and point out those operations that are unnecessary if a lower standard of result is sufficient.

Types of level instrument
The various types in common use have been described in Chapter 3. It should not be forgotten that a theodolite may be used as a level on occasion.

Types of staff
A large number of different types of staff are available to the surveyor. Folding or telescopic staves are easier to carry whereas

rigid staves are more accurate since they maintain their length better over a period of time. Geodetic levelling staves are constructed of wood with the graduations on a strip of invar held securely at the bottom of the staff but which is free to expand elsewhere along its length. The coefficient of expansion of invar is so small that temperature corrections need not be applied as a rule. The figuring of staves also varies from type to type. Modern staves with bar-coded scales are revolutionizing levelling, by making it consistently more accurate and cost effective.

Some geodetic staves have two different scales side by side so that a gross error in reading the staff may be detected at once and the reading repeated. To ensure that the staff height at a change point is not altered when it is turned around, a special steel foot with a dome-shaped top is employed.

Booking of readings

The process of levelling requires a very orderly approach to recording data because so many readings are taken. Data recorders are widely used with robust software to detect omissions and assess quality as the work proceeds. However the principles are the same for hand booking which is now outlined with respect to the example shown in Figure 9.5 and Table 9.1.

Figure 9.5 shows a cross section and plan along the route taken by a surveyor levelling from point 1 to point 11. The instrument positions are at A, B, C, and D, and the change points are 5, 9 and 10. Table 9.1(B) shows a typical set of readings for the levelling illustrated in Figure 9.5. The algebraic version given in Table 9.1 (A) is to assist in the explanation.

The instrument is set up at A and carefully levelled. A staff is held at point 1 and the reading R1 is taken. The staffman then moves in turn to points 2, 3, 4 and 5 at which readings R2, R3, R4 and R5 are made.

The intermediate sights are booked in the second column of the field page, with the change points in columns 1 and 3 for backsights and foresights respectively. The differences in height between each point are booked in the 'rise' and 'fall' columns where appropriate.

The instrument is then changed to position B while the staff is kept at point 5. Readings from position B are denoted by r5, r6, r7, r8 and r9 with the staff at points 5, 6, 7, 8, and 9 respectively. The backsight reading is r5 and the foresight reading is r9.

Note that only the change points (CP) have two readings taken to them. The reduced level of each point in turn is obtained by adding the rise or subtracting the fall from the height of the point to which the difference of height was related, i.e. the point above it in the field book. At the foot of each field book an arithmetical check is made.

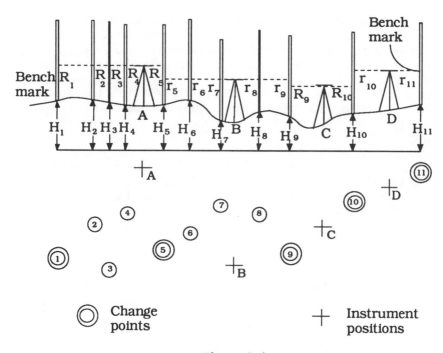

Figure 9.5

The following four columns are summed separately and the totals entered at the foot of each column:

(a) The sum of backsight readings.
(b) The sum of foresight readings.
(c) The sum of rises.
(d) The sum of falls.

The difference in height between points 1 and 11 in this example is

$$H_{11} - H_1 = 0.473$$

The various checks then are:

$H_{11} - H_1$ = sum of the backsights – sum of the foresights
$H_{11} - H_1$ = sum of the rises – sum of the falls

This can easily be proved from the algebraic example.

The totals at the foot of one levelling page are carried forward to the top of the next page, taking great care not to make a mistake in the transfer. It must be stressed that these checks in no way verify the readings but merely their abstraction into heights.

Normally the line of levels will start and close on a benchmark, though a loop may be run which begins and ends on the same point. The adjustment of any misclosure is considered below. The method of booking given above is the 'rise and fall' method.

Table 9.1 A
Rise and Fall Method : algebraic version

Back sight	Intermed–iate sight	Fore sight	Rise	Fall	RL
R1					H_1
	R2		R1–R2		H_2
	R3			R2–R3	H_3
	R4		R3–R4		H_4
r5		R5		R4–R5	H_5
	r6		r5–r6		H_6
	r7			r6–r7	H_7
	r8		r7–r8		H_8
R9		r9		r8–r9	H_9
r10		R10	R9–R10		H_{10}
		r11	R10–R11		H_{11}

Table 9.1 B
Rise and Fall Method : numerical version

Back sight	Intermed–iate sight	Fore sight	Rise	Fall	RL	Notes
1.198					36.469	BM1
	0.811		0.387		36.856	
	1.305			0.494	36.362	
	1.283		0.022		36.384	
0.920		1.588		0.305	36.079	CP
	0.884		0.036		36.115	
	1.109			0.225	35.890	
	0.655		0.454		36.344	
1.393		2.097		1.442	34.902	CP
1.603		1.015	0.378		35.280	CP
		0.887	0.716		35.996	BM2
Checks						
5.114		5.587	2.193	2.466	0.473	
		0.473		0.473		

One other worthy of mention is the 'height of collimation' method. In this method we book the height of the line of sight and from it subtract the individual readings to obtain each ground height. For example, in Figure 9.5 the line of sight of the instrument in position A is $H_1 + R_1$.

A typical booking by height of collimation of the readings of Table 9.1 is given in Table 9.2. The sole arithmetical check on the booking is

$$H_{11} - H_1 = \text{sum of the backsights} - \text{sum of the foresights}$$

There is no arithmetical check on calculating the heights of the intermediate points. For this reason this method is less favoured than the rise and fall method. It is however more convenient in the setting out of height points on the ground, and in recording heights below roofs or arches in engineering work.

Table 9.2
Height of Collimation Method

Back-sight	Intermed-iate sight	Fore-sight	Height of collimation	RL	Notes
1.198			37.667	36.469	BM1
	0.811			36.856	
	1.305			36.362	
	1.283			36.384	
0.920		1.588	36.999	36.079	CP
	0.884			36.115	
	1.109			35.890	
	0.655			36.344	
1.393		2.097	36.295	34.902	CP
1.603		1.015	36.883	35.280	CP
		0.887		35.996	BM2
Checks					
5.114		5.587		0.473	
		0.473			

9.3 Errors in levelling

Because levelling is a chain process in which small errors accumulate attention has to be paid to errors which, though small in themselves, are significant in total. The following are the various sources of error in levelling

Gross reading error
In precise work each staff has two scales engraved upon it so that a mistake will be detected by the booker at once. Alternatively the

readings to the stadia hairs are used for the same purposes. Bar-coded staff reading levels result in few mistakes, if any.

In all levelling it is standard practice to close the work, either out and back, or round in loops, to detect mistakes. The prudent surveyor identifies semi-permanent change points to save having to re-level too much work.

Staff datum error

A staff may have a small datum or zero error. This means that the complete scale on the staff is moved through a small amount with respect to the base of the staff. If a staff becomes badly worn at the base, such an error is introduced. If only one staff is used to height two points, the staff error from this source will not affect the difference in height since it is common to both readings. If two staves are used for speed, as in geodetic levelling, the error in the difference in height of one set-up is the difference between the two datum errors of the staves. After every pair of set-ups the error is eliminated if the back staff is leap-frogged to the forward position. Hence if two staves are used, there should be an *even number of set-ups* between each bench mark.

Graduation error

In precise work the errors in the graduations of each staff are determined and applied to the staff readings where applicable. Each staff is placed upon a specially designed bench along which is a very accurately divided tape or scale and against whose graduations those of the staff are compared, with the assistance of a travelling microscope. For best results a polynomial is fitted to the calibration data and corrections are applied to every reading: not an arduous task if the bookings are originally by datalogger.

Non-verticality of the staff

If a staff is held at an angle θ to the vertical the reading is $S \sec \theta$ instead of S. Hence the reading error is

$$S \sec \theta - S = \frac{1}{2} S \theta^2 \quad \text{approx}$$

This is a serious source of error in all types of levelling. In ordinary work, the staffman swings his staff backwards and forwards about its base so that it moves through the vertical position at some stage. The observer then obtains a correct minimum reading at this point. In some staves a bubble is fitted to the back of the staff to enable the staffman to hold it vertically. The adjustment of a staff bubble should be checked every week against a plumb line. A geodetic levelling stave is provided with two rods which enable the staff man to hold it vertically for some time. Some surveyors prefer a staff clamped to an adapted survey tripod for precise work.

Warpage of the staff

If the staff becomes warped the effect is similar to non-verticality: a reading becomes too high by an amount proportional to the reading itself.

Temperature errors

In refined work, the graduations are engraved or painted on a strip of invar attached to the bottom of the staff. Thus the scale is free to expand and contract without stresses being set up. Since the coefficient of linear expansion of invar is very small (of the order of 14×10^{-7} / deg C) temperature corrections need not be applied as a rule, although in the tropics they may be significant.

Staff illumination

It has been found that engraved staves may exhibit a small systematic error under certain conditions. If one staff is constantly illuminated by the sun, whilst the other is constantly shaded, a small error of about 60 μm per shot can arise. The reading on the illuminated staff is too low because the bottom portion of the graduation is not seen. If a line of levelling is run from south to north, the south facing staves (the foresight staves) are consistently read too low and the difference in height (back − fore) is therefore too large if positive and too small if negative. Thus the northern benchmarks are give too great a height. The same effect will be obtained when running a line of levels east to west, although the effect will tend to be balanced if the line is levelled in one direction throughout the day. Painted staves do not exhibit this effect to any marked extent.

Bubble and compensator sensitivity

In a level, the bubble should be sufficiently sensitive to permit staff readings of the required accuracy to be read, whilst at the same time they should not be so sensitive that much time is wasted in levelling the bubble. In a geodetic level the bubble is capable of being levelled with a standard error of 0".25, which is equivalent to 0.1 mm at about 100 m.

Temperature effects

Most good instruments are designed so that temperature does not affect their performance to any extent. However in precise work, the instrument is always shaded from the sun by a survey umbrella to prevent errors arising from differential heating and twisting of the tripod.

Collimation errors

If a level has a collimation error of θ the reading error on a staff will depend on the length of the line of sight. In Figure 9.6 the backsight distance B is less than the foresight distance F with the result that the staff errors in reading are B θ and F θ.

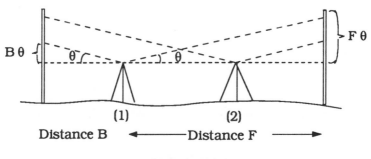

Figure 9.6

The error dH in the height difference is given by

$$dH = \theta\,(B - F)$$

If the lengths of the back and foresights are equal there is no error in the height difference. Hence in all types of levelling these sights should be equated where possible, and the collimation error should be made as small as possible by adjusting the level by the two peg test described in Chapter 3.

In very precise work, since it is impossible to eliminate the collimation error completely, its value may be determined and a correction applied to the difference in the sums of the backsights and foresights for the whole line. In precise work this difference is kept to 1 m if possible so that any correction will be negligible. But in levelling up steep slopes it can be impossible or very difficult to equate backsight and foresight distances.

The collimation error of the level may be determined in the following manner. The instrument is set up, as in Figure 9.6, close to the back staff at A. Readings are taken to both staves together with the stadia readings u and l, from which the distances B and F are calculated from 100 x (u – l). The error is as before

$$dH_1 = \theta\,(B - F)_1$$

The instrument is set up again in a second position close to the forestaff B and the same readings repeated giving a second expession for the error:

$$dH_2 = \theta (B - F)_2$$

Equating the true difference in height gives

$$B_1 - F_1 + \theta (B - F)_1 = B_2 - F_2 + \theta (B - F)_2$$

Which re-arranges to the rule

$$\theta = \frac{\text{sum of near staff readings } - \text{ sum of far staff readings}}{\text{sum of near stadia intercepts} - \text{ sum of far stadia intercepts}}$$

Once θ has been found, a correction dH may be applied to the observed difference in height over a line of levels of

$$dH = -\theta (B - S)$$

where B – S is the difference between the total sum of the backsight distances and the total sum of the foresight distances between the bench marks over which the levelling is carried out.

The distances are not measured by tape because the stadia distances are used in the levelling itself: so any error due to the tacheometric multiplier will cancel out.

Example
Table 9.3 shows a set of readings taken to establish the collimation error of a geodetic level, together with a set of readings taken at the mid-point of the line as a check. The stadia constant of 100 can be neglected throughout the calculation.

Table 9.3

	First set-up	Second set-up
B	0.061	0.853
F	0.853	0.061
B–F	– 0.792	+0.792
RB	1.4615	0.6230
RF	2.3360	1.4886
RB–RF	– 0.8745	– 0.8656

$$\theta = \frac{-0.8745 + 0.8656}{-0.792 - 0.792} = 5.6 \times 10^{-3}$$

Applying this collimation correction of $- 0.0056\,(B - S) = + 0.0044$ to the first set-up we have the corrected difference in height of

$$-0.8745 + 0.0044 = -0.8701$$

and to the second set-up a corrected height difference of

$$- 0.8656 - 0.0044 = - 0.8700$$

As expected, the mean of the two differences gives the same result.

Sinking or rising of the staff
If the staff is set up on soft ground it will gradually sink lower whilst the observer is working. In Figure 9.7 the levelling staff was at a position F when the reading R_B was taken from position I, but it had sunk to position F' by the time that the reading was taken from position i to give a reading r_B which is in error by e.

There is therefore a strong argument for reading on to a staff in quick succession to minimize the time during which the staff may sink. As will be seen below, this conflicts with the effect of the sinkage of the instrument. Both effects can be avoided if all precise work is carried out along a hard surface such as the kerb stones of a

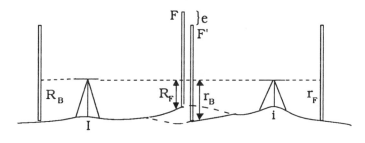

Figure 9.7

road. In hot weather a tarred surface will permit the staff to sink, whilst pegs driven into the ground will rise for a time after being driven, producing the opposite effect.

Sinking of the instrument
In a similar manner to staff sinkage, the instrument may sink into the ground whilst the observations are being made. In Figure 9.8 the

instrument sinks a distance e_1 in position 1 and e_2 in position 2. If the same degree of sinking occurs at both positions and the observer takes the same time to carry out his reading, there is a strong argument for reading first to the backstaff at A then to the forestaff at B, during which time the error e_1 has occurred. After changing to position 2, the first reading should be to the forestaff at C and finally to the backstaff at B, during which the error e_2 has occurred. The chances are that these errors e_1 and e_2 will be nearly equal and therefore the height difference between A and C will be correct.

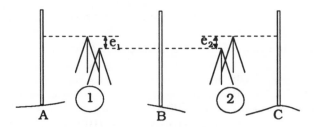

Figure 9.8

Again this means that there should be an even number of set-ups between the bench marks. Since it is argued that the instrument is more likely than the staff to sink on account of its greater weight, the procedure is to read alternately to back and fore staves first in precise work, or which is the same thing, always read to the same staff first, the staves being leap-frogged forward.

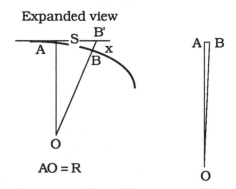

Figure 9.9

Effect of the Earth's curvature
In Figure 9.9, AB' is a horizontal line of sight at A. Due to the curvature of the surface of the Earth, a staff reading at B will be too great by BB'.

This is a very small amount compared with R, the radius of the Earth. To explain the geometry of the figure, refer to Figure 9.9, but remember that the true dimensions are those shown on the right. Let AB = AB' = S and BB' = x. Then in triangle AB'O

$$B'O^2 = AO^2 + S^2$$
$$(R+x)^2 = R^2 + S^2$$
$$R^2 + 2Rx + x^2 = R^2 + S^2$$

then

$$x = \frac{S^2}{2R} \quad \text{approx}$$

since x^2 is negligible.

It will be obvious that if the instrument is set up halfway between the two staves both the backsight and foresight readings will be in error by the same amount, and therefore the height difference is free from error. However, the instrument need not be set up each time at the mid-point between the staves to eliminate this error; if the total sum of the backsights equals the total sum of the foresights, the total curvature error over the whole line of levelling will be nil.

Effect of refraction

Two points B and F in Figure 9.10 are at the same height for simplicity, and R_B, S_B etc are the staff readings and distances. We neglect the curvature of the Earth. Since the light ray from the instrument to the staff is refracted as it passes through the atmosphere, a recorded staff reading is too low. The errors on the backsight and foresight due to refraction are e_B and e_F respectively.

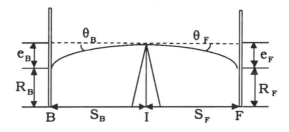

Figure 9.10

If α is the angle subtended by AB = S at the centre of the Earth, whose radius is R (Figure 9.9), and the coefficient of refraction is K, the angle of refraction θ_B is given by

$$\theta_B = K \alpha = KS^2/R$$

The coefficient of refraction K, about 0.07, varies considerably. The main factor affecting it is the rate of change of temperature with height above the ground; the greater this rate of change, the smaller the refraction. If K is constant during the time that both sights are observed, the effect on the difference in height will be nil, if the instrument is positioned mid-way between the staves. But in many cases K is not constant over a given period, hence a small systematic error will be introduced into the line of levels. If a line is levelled from early morning to noon, the refraction becomes progressively less. In this case, always reading to the same staff first will help to eliminate the refraction error since the change in refraction will alternately affect each height difference with an opposite sign, each even number of height differences being relatively free of error.

When a line is levelled up a steady slope, the foresight is always closer to the ground and refraction is again less there. In this case, the effect will be systematic and the uphill bench mark will be too low. Wherever possible, the line should be run in the same conditions for both the forward and backward levelling so that the mean differences in height will be almost free from error. If the total sum of the backsights equals the total sum of the foresights, but the individual set-ups are not mid-way between the staff positions, the error due to refraction will not be eliminated, as a rule, because of the change in refraction with time. The average value of refraction for the backsights will not be equal to the average value for the foresights unless these sights are equated as closely as possible in time, i.e. at every set-up.

Combined curvature and refraction
It is usual to combine the curvature and refraction corrections into the one formula

$$C + r = \frac{S^2}{2R}\,(1 - 2K) \qquad (1)$$

Taking average values and S in km this correction is

$$67\,S^2\ \text{mm}$$

Normally it does not need to be applied because backsights and foresights are equated. In some special cases and in trigonometrical heighting it will be applied.

Practical levelling
As a result of considering the above sources of errors in levelling, it is apparent that for all types of work most errors will be avoided if the following three rules are adopted.

(a) To detect gross errors, a line of levels should be run between two points of known height, or twice over the same line, or using a staff with two scales.

(b) The staff should be vertical when a reading is taken; this is achieved either by swinging it through the vertical position, or with the aid of a bubble, which should be tested.

(c) The length of the backsight should equal the length of the foresight at every set-up.

In ordinary work on engineering sites it is not feasible to have backsights and foresights exactly equal, and in intermediate sights no compensation for curvature, refraction, and collimation takes place. Hence it is imperative that the level has very little collimation error, otherwise tedious corrections will have to be applied.

Errors due to curvature and refraction are negligible for most work with sights not exceeding 100 m. Again it is important to avoid soft muddy ground, but for obvious reasons this is not possible on an engineering site. Remember, pegs knocked into the ground will rise up for several minutes afterwards.

9.4 Geodetic levelling

In geodetic levelling all the above sources of error have to be taken into account and a field procedure adopted which will eliminate their effect as much as possible. Usually, very long lines of up to 50 km are run, so that small systematic errors can accumulate into quite large proportions. Where possible the work is repeated under varying conditions to change systematic errors into random errors and thus bring about some degree of compensation. Lines of geodetic levelling are always run twice, back and fore along the line K km, within an acceptable tolerance of 2 $\sqrt{K}$ mm between the two measures. If the average length of sight is 40 m, and there are about 25 set-ups per km, the permissible difference between two measures of the height difference at each set-up is 1.2 mm, (based on a 3 σ statistic).

There are no intermediate sights in geodetic levelling. To avoid misreading or booking in the wrong column, each staffman should have some distinctive dress. This enables the observer to point first at the same man, and therefore at the same staff at each set-up. Before booking the first reading of a set-up, the booker will enter an arrow pointing up the page if the distinctive man is being observed. Since alternate back and foresights should be read, the arrows will appear alternately in the back and fore columns. If this pattern is broken, the booker draws the attention of the observer to his mistake.

To avoid errors arising from differential heating of the instrument, it should be shaded by an umbrella. Since two staves are invariably used in geodetic work, and because the effects of sinking of the instrument and changes in refraction will be partially eliminated, there must always be an even number of set-ups between bench marks.

All bench marks and section points have to be well planned before work commences.

The future of geodetic levelling

It is difficult to predict how much geodetic levelling will be carried out in the future. On the one hand the GPS system, which can be used to interpolate between bench marks, and trigonometrical heighting can achieve equal standards: while on the other hand, the new bar-coded staff reading levels are so quick and accurate that they must be given serious consideration. The likely outcome is a hybrid of all three methods.

Booking and recording readings

With conventional levels and staves, modern practice is to use data loggers to booking readings. These loggers process the data for statistical rejection in real-time to alert the observer. Generally several readings are taken, following an orderly scheme, to produce sufficient redundancy for these tests.

Levelling across a water gap

It sometimes happens that a line of levels has to be carried across a wide water gap of from 100 m to 10 km. Under these circumstances it is impossible to read the staff from the instrument end with the required degree of precision. One method of overcoming the problem is to fit special targets to the staff which are moved up and down until the observer signals that the target is bisected by his cross hair. The staffman then reads the staff. A successful target has been a band of brown paper tightly wrapped round the staff. This method is less precise than those now described.

By gradienter

The line of sight of a level may be tilted through a small angle by a micrometer tilting screw or gradienter which is calibrated in terms of the tangent of the angle through which it is tilted. Readings are in parts per thousand, i.e. a reading r represents an angle whose tangent is $r/1000$. A target is placed at some convenient mark on the staff and the gradienter reading r taken. A reading on the gradienter r_0 is also taken when the bubble is level.

If possible, a second reading should be taken to a target below the horizontal line from the level to the staff so that a negative gradienter reading is obtained. The distance s is obtained by EDM and radio communication is necessary. It is clear that the backsight and foresight distances cannot be equated and therefore the various errors arising from this lack of equality will be considerable. If the level is now moved to a position close to B and the total backsights and foresights thus equated, all errors other than refraction are eliminated by accepting the mean of the two height determinations.

Refraction is not eliminated because of the time that elapses in moving the instrument from one bank to the other. To eliminate this effect, two levels are used to take simultaneous reciprocal observations. Then all equipment is interchanged and the process repeated to eliminate the effect of two observers and instruments. As many as twenty complete determinations of the height difference may be required to achieve the desired precision. The effect of refraction is not entirely eliminated since the propagation curve is asymmetrical. However the procedure of simultaneous reciprocal levelling will eliminate most of the error.

By precise theodolite observations

As a result of a very comprehensive series of tests involving height determinations over some 45 lines varying in length up to 8 km, the Ordnance Survey of Great Britain has produced results of the same accuracy as geodetic levelling for height transfers by observation of truly simultaneous reciprocal vertical angles using geodetic theodolites at either end of the line. Recent experience of the author in California measuring at high temperatures throughout the day by similar methods gave good results (2 mm/ km) over 200 m lines.

9.5 Orthometric and dynamic heights

Early in this chapter we defined the height of a point to be the linear distance from a datum surface at mean sea level to the point. However, the concept of 'height' may be considered from a different point of view in terms of the force of gravity acting on a mass situated near the surface of the Earth.

The gravitational potential at a point at a distance D from the centre of the Earth is given by $dP = g\,dD$, where g is the acceleration due to gravity. An equipotential surface is one on which all the points have the same gravitational potential, i.e. the potential P is constant. Since g varies with latitude, being greater at the poles than at the equator, an equipotential surface is closer to the centre of the Earth at the poles than at the equator.

Figure 9.11 shows the shape of two equipotential surfaces defined by their potentials P_1 and P_2. If H_0, H_{45} and H_{90} are the linear separations between these equipotential surfaces at the latitudes of 0°, 45°, and 90°

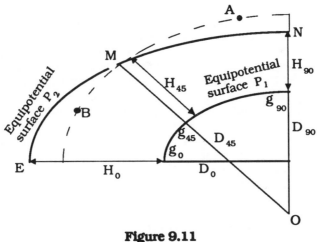

Figure 9.11

respectively; g_0, g_{45} and g_{90} their respective gravities we can define a constant "height" difference by

$$dP = \text{constant}$$

$$g_0\, H_0 = g_{45}\, H_{45} = g_{90}\, H_{90}$$

Since no work is done against gravity, if a point moves over an equipotential surface, such a surface is dynamically flat. For example if a pipe line followed P_1 in Figure 9.11, no water would flow along it under the action of gravity. But if the pipe line took the path of the broken line which is equidistant from P_1, water would flow from A to B under the force of gravity due to the difference of potential at these points. Thus, although A and B are at the same *linear* height above P_1 in a *dynamic sense* there is a slope from A to B. Points defined by linear distances along the vertical are called *orthometric heights*, and those defined in terms of potential are called *dynamic heights*.

To relate the concept of potential to a linear system of height, a standard latitude is chosen (45˚ for a world system, 53˚ for England and Wales) at which both types of height are defined to be equal; e.g. in Figure 9.11 the linear separation h between P_1 and P_2 at latitude 45˚ is used to define the potential surface P_1 with respect to P_2 as datum in linear units. Hence the *dynamic heights* of points E, M, and N all equal h, the orthometric height of M above P_1 at the standard latitude 45˚. The *orthometric height* of E is H_0 and of N is H_{90}. The orthometric height at the equator is greater than the dynamic height, and vice versa at the poles, the two being equal at 45˚.

Relation between orthometric and dynamic heights

To relate the two concepts we use the fact that

$$g_0 \, H_0 = g_{45} \, H_{45} = g_{90} \, H_{90}$$

Then

$$h = H_{45} = g_0 \, \frac{H_0}{g_{45}}$$

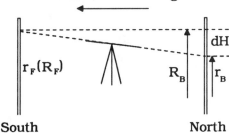

Direction of levelling

South   North

Northern Hemisphere

Figure 9.12

It can be shown that theoretical gravity g_ϕ at latitude ϕ, is related to the gravity g_0 at the equator by the simplified expression

$$g_\phi \approx g_0 \, (1 + 0.0053 \sin^2 \phi)$$

therefore

$$h \approx H_0 / 1.00265$$

and for any point at an intermediate latitude

$$h \approx H_\phi \, (1 + 0.0053 \sin^2 \phi) / 1.00265 \tag{2}$$

If the orthometric height H_ϕ is known, its equivalent dynamic height h may be derived from this expression (2).

Example

If the orthometric height of a point on the equator is 1000 m its dynamic height is 997.4 m, i.e. there is a difference of 2.6 m.

Orthometric heights from levelling

In the actual process of levelling, the bubble sets itself tangential to the equipotential surface through the instrument, whilst the readings on the staves are orthometric operations.

In Figure 9.12 the divergence between the orthometric and dynamic surfaces through the instrument gives rise to an error dH in the backsight reading r_B such that the correct reading is

$$R_B = r_B + dH$$

when levelling from north to south in the northern hemisphere.

Defining the difference in latitude over the line $d\phi = \phi_B - \phi_F$ i.e. in the same sense as dH, $d\phi$ is positive in this case, and for a line run from south to north it is negative. Over the equipotential surface, gH is constant, therefore

$$H\,dg + g\,dH = 0$$

$$dH = -H\,dg\,/g$$

$$= + 0.0053\ H\ \sin 2\,\phi\,d\phi$$

Thus to obtain the orthometric height difference from the levelled height difference, a correction of + dH is applied. Because this correction is very small for one set-up, it is normally applied to lines of levels at about every 10 km or at convenient section points along the line. The total effect for the line is the summation of all the individual corrections, i.e. the corrections are applied by numerical integration over the whole line.

Example
If the average height of a section of levels 10 km long run from south to north at latitude 10° N is 1000 m, and the indicated height of the northern bench mark was 1026.9384 m, calculate its orthometric and dynamic heights, assuming that the height of the southern bench mark was orthometric. In this case d ϕ is negative and

$$dH = -1000 \times 0.0053 \times 0.3420 \times 10/\ 6400 = -0.0028\ m$$

Hence the orthometric height of the required bench mark is 1026.9356 m, and the dynamic height is given by

$$h = 1026.9356\ (1 + 0.0053 \times 0.03014)/\ 1.00265 = 1024.3850$$

Geopotential numbers
Since the acceleration due to gravity varies not only with latitude ϕ, but with height above sea level h and with the composition of the Earth, allowance is made for these factors, and a system of heights, called *geopotential numbers*, has been introduced. The system,

devised by the International Association of Geodesy, defines the geopotential number of a point A to be

$$C_A = \int_0^A g\,\Delta h$$

The units to be used are the kilogal/metre. i.e. g is in kilogals and Δh in metres above sea level. (One gal is an acceleration of 1 cm s^{-2}).

It is desirable that observed values of g are used, though theoretical values based on some hypothesis concerning the structure of the Earth may be used for want of observational data.

9.6 Trigonometrical heighting

A common method of determining height difference ΔH is to observe the vertical angle θ to a target or object and compute the difference in height between the theodolite and the target by the relationship

$$\Delta H = S\tan\theta$$

where S is the horizontal distance between the stations; or from

$$\Delta H = S'\sin\theta$$

where S ' is the slant range as measured by EDM.

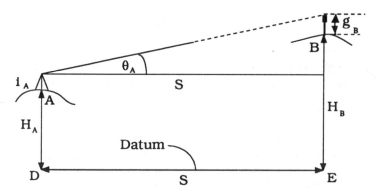

Figure 9.13

In Figure 9.13, H_A and H_B are the heights of the stations A and B above a horizontal datum line DE, which is considered straight meantime, i_A is the height of the trunnion axis of a theodolite at A referred to some reference mark on the station at A, g_B the height of the target at station B, the vertical angle observed at A is θ_A, and the horizontal distance DE is S. These quantities are related by the expression

$$H_A + i_A + S \tan \theta_A = g_B + H_B \qquad (3)$$

or putting $\Delta H = H_B - H_A$

$$\Delta H = i_A + S \tan \theta_A - g_B$$

Note that this equation holds for all values provided the following conventions are adopted.

(a) Heights above a station are considered positive, those below are negative.
(b) The height difference ΔH from A to B is defined as $H_B - H_A$.
(c) An angle of elevation is positive, and a depression is negative.

These conventions are commonly used in everyday life and should not cause any difficulty. No diagrams are necessary when computing a height difference, on the contrary, a diagram will often confuse the issue.

Example

If $H_A = 100$ m, $S = 1000$ m, $\theta_A = +1°25'57''$, i.e. an elevation, $i_A = 1.5$ m and $g_B = 10.00$ m, the height H_B is derived as follows:

$$H_B = 100 + 1.5 + 1000 \times 0.02501 - 10.00 = 116.51$$

It should be noted that although i_A and g_B are normally positive they can have negative values, for example if an instrument or beacon is erected below the reference mark on a ceiling or tunnel roof.

It is usual to observe a vertical angle on face left and on face right so that the effect of the vertical index error of the theodolite is eliminated on taking the mean. Whenever possible, observation should be taken from A to B, and from B to A. It is always better to compute the height difference by two separate computations using equation (3), rather than a formula which combines the observations taken at both ends. The reason for this is that a gross error may be detected from any discrepancy between the two values. Great care should be taken to avoid mistakes in recording beacon and instrument heights.

Earth curvature

The reference datum line DF is not straight but is curved as in Figure 9.14. The linear amount of the Earth's curvature FE = x is given by

$$x \approx \frac{S^2}{2R}$$

where R is the radius of the Earth. This curvature effect may also considered as an angular correction α to the observed angle of

$$\alpha \approx \frac{x}{S} \approx \frac{S}{2R}$$

$$\alpha'' \approx 206265 \frac{S}{2R}$$

We then use the effective angle $(\theta + \alpha)$ in calculations to allow for Earth curvature. Notice α is always positive whereas θ may be either positive or negative.

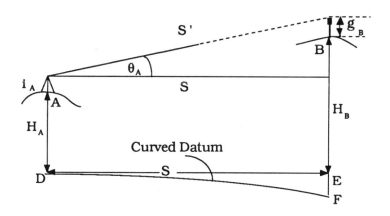

Figure 9.14

Refraction effect

In Figure 9.15 the light path from theodolite to target J is shown as a curved line instead of a straight line. This is so in practice, due to the refractive effect of the atmosphere. The telescope points to G in the direction of the tangent to the curved refracted ray instead of along the direction of the chord, and therefore records an angle which is greater than that used in the computation. Hence to obtain the equivalent angle free from refraction, the angle β has to be *subtracted* from the observed angle θ. The refraction angle β is given by

$$\beta \approx \frac{KS}{R}$$

The coefficient of refraction K is about 0.07, though it varies according to the density of the air through which the light passes, and should be determined for the average conditions of the survey area.

Combined curvature and refraction

To reduce an observed angle θ to its equivalent plane angle φ we have :
Corrected plane angle = observed angle + curvature – refraction

$$\phi \approx \theta + \alpha - \beta$$

or

$$\phi \approx \theta + 206265\,(1 - 2K)\,\frac{S}{2R} \qquad (4)$$

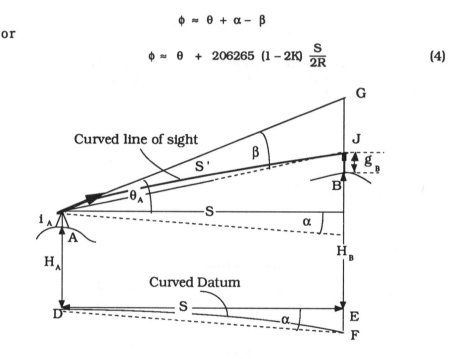

Figure 9.15

The combined correction works out at about 14" per km, putting K = 0.07 and R = 6400 km. This combined curvature and refraction correction is always positive because the curvature correction exceeds the refraction correction. Since the latter varies with the prevalent air conditions, the combined correction will also vary from place to place and with time of day. Since refraction is least at the mid-day period, usually taken from 12.00 hrs to about 15.00 hrs, and since its value during this period varies little from one day to another, vertical angles are best observed during these hours. However, if two observers are situated at each end of the line, truly simultaneous vertical angles may be observed, which enables the mean value to be almost free of refraction error. Such observations may be taken at almost any time of the day.

In practical computation of trigonometrical heights a few reliable observations, taken over several long lines, are used to compute the combined curvature and refraction correction in seconds of arc per kilometre, and thereafter this value is used to compute all heights by single rays.

Example

A line EG observed between two pillars is chosen from which to compute the combined curvature and refraction correction. The height difference from E to G is computed without applying a correction for curvature and refraction. This is repeated for the direction G to E. See Table 9.4.

Table 9.4

Line	Vertical angle θ	Distance Sm	i	g	ΔH
EG	+ 00° 07′ 39″	8024.90	0.207	0	+ 18.065
GE	− 00° 11′ 33″		0.253	0	− 26.709

The difference these two values for the difference in height is due to twice the effect of curvature and refraction, which is therefore calculated from

$$\frac{8.644 \times 206265}{2 \times 8024.90} = 111''$$

This is the correction to be applied to a single ray observation over the line. The value of the correction to be used for the area of the survey is therefore

$$13.843 \text{ '' per } 1000 \text{ m}$$

As a check on the arithmetic and to illustrate the method of single ray calculation, these two observations are recomputed separately using this value for the combined correction (see Table 9.5). The combined correction is

$$13.843 \times 8024.90 = 111'' = 01' \; 51''$$

The accepted mean difference in height is 22.387m. It will be noticed that this same value is obtained from the mean of the two values 18.065 and 26.709 computed without recourse to the correction at all.

To adopt this latter method of computing is dangerous because there is no clear indication that the results are free from gross error. Thus the method of single ray computation, with an applied curvature and refraction correction, allows the quality to be assured.

Table 9.5

Line	Vertical angle θ	Distance S m	i	g	ΔH
EG	+ 00° 07′ 39″		0.207	0	
	+ 00° 01′ 51″				
Cn	+ 00° 09′ 30″	8024.90			22.383
GE	− 00° 11′ 33″		0.253	0	
	+ 00° 01′ 51″				
Cn	− 00° 09′ 42″	8024.90			22.390

Linear value of the curvature and refraction correction
It is useful to evaluate the linear effect of the combined curvature and refraction over the distance S. The linear corrrection is

$$S(\alpha - \beta) = S^2 \frac{(1 - 2K)}{2R} = 0.0672 \, S^2 \, m \tag{5}$$

where S is in km.

This formula is useful in working out intervisibility problems. For example, if the distance S is 8 km, the combined correction is approximately 4.3 m.

Precision of measurements
We now consider briefly the precision required in the various measured quantities to achieve a given result for the difference in height. Denote $\tan \theta$ by t, the distance by S, and the height difference ΔH by h, then

$$h = St$$

Since the height of instrument and beacon enter directly into the value obtained for h they must be measured to at least the same precision as is desired in the final result. In normal work, these quantities are measured to 0.005 m by taking readings to a tape held vertically as with a level staff, and by direct measurement of the height of theodolite from the secondary axis, applying the small alignment offset if necessary.

Consider small changes dt in t, dS in S, and dh in h, related by

$$dh = S \, dt + t \, dS$$

If dt = 0, and θ = 1', and if dh is to be 0.01 m,

$$0.01 = dS/60 \qquad dS = 0.6 \, m$$

Thus distances are not required to any great accuracy.

Again, if dS = 0, and if S = 1000 m

$$d \, \theta'' = 0.01 \times 206265 / 1000 = 2 \, ''$$

To guard against mistakes in reading the vertical angle, one reading should be taken on each of the stadia hairs, thus obtaining an independent check from the mean.

9.7 Least squares estimation of heights

The estimation of trigonometrical heights and of level networks by the method of least squares are similar. The difference in height brought through a line of levels is treated in the same manner as a height difference computed for a long line from vertical angle observations. The ends of such a line of levels are called 'junction points'. The method is to obtain best estimates for these junction points or trig stations, and in the case of a line of levels, to adjust the individual points along the line by proportion if required.

Dispersion matrix or weights

Some thought has to be given to an estimation of the quality of the observations, or to the *a priori* dispersion matrix of the height differences. For each line of levels or trig height difference, expectations are taken of the error equation

$$dh = S \, dt + t \, dS$$

giving variance estimates

$$\sigma_h^2 = S^2 \, \sigma_t^2 + t^2 \, \sigma_S^2 \tag{6}$$

Assigning reasonable values for the variances in the angles and distances yields the initial dispersion matrix elements. In levelling the distance error does not matter because t = 0, so we have

$$\sigma_h^2 = S^2 \, \sigma_t^2$$

The degree of sophistication chosen for the error model depends on circumstances, but normal practice is to work with a reasonable estimation per set-up between staves of say 2 mm in ordinary work. It is further assumed that the length of sight is kept more or less constant for the line which has n set-ups thus yielding a line estimate of variance of $2\sqrt{n}$ mm giving, for 25 set-ups per km, about 10 mm per km for ordinary work. Geodetic levelling can achieve 1.5 mm per km. In trigonometrical work the side length S is crucial. It is usual to calculate variances for every line according to the variance formula (6).

In Chapter 5, section 5.12, the least squares treatment of a small network of levels by the method *observation equations* is given. Although these problems can be handled very simply by the method of *condition equations*, (see Appendix 5), observation equations have been preferred because of their generality and the ease of computer programming.

After the heights of junction points have been estimated, the heights of individual set-up points, if required, are obtained by adjusting the change to the observed line height difference in proportion to the number of set-ups in the line.

9.8 Intervisibility problems

Initially we adopt the simple model which neglects curvature and refraction. Figure 9.16 shows two stations A and C whose heights above a horizontal straight line datum are H_A and H_C respectively. Considering for the moment that the line of sight from A to C is straight, we wish to establish the height H_B of an intermediate point B lying on this line.

This problem arises when a survey reconnaissance is being carried out or when a tower has to be erected to a computed height above datum so that observations may be made along a line, such as a traverse. Equating values of tan θ

$$\frac{H_A - H_C}{A'C} = \frac{H_A - H_B}{A'B'} = \frac{H_B - H_C}{B'C}$$

Example
If $H_A = 100$ m, $H_C = 50$ m, AC = 10 000 m, and BC = 8 000 m, the height of B is given by

$$H_B - H_C = \frac{H_A - H_C}{A'C'} \, B'C$$

$$= \frac{50 \times 8\,000}{10\,000} = 40$$

Therefore H_B is 90 m above datum. If the hill at B' lies below this height, say at 85 m, line of sight will pass from A to C, and vice versa. Or to erect a tower on line AC its height has to be BB' = 5 m.

In practice, because of refraction of the line of sight and the curvature of the datum surface, the simple case above has to be modified.

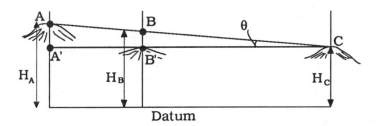

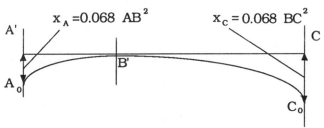

Figure 9.16

In Figure 9.16 it will be seen the net effect of these factors, curvature and refraction, is to lower the datum at A and C with respect to B by x_A and x_C. Hence if the heights of A and C are decreased by their respective corrections, the problem is reduced to the simple model just treated above.

Example
If the heights of A,B and C are respectively 417.27 m, 273.10 m, and 214.58 m; and AB = 24.30 km, BC = 12.23 km, what height of tower should be erected at C so that the line of sight from a theodolite at A, 1.2 m above ground, will clear the ground at B by 6.1 m ?

The effective height at B is 279.2 m. The curvature and refraction corrections to be applied to H_A and H_C are respectively 40.15 m and 10.17 m.

The effective height at A is then 417.27 + 1.2 − 40.15 = 378.32. The effective height C is then given by

$$H_B - H_C = \frac{H_A - H_B}{A'B'} \quad B'C = \frac{99.12 \times 12.23}{24.30} = 49.89$$

$$H_C = 223.21$$

Since the ground height of C is 214.58 m above datum and x_C is equal to 10.17 m, the tower will be required to be 18.8 m tall.

Practical applications

In practice, a section has to be drawn from map contours and a likely position for a station is chosen. Any doubtful intervisibility problems are solved as above. Intervisibility problems, now tackled by computer software, are of some interest in the production maps of visibility and obscurity required for optical communications, flight paths into airports, planning applications and military purposes.

Theodolite height traversing

Sometimes approximate bench marks to about 0.25 m accuracy are required for water table studies and altimeter bases. Such stations are usually required at the bottom of a valley and in thick impenetrable forest or other inaccessible terrain. A rapid accurate height may be obtained by running a traverse from a trigonometrical station often over distances of several km. No horizontal angles are measured, only verticals are recorded at every other station. Distances may be taken with a tape or by tacheometry or EDM. To safeguard against gross errors in reading the vertical angles, stadia hairs must be used, or failing that, observations should be taken to a second point at some set distance (say a metre) above or below the standard target, usually a mark on a ranging pole, or a particular staff graduation. Results are reduced immediately to locate gross error and the traverse should close on another bench mark if possible, or at least back on the original point.

Heighting from air photo distances

In good open country, such as plateau Africa with kopjes, the surveyor can identify hills from afar. Vertical and horizontal angles are taken to these points, and later the distances are scaled from a photogrammetric plot. Thus the heights of these points may be computed and used to control photogrammetric contouring. Alternatively, and more accurately, the surveyor takes a vertical observation to a trig beacon from some point which he has identified on the photography. Only one trig need be observed and the distance scaled from the photoplot. Needless to say more than one trig would be observed if possible.

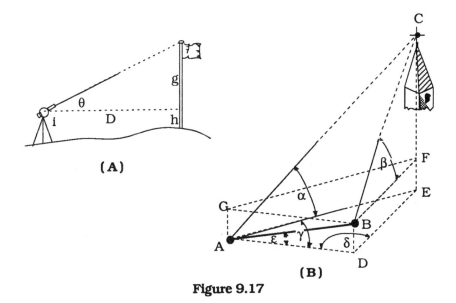

Figure 9.17

Observations to inaccessible stations

When the height of a survey beacon cannot be measured directly by tape, an arrangement such as that shown in Figure 9.17 (A), in which D and h are measured, will enable the beacon height g to be computed.

If the beacon is a tall object such as a church spire whose top has been intersected, a baseline may have to be set out some distance back from the church and angles α, β, γ, δ and ε are observed, together with the length AB (Figure 9.17 B). Triangles AED, ACE and BCF are solved. CE gives the height of the spire relative to A, with CF + BD as a check. In these determinations the various angles should give well conditioned figures for accurate results. The solution is much simpler if A, B, and C are selected to lie in the same plane.

9.9 Barometric heighting

The surveying altimeter was much used in past times to provide heights accurate to about 1.5 m to control aerial mapping. It may still have limited applications to reduce EDM lines for slope in seismic and other surveys. (Pressure is of course still required to calculate the refractive index of electromagnetic waves through the astmosphere). Modern technques such as GPS, Radar altimetry, EDM and analytical aerial triangulation have made the technique of altimeter heighting virtually obsolete.

It must be pointed out that one altimeter is virtually useless for surveying. At least two instruments are needed: one to remain at a base station of known height to monitor changes of pressure, while the other moves to the unknown points. Recorded height differences, allowing for relative index corrections to instruments, are then scaled by a meteorological factor which depends on air temperature and humidity. This factor can be determined empirically from two fixed instruments remaining at bases of known height. Therefore three calibrated instruments are the minimum for serious work. These complications are occasioned by the need to allow for a variation in pressure which can amount to more than 100 metres of indicated height during a day. Careful time keeping is necessary to reduce readings to common instants for comparison of heights.

The true height difference, ΔH_t at time t, is obtained from the indicated altimeter height difference, Δh_t, by

$$\Delta H_t = F_t \ \Delta h_t$$

The factor F_t is otained from simultaneous readings between two known bases whose height difference is B_t and the indicated height difference is b_t giving

$$F_t = \frac{B_t}{b_t}$$

The undoubted attractions of the altimetric method are its simplicity speed and convenience, with the illusion of potential accuracy as a goal. Readers should consult reference (8) for more information.

Comment on heights and levels
The importance of very precise heights and levels to engineering and hydrologicial works has to be stressed.

Geometrical methods such as satellite, photogrammetric and theodolite systems cannot compete directly for accuracy in the last analysis , because of uncertain knowledge of the Earth's gravity field at the millimetre level. Combinations of systems can be effective in which the geometrical systems are used to interpolate values controlled by levelling.

The two-plus-one coordinate system will continue to be of continued importance, while some purely geometrical applications in engineering and construction may be treated in a three dimensional cartesian coordinate system which ignores the vertical altogether.

10 Detail mapping and processing

Chapter Summary

In this chapter a summary is given of the methods used in the surveying, the presentation and the completion of map detail. It also deals with some miscellaneous topics closely related to cartography and photogrammetry, which are not included in this book.

We discuss the following topics in order: map detail; the plotting of grids; the construction of contours; and some allied processing of map information, such as area computation and subdivision of plots.

10.1 Reconnaissance

The first, and perhaps the most important stage of survey is the reconnaissance, or recce. Experience of the full technical and administrative factors of surveying is required to carry out an effective recce. This experience includes the whole production chain subsequent to field work so that potential problems and difficulties can be avoided.

A pre-analysis of any information related to the task or its specification is valuable, such as knowledge of any previous surveys, availability of controls and datum information, copies of old maps and aerial photography etc.

The site has to be visited to decide how to carry out the work, where to located points and marks, the availability of accommodation, possible labour and services etc.

A computer simulation of the technical proposal may also be feasible if software is available. A final decision has to be taken on how the task will be executed, by whom, and with what equipment. Suitable documentation has to be prepared for the surveyors who will actually carry out the task.

Documentation

Proper clear documentation of surveys is required at all times. This includes station descriptions, field books or records of data loggers used, any changes to original objectives, and so on. The whole data set should be clearly cross-referenced, and kept in a clear neat form. This documentation may be the only evidence that the work was properly carried out, and may have to be submitted to a court of law as evidence. The copying of field notes without retention of the originals is tantamout to fraud.

Station marking

There is much wisdom in marking stations, even though it may not be part of the specification to do so. Certainly while the task is in progress, marks should be monumented if only with wooden pegs. It is also wise to reference any station marks by *repere* or witness marks. This will enable the station point to be recovered should it be damaged or removed, as happens frequently on engineering sites as construction work progresses.

The basic purpose of a *repere* is the location of a point either from intersecting lines of sight or from taped distances, or by a combination of both. Examples are to be found in all branches of surveying.

A simple case arises when selecting the site of an instrument station in the middle of an open field. See Figure 10.1. A little inspection of the surrounding fences will reveal pairs of recognisable points which can be used to line-in the point. Better still, orientation can usually be achieved from lines established to distant points on the horizon. Identical methods are employed to check or field-complete a survey. The optical square is a useful tool for placing oneself on line to such reperes. It is surprisingly accurate to locate one's position in a field, such as at point F, by holding the arms outstretched at 180° to each other while sighting on A and D in turn, then establishing the perpendicular direction to B by clapping the hands together at speed. The distance FP can then be taped or paced. The optical square makes the procedure even more accurate.

A record of the procedure can be used to relocate the station peg without difficulty.

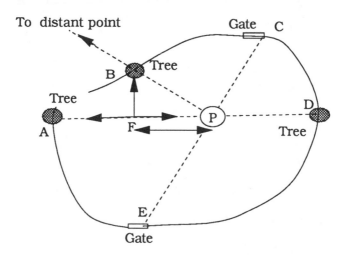

Figure 10.1

10.2 Introduction to detail surveying

The methods described here are those used to make plans at large scales from given control, to revise existing plans and to complete plans made in part by air survey methods. Every survey is based on a rigid framework which is regarded as correct until proved to be in error. Generally speaking all details such as structures and natural features can be *shown to scale* on a *plan* but not to scale on a *map.*

This rule is often broken in special cases as for example, when a forester wishes named trees, which are too small to plot to scale, to be located in position. They have to depicted by *conventional signs* . Map detail can be classified under the following headings.

(a) *Soft detail:* has an outline incapable of exact definition, or is likely to change; such as vegetation and crops.

(b) *Hard detail:* is clearly defined, such as a *building line* where the building enters the ground. A useful rule is that anything that would obstruct the normal passage of a bicycle is shown as a full line, while that which does not obstruct, such as a road kerb, is shown by a pecked line.

(c) *Overhead detail:* constitutes no obstruction at ground level, such as power lines or tree canopies; shown by conventional signs.

(d) *Interior detail:* is surveyed if it is of special interest, such as a party boundary delimiting ownership. Building interiors have separate plans for each floor linked together by vertical correlation.

Depiction of detail

The manner of depicting features varies a great deal according to map scale and purpose: for example, from topographic maps, building interiors and archaeological maps, to thematic plans of a rail network. The reader should inspect a range of these plans to see the symbols used, many of which are obvious. Special plans showing street furniture (lamps, gulleys, man-holes etc) are drawn to the specification of the client. Every user has his special wishes which must be discussed and agreed, particularly whether the item is to be shown to scale or not.

10.3 Detail surveying

In surveying detail point positions we need to record sufficient information to enable a map to be plotted from them. Because the dot plot of Figure 10.2 is the only information from which to plot the map it needs to be completely comprehensive and intelligible. When operating in the field, the plotting process must be kept in mind, so that points can be plotted from a minimum of two position lines. The commonest method of data capture is by radiation.

Points need to be classified as well as positioned, so that their common identity can be used to delineate special features, such as house corners. It is important to know how the map is to be plotted, before embarking on the capture of field data.

If the plotting is to be manual, the collection of data can be flexible and inconsistent. If the data collection serves a computer software system, a rigorously ordered approach is essential.

Figure 10.2 shows a typical part of an urban area to be mapped by radiation from a traverse station B. It will be seen that *nodes*, where two or more lines meet, are recorded by a different key from single point features which lie on a *string*, such as marks the road kerbs. Some points will be both nodes and string points. The curved corners of the road kerb, to be drawn by a circular arc, are marked by the curve end points, which are shown as special nodes. Even when a computer plotting system is used, most surveyors prefer to draw field sketches to label the points and identify their types.

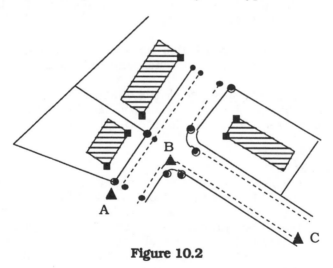

Figure 10.2

Not all details need be picked up by radiation. The enlarged diagram to the right of Figure 10.3 shows much detail, such as lamp-posts, which can be tied in by taking rectangular off-sets to a straight line, such as the kerb. This *line and off-set* method uses a local x,y coordinate system based on the kerb.

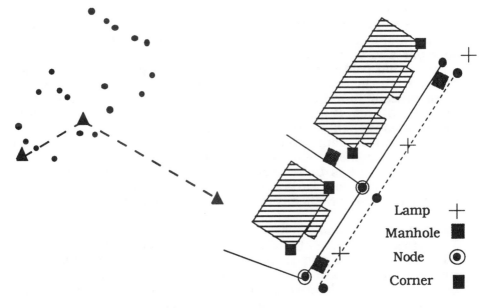

Figure 10.3

Ancillary methods of detail fixing
In addition to the radiation and line and off-set methods of fixing details, other geometrical principles can be used to advantage, particularly as check measurements. These methods include the following.

(a) Theodolite intersection of inaccessible points like trees, and corners of buildings.

(b) Sighting tangents to irregular or rounded shapes like tanks, ponds and clumps of trees.

(c) Off-sets from radiation: to the left, or to the right, or plus on-line measurements to reach inaccessible points such as hedge centres.

(d) Intersecting rays to already established lines, such as building frontages: useful when mapping building elevations.

(e) Direct taped measurements of key dimensions, like buildings, spaces between buildings and gates.

(f) Dimensions to points where face-lines cut out on others.

Field completion

No matter how much care is taken to avoid mistakes and omissions in map making, some do escape the checking systems used by the surveyor and cartographer. It is important that a map should be given a final field check, mounted either on a plane-table or small board such as that used for map revision. The checker works systematically round the horizon at selected key stations, noting all plotted features and looking for those that are omitted. Graphic methods usually suffice to make corrections.

Plots compiled from aerial photographs show the roof line instead of the building line. A tedious but necessary process of measuring the roof overhangs has to be carried out to complete the map by off-setting from the plot. Parts of ground not visible from the air, such as forest paths, also have to be added by ground methods.

Graphic map revision

The principle of lining in and cutting back from established points is basic to the simple graphic method of completing or revising maps. Instruments used are an optical square, a 30 m plastic tape with a hook at one end, a light-weight field drawing board, and a set square and ruler. The reviser works alone on the original map, or part of it, fixed to a small board. Although the following explanation is complicated, the actual process is rather easy and interesting to carry out once the idea has been grasped. The technique is best learned by experiment, rather than from a book.

Suppose the new building A of Figure 10.4 has to be drawn on the map which already shows buildings B,C,D, and E and fence lines BF and FG. The reviser checks the original map to ensure that these buildings are as plotted. Suppose it is found that building E has been demolished. It will be recorded as missing. Otherwise the map is correct. To fix building A, lines of sight to others will be used with the minimum of taping. Points which can be occupied on the ground are established by the intersection of identifiable lines. For example, the surveyor can occupy point P by cutting in from C to the line BD using the optical square. By moving to Q, the perpendicular off-set to corner R can be drawn after taping the short distance PQ shown with a heavy line. A similar process allows a line to be drawn to S. If the distance QR is more than a tape length it will not be measured to fix R. However, before leaving this place it was noticed that the building face line VR cuts out on BD at W, so QW is taped.

A perpendicular off-set from T to R will be used to fix R, after locating T on the line BF produced and FT measured. To fix the direction of RS, the point at which this line cuts the fence BF in U is established and FU taped. Then S is fixed by UR produced. Usually the building dimension RS will be taped to check. The direction of RV was already noted whilst at Q.

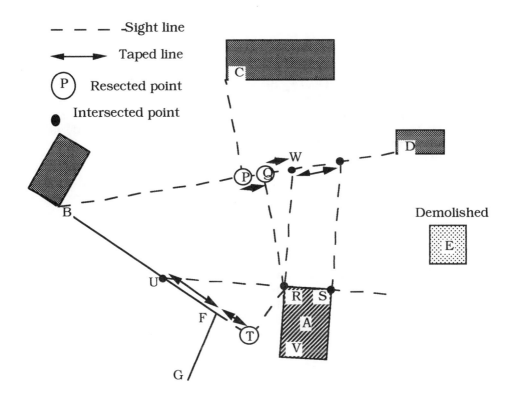

Figure 10.4

10.4 Constructing a map or site grid

Although most map grids are plotted by coordinatograph, or by automatic plotter, the surveyor sometimes has to plot a grid by hand, or has to check one for fidelity. Hand plotting instructs in the principles of construction of all grids, be they on a map or on the ground for engineering works.

A grid has to be formed by *straight lines*, which have to be *perpendicular* to each other, that is *orthogonal*; its *size* has to be correct, and its *position* and *orientation* have also to be correct.

Straightness
Straightness is achieved mechanically by a *straight edge*. This can be checked by drawing a line twice, first with the edge in the normal position, and again with it rotated through 180°. Any gap between the two lines indicates non-straightness.

Orthogonality

To achieve orthogonality, we use the property of the angle in a semicircle being 90'. Consider Figure 10.5, which shows a sheet of paper and various stages in the construction of a grid. The centre of the page is located by drawing its diagonals to meet at O. With centre O measure four equal distances to fix points A, B, C and D. These distances may be any convenient length if the position of the final grid on the page is not important. (In cartography their length will be calculated in order to position the grid in the correct place on the map sheet.)

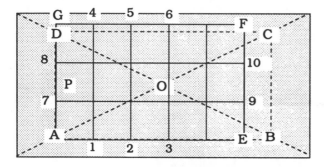 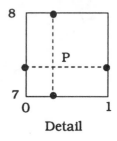

Figure 10.5

The figure ABCD should be an exact rectangle. This is checked to see that the diagonals AC and BD are equal, and so too are the opposite sides. Drawing accuracy is usually taken to be 0.2 mm for this check.

(It should be appreciated that exactly the same procedure can be adopted when setting out a grid on the ground, straightness being achieved by line of sight, and distances by tape or EDM.)

Scale

After the rectangle has been drawn it is used to control the drawing of the grid itself. All lengths are scaled down to the correct size of the map. For map construction a properly calibrated scale should be used. Interpolation of fractional parts is assisted by a diagonal scale which works on the wedge principle.

Plotting the grid

Suppose we take point A as the grid origin. Along AB and DC the points 1,2,3, and 4,5,6, are marked out. Notice that this is done by setting out the lengths A–1, A–2, A–3 etc and not by the accumulation of A–1,1–2, 2–3 , which would accumulated systematic error very quickly. Joining pairs of opposite points 1–4 etc gives the north-south grid lines. The same procedure uses points 7–9 etc to give the east-west grid lines. Notice that the last grid lines EF and GF will not fall on BC and

DC unless the lengths of OA etc had been calculated for this to happen. The grid intersections can now be checked by laying the straight edge across *all* square diagonals. Sizes are also checked by *measuring* G–7, F–9 etc. (Again, on the ground this check is to line-in *all* pegs across *all* grid diagonals.)

Emphasis has been put on checking *all points* (pegs). Because the grid controls the whole map or construction, it must be accurate and it must be checked for reliability. Clearly it might not remain so because of paper distortion (Pegs being knocked). Thus any later measurement from the map could be wrong. If the grid is checked at intervals, corrections for distortion can be applied.

Plotting control points
The control points of the survey have to be plotted within the nearest grid square by measuring fractional distances along pairs of grid lines, and joining opposite pairs of lines. See detail insert of Figure 10.5. After all points have been plotted, some selected distances between pairs of controls are measured and compared with their known distances converted for scale. It is prudent to compare with the surveyor's diagram to check the plotting. All this work must be done with the greatest care.

Checking computer plotted grids
It is not unknown for an automated plotting system to plot incorrectly. The way to check a grid for orthogonality is to make two plots and see if they fit, on overlay and after a rotation of 90° with respect to each other. Size has to be verified by a calibrated scale.

Position and orientation
Often the grid is required to be located in a special position; for example a map may have to be positioned central to a page which also contains other items such as a title, a map key, a coordinate list and so on. The map may even have two different grids on it. In engineering, the site grid will need to be placed correctly in relation to the ground. In these cases, a coordinate transformation may be required according to the methods of Chapter 2. Often some simple off-sets from known points, such as page corners (or existing buildings) will suffice to locate the grid. Clearly these must be checked most carefully.

Coordinate references
Points are located uniquely by map reference to coordinates. Often many references have to be made to illustrate reports etc. A quick, practical system for reading off coordinates is by Romer. This little device for interpolating fractional parts of a grid is illustrated in Figure 10. 6. It can be drawn easily on a piece of card for each map scale used. The Romer scale is marked off in the opposite direction to

the grid coordinates. With the Romer parallel to the grid, the top right edge is place against the point P whose coordinates are read off directly.

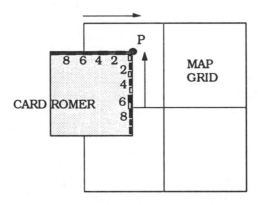

Figure 10.6

Paper distortion

Although all base maps should be plotted on stable plastic, their paper copies are subject to distortion. Usually paper shrinks or expands differentially due to the grain of the paper. This gives rise to different scale distortions in two orthogonal directions aligned to the direction of the paper grain. A suitable mathematical model to correct for this distortion can be set up.

The distortion in a grid direction is found by comparing a measured grid interval with its correct value. Thus we have, for the x and y axes, the two known distortions m and n, usually expressed as a percentage. Also we have the various relationships.

$$x = S \cos t \qquad y = S \sin t \qquad m = \frac{dx}{x} \qquad n = \frac{dy}{y}$$

The problem is to find the distortion p at any bearing t where

$$p = \frac{dS}{S}$$

Now

$$dS = dx \cos t + dy \sin t$$

therefore

$$p = \frac{dS}{S} = \frac{dx}{S} \cos t + \frac{dy}{S} \sin t = \frac{dx}{x} \cos^2 t + \frac{dy}{y} \sin^2 t$$

$$p = m \cos^2 t + n \sin^2 t \qquad\qquad (1)$$

Thus the scale distortion of a line S at bearing t can be evaluated with reference to the grid distortions m and n, and applied as a correction.

The factor for correcting an area A is q^2 where q is the *mean distortion* at bearing t given by

$$q = \frac{1}{2\pi} \int p \, dt = \sqrt{(mn)} \qquad q^2 = mn \qquad (2)$$

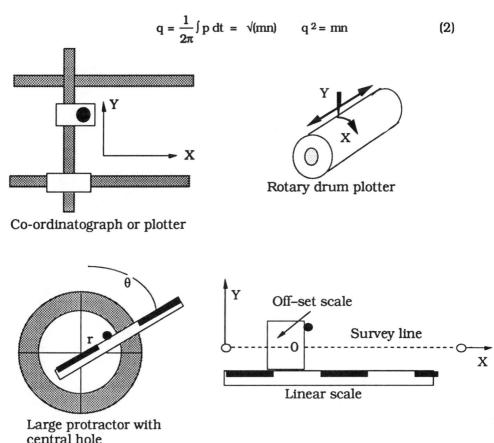

Co-ordinatograph or plotter

Rotary drum plotter

Large protractor with central hole

Linear scale

Figure 10.7

10.5 Plotting details

If the details points are plotted by computer, their coordinates are converted into rectangular cartesian form, for output to a coordinate plotter. The most accurate plotters employ a flat-bed system in which the x–y slides move over a table at great speed responding to the commands of the software.

Other plotters use a rotating drum carrying the paper to create the abscissa x, and a fixed slide for the ordinate y. Although satisfactory for most work, these plotters are not good enough for multicoloured map making , for which accurate colour registration is essential.

Some surveyors use the automatic system to plot points only. These are used as the basis for hand finishing. In this system a dot-plot is made, possibly with the addition of point numbers. By reference to the field sketch appropriate points are joined up to form the map.

Automated systems require rigorous field coding which accords with the demands of the software.

Hand plotting usually matches the mode of data capture. A large protractor and scale for radiation: a linear scale and off-set-scale for line and off-set. Figure 10.7 shows these plotting devices in schematic form. Manual plotting is usually in pencil on paper; with the final plot traced in permanent ink on transparent stable plastic.

Map digitizing

Much input to a geographical information system (GIS) is from maps digitized on special tables which operate like plotters in reverse. A cursor is worked over the map to digitize selected discrete points in much the same way as a surveyor would capture the original points and features. A coding system is also adopted with nodes and points etc, as well as any attributes such as tree types.

A key procedure in digitizing is the control of scale and distortion by digitizing four grid corner points; or more if necessary. Within the software, transformation formulae (see Chapter 2) are set up. In this way the digitized shapes can be corrected geometrically.

Mapping at small scales

Because many important features such as roads, buildings, etc cannot be plotted at the scale of the survey, *conventional signs* have to be used to depict them. Wherever possible, the centre of the sign is taken as the exact map position of the feature. For example, if a road is represented on a map at a scale of 1: 25 000 by two lines 1.5 mm apart, the corresponding ground width is 37.5 m. The centre line is in the correct position. However there is a knock-on effect. A house at the side of the road has to be misplaced laterally with respect to the road centre line. In urban areas, this process is impossible, so the whole set of buildings has to be conventionalized into block.

Plan information

All maps and plans should bear the following information.

(a) Title.
(b) Scale.
(c) Grid.
(d) Key to map symbols.
(e) Datum information: plan and height.
(f) Sheet history; date sources etc.

(g)   Projection details if relevant: type and convergence.
(h)   Orientation: possibly with magnetic north.
(i)   Graticule if relevant.

Although it has been customary in the past to write the map scale in words or as a representative fraction, there is a trend away from this practice because of the dangers arising from the widespread use of cheap photocopying devices. Grids show the correct scale irrespective of the distortion and scale change, while line scales give some indication of the effect in one direction only.

10.6 Heights and contours

A *contour line* is a line drawn on a map through points at equal heights above some datum or reference point. If we imagine the land surface to be flooded to some height above sea level, say 100 m, the shore line of the flooded area will trace out the 100 m contour line. The task of the surveyor is to fix enough points on this line to enable it to be drawn on a map. The mathematical modelling process consists of selecting as few discrete points as possible to carry out this mapping. The density of points will depend on how complex the ground is, and how accurate a result is required.

The specification for contours has to be drawn up in discussion with the client. Matters of importance are

(a) The datum to be adopted: national levelling or arbitrary.
(b) The vertical interval between contours.
(c) The accuracy of contours, often expressed as a fraction of the vertical interval.

Contouring methods
There are four main ways of producing contours.

(a) Tracing out the contour by direct levelling.

(b) Interpolating contours between *spot heights* at the corners of a grid.

(c) Interpolation from spot heights at positions carefully chosen at changes of slope and direction.

(d) Plotting by stereo-photogrammetry.

The first method (a) is used in those engineering works where great accuracy is required, such as in irrigation and other water-related problems. A point on the contour is established by levelling from a *bench mark* (reference height point) and thereafter the staff is located by trial and error on the contour. The contour is pegged out on the ground. In forested areas sometimes a hose-pipe is used to augment levelling and to avoid excessive tree cutting.

In the second method (b), a grid of *spot heights* is set out on the ground at intervals according to the complexity of the surface and accuracy required, and the grid intersections are levelled or heighted. Figure 10.8 shows such points at the numbered grid intersections. Since both plan and heights are known, the contours can be interpolated within the grid.

In the third method (c), the surveyor anticipates how the interpolation from a dot plot will be carried out, and supplies spot heights, by total station or stadia tacheometry, at carefully selected locations. This is a more efficient process than the grid method, which requires more skill from the surveyor.

The photogrammetric technique (d) usually requires some height control points fixed by ground methods to relate the stereo-models to the ground. As with the first method, it has the great advantage of tracing out contours directly, thus giving much better shape. The practical limit of accuracy by this method is about 0.1 m.

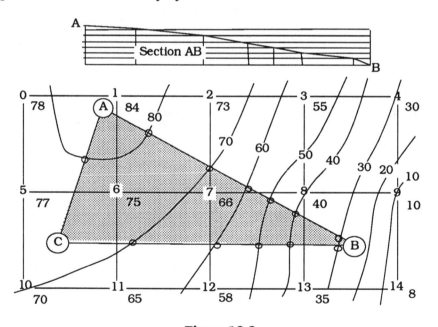

Figure 10.8

Contour drawing

Except in photogrammetry or direct surveying, contours have to be interpolated from discrete points. This can be done manually or by computer software. To draw the contours, the approach is to join up the spot heights, such as A, B and C of Figure 10.8, by a series of non-overlapping triangles and interpolate linearly along all sides to give a series of points on the required contours, marked by dots in the figure. When plotting by hand, the position of a contour on a sectional line, between two points not on contours (shown by the open circles in Figure 10. 8), can be interpolated with the assistance of an elastic band marked at even intervals to act as a convenient analogue computer. Otherwise the process is carried out by linear or non-linear numerical interpolation.

These discrete points on contours are joined up by curves, paying due regard to the actual shape of the ground, which will have been recorded in a field sketch.

Because it is much easier for computer software to operate on a grid, heights may be captured in grid form; alternatively, a grid of points may be interpolated from the points of the selective method. The selection of most nearly equilateral triangles to fit the surveyed points can be made automatically by the Delaunay triangulation algorithm. Thereafter the contours are interpolated according to various schemes of varying complexity. An important test of computer software is that the contours must preserve the integrity of the original heights from which they have been derived. In other words, a spot height must be on the correct side of a contour line.

There is much to commend a hybrid approach to contour plotting, part-computer part-manual. The computer is used to produce a *dot-plot* with the spot heights as a series of dots in their correct plan position. The contours are then finished off manually. This procedure also gives the surveyor an opportunity to inspect the work for any mistakes, as part of a quality control scheme.

10.7 Calculation of areas

The area contained within a closed figure is of interest in engineering and cadastral surveying. Basic to most methods is the calculation of Δ, the area of a triangle ABC, from one of the following formulae

$$\Delta = \frac{1}{2} bc \sin A \tag{3}$$

$$\Delta = \sqrt{s(s-a)(s-b)(s-c)} \tag{4}$$

$$2\Delta = \begin{vmatrix} E_A & N_A & 1 \\ E_B & N_B & 1 \\ E_C & N_C & 1 \end{vmatrix} \tag{5}$$

The area of a polygon is evaluated as the summation of a series of adjacent triangles, as indicated in Figure 10.9.

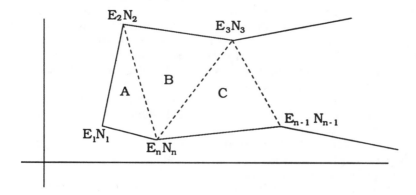

Figure 10.9

Instead of summing up the determinants of each triangle individually, we use the *Rule of Sarrus* to cross-multiply the coordinates taken in clockwise order round the figure according to the scheme

$$2\Delta = -\sum_{i=1}^{n} E_i \sum_{i=2}^{1} N_i + \sum_{i=2}^{1} E_i \sum_{i=1}^{n} N_i \qquad (6)$$

To minimize the numbers in calculation it may be necessary to transform all coordinates to the centroid before evaluating the area.

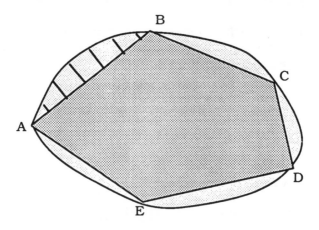

Figure 10.10

The areas of plots bounded by curved boundaries are calculated by combining the areas of polygons and of irregular shapes. Off-sets from the polygon sides are taken at regular intervals to compute the excess areas by Simpson's rule for approximate integration. Figure 10.10 illustrates a method of calculating the area of a typical field. To the area of polygon ABCDE are added the extra small areas measured by regular off-sets y from the perimeter lines such as AB.

Successive application of Simpson's rule to the n off-sets y separated by even intervals x gives the area A as

$$A = \tfrac{1}{6} \, x \, (y_0 + 4 \, y_1 + 2 \, y_3 + \text{and so on to the end of the line)}$$

If n is odd, the last term is y_n : but if n is even, the area of the last figure, a trapesium, is $\tfrac{1}{2} x \, (y_{n-1} + y_n)$

10.8 Subdivision of land parcels

In cadastral and planning surveys, problems arise with the subdivision of plots. These problems vary from simple to very complex. A general approach is to obtain a solution by variation of coordinates.

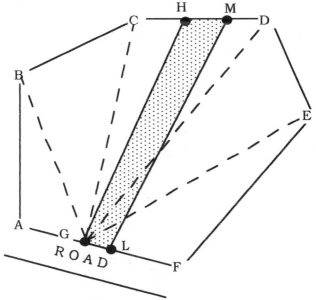

Figure 10.11

At its simplest, the problem might be to divide the plot ABCDEF of Figure 10.11 into two agreed portions and the road frontage AF equally at G. Suppose the portions have to be equal. The total area is calculated, call this A_T. Since point G is given, the variables are the coordinates of H. The first stage is to find which line of the plot is cut by GH. This is done by calculating the areas of the triangles with sides radiating from G. Suppose we find that area GABC is less than 0.5 A_T and that area GABCD is greater than 0.5 A_T. In this case we know that H lies on CD. This process of selection can be programmed.

Although this simple problem has a particular direct solution obtained by remembering that H lies on the line CD and that the area GABCH = 0.5 $A_{T,}$, a more general approach to the solution of such problems is needed. If we assign approximate coordinates to H, selected perhaps as the mid point of CD, the general method will proceed as follows.

Suppose the equation of the line CD, in which both m and c are known, is

$$E - m\,N - c = 0$$

Then we have

$$dE - m\,d\,N = 0 \tag{7}$$

The provisional computed area of figure GABCH subtracted from the known required value gives the absolute term k of the differential area equation

$$P\,dE + Q\,dN = k \tag{8}$$

where

$$P = -(N_G - N_C) \qquad \text{and} \qquad Q = (E_G - E_C)$$

Equations (7) and (8) give the solution dE and dN. If the changes to the provisional coordinates are very large, iteration may be needed.

Complex example

In a more complex example, where the sub-division may include a new road (2 units wide) occupying an unknown area A_R, the solution is more difficult. Consider the coordinate data of Table 10.1. We will assume that the subdivision has to be made from GH and LM preserving these as parallel road sides (at a provisional bearing of zero), and that the areas of the triangles GCH and LMD have to be equal. From the initial coordinates of G and H these areas are 40 and 26.4 square units respectively, giving a difference of 13.6.

The solution for G and H is given, together with the checks that the road sides have the same gradient, that M and H lie on CD (whose gradient is 1.25) and that the required areas are equal. See Table 10.1.

The method is to form the observation equation for area difference, and to link it with the constraints of the problem by Lagrange's method, explained in Appendix 5.

Table 10.1

Point	E	N
G	5.00	0.00
C	0.00	12.00
D	10.00	20.00
L	7.00	0.00
H*	5.00	16.00
M*	7.00	17.60
	Solution	
H	4.24	15.39
M	6.16	16.93
	Check bearings	
HG	ML	MH
-0.05	-0.05	1.25
	Check areas	
GCH	33.85	
LMD	33.85	

The observation equation is formed by considering the changes to be made to the areas of the provisional triangles GCH and LMD to equate them, that is

$$(N_G - N_C) \, d \, E_H - (E_G - E_C) \, d \, N_H + (N_L - N_D) \, d \, E_M - (E_L - E_D) \, d \, N_M$$

$$= 2 \, (A_{GCH} - A_{LMD})$$

Two constraint equations ensure that the points H and M lie on CD that is

$$d \, E_H - 1.25 \, d \, N_H = 0$$
$$d \, E_M - 1.25 \, d \, N_M = 0$$

And two more preserve equal changes dm to the gradients of GC and LM originally assumed zero. That is

$$d E_H - m d N_H + N_H \, dm = 0$$
$$d E_M - m d N_M + N_M \, dm = 0$$

The numerical values of the coefficients and absolute term are given in Table 10.2.

Table 10.2

	Observation equation				ABS
−12	−5	−20	3	0	27.2
	Constraint equations				
1	0	0	0	16	0
0	0	1	0	17.6	0
1	−1.25	0	0	0	0
0	0	1	−1.25	0	0

To obtain a solution we introduce Lagrange's method giving the hypermatrix

$$\begin{bmatrix} A^TA & C^T \\ C & 0 \end{bmatrix} \begin{matrix} x= \\ \lambda \end{matrix} \begin{matrix} b \\ d \end{matrix}$$

The numerical values are given in Table 10.3 together with the solution. Because of lack of space, the symmetric hypermatrix has been split into two parts ; one printed under the other.

This example has been given to illustrate a general method of approaching quite difficult problems, such as when one of the boundaries is curved. An equation is written for this curve (possibly a cubic spline) and appropriate constraint equations formed. Provided a solution actually exists, the variation of coordinates method will arive at a solution, possibly with iteration.

Table 10.3

Hypermatrix

144	60	240	−36	0
	25	100	−15	0
		400	−60	0
			9	0
				0

					ABS	SOLN
1	0	1	0	−326.4	−0.7692	
0	0	−1.25	0	−136.0	−0.6154	
0	1	0	1	−544.0	−0.8462	
0	0	0	−1.25	81.6	−0.6769	
16	17.6	0	0	0	0.0481	
0	0	0	0	0	0	
	0	0	0	0	0	
		0	0	0	0	
			0	0	0	

10.9 Engineering surveys

On completion of the site plan for an engineering survey, the design work has to be completed. Eventually the design has to be implemented in the field. If possible the same control points are used to maintain consistency and design fidelity.

Before beginning the setting out, a field check is essential. In particular it is vital to see that new works, such as road alignments, match up with the existing terrain. For instance sometimes the alignment of an existing road has to be established. This is done by marking several centres of the road over a distance of about 100 m, then eye-balling a best mean fit line to the centre marks. Similar techniques are used for off-sets from walls, directions of power lines etc. These alignments are compared with the setting out data and plans for accuracy.

Setting out markers or reperes

The points of the design are set out usually by a trial and error process (see Chapter 11). Because these key positions are likely to be disturbed from the very nature of the task, a complex system of repere marks is adopted to recover them throughout the life-time of the task. The technique is used extensively in setting out works to control such operations as the excavation of trenches for the footings of buildings, for pipe laying, pile driving, tunnel alignment etc.

A typical arrangement of reperes to control trench digging for the footings of a building is shown in Figure 10.12. The cross rails are marked by saw-cuts which are easy to position, and difficult to remove. When needed, the lines are established by strings between the boards. Excavation depths are also controlled from these strings, although this may be controlled separately by levelling if the ground is steep. All dimensions are measured by tape giving many checks.

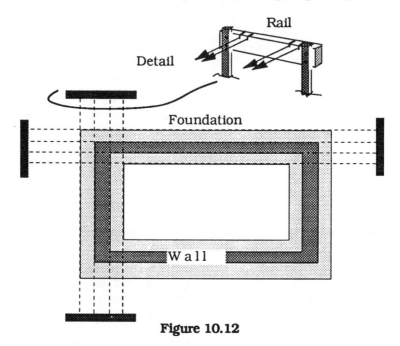

Figure 10.12

Much use is also made of reperes in a vertical plane to control the slopes of cuttings and embankments in road works, and the fall of a pipe. The latter has to be particularly accurate to preserve the direction of flow. Typical arrangements using fixtures, variously called *batter boards, sight rails, boning rods* and *travellers*, are shown in Figures 10. 13 and 14. The sight rails define the slope of the pipe and its trench, which is different from the ground slope. The traveller acts as a template to control the depths of the trench and of the inside of the bottom of the pipe. The rails are set out with a level using a height of collimation carried in from a bench mark.

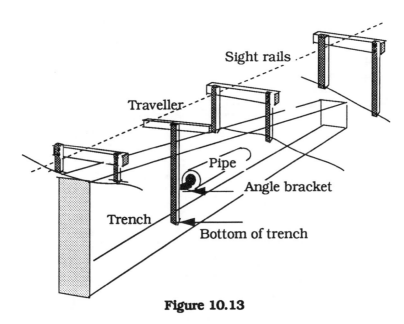

Figure 10.13

Where practicable more than the minimum number of reperes should be used so that the work can be verified as it proceeds. As they can easily be disturbed, reference back to semi-permanent control marks is essential. Such marks will be concrete blocks cast in situ, or positioned after casting. Important primary blocks are best guarded on site by massive protecting fences. It is not always easy to persuade a site engineer to co-operate in their construction.

To control slopes , the arrangements of Figure 10.14 are typical. The upper diagram shows how the slope is controlled during the excavation of construction. The required line of sight is established by a rail on which the traveller is lined. The lower diagram indicates how the rails are set in position using a level. The holding nails should be hammered through the sloping member before trying to locate them on the uprights. See Chapter 12 for more details.

Lines may also be established by visible lasers, or by invisible systems using special detectors. Laser safety on site needs to be considered when using these devices.

A similar procedure arises in industrial surveying in which the required alignment is established by a collimator, or components are located on-line by auto-collimation or auto-reflection to a mirror mounted on a component to be positioned. See Chapter13 for more details.

The floor plans of buildings have to be correlated vertically using any of the vertical alignment methods described in Chapter 13.

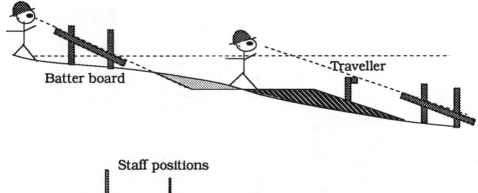

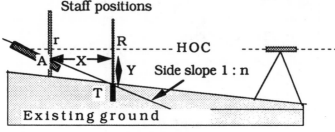

Figure 10.14

Comment
It is vital that good records are kept describing the location of points and the nature of any reperes used in a construction process. These will enable work to be checked independently, or to be presented as evidence in the case of dispute or litigation.

11 Curve surveying

Chapter Summary

This chapter is concerned with the use of ground methods to locate points in predesigned positions on engineering structures and works, with special concentration on road and rail curves. The geometrical part of the task usually follows three stages: site survey; planning and design; and setting out or building. At all stages the tolerances of measurement have to be assessed and controlled. This is now possible with on-line data processing from electronic total stations.

11.1 Engineering and industrial surveying

Almost all standard surveying techniques are applicable to engineering and industrial surveying; hence almost all chapters of this book are relevant. Chapter 13, which treats aspects of industrial surveying, is particularly relevant.

The distinction between *Industrial* and *Engineering* surveying is somewhat arbitrary, although the former is usually attempting to achieve accuracies at the micrometre (0.001 mm) level, and the latter seeks tolerances of from 0.05 to 0.002 m.

It is common practice to base all dimensional problems on a numerical coordinate system developed and processed by computer software, and to present results in digital form for future use. Methods of working have become very versatile as a result. It is not possible to describe particular systems here. We deal only with the essential principles which enable the surveyor to apply common sense to a working medium which is dominated by computers and a very special close-knit relationship with the construction industry.

The chapter deals with the surveying and setting out of typical curves in the horizontal and vertical planes.

Because the metre and the foot are still commonly used as units of length in engineering throughout the world, many of the examples merely refer to 'units', without specifying which. The principles involved are of main interest. However, where realistic tolerances have to be quoted specific units have to be mentioned.

Error analysis and instrumental techniques are discussed in Chapters 7 and 13.

Setting out works

Setting out engineering works involves the siting on the ground of the various elements of the works in accordance with the dimensioned plans and drawings supplied by the designer. Most significant dimensions are readily available as are coordinates from a computer model, although on-the-spot calculations are not unheard of. These too are performed by field computers with great computational capability.

Since it is often not possible to set out the whole of the works before construction commences, the accurate positioning of each element independently is highly important and errors or mistakes can be very expensive.

It is important to remember that inaccurate preliminary surveys are not unknown, nor are dimensions scaled from distorted drawings or prints, so that it is advisable to check all leading dimensions on the site before commencing any setting out whatsoever. Calibrated EDM and tapes are essential for these checks.

Normally the dimensions of individual elements (buildings, roads, bridges, etc.) will be fully figured on drawings, and their relative positions given in computer output. However, it is advisable to look for any controlling factors that will influence the actual positioning of the element, e.g. if space is to be left between two houses for two prefabricated garages, each 2.5 m wide, then these houses must be set out at 5.0 m apart against a possible scaled dimension of 4.9 m.

Marks

The usual practice is to mark key points (corners of buildings, centre-lines, or kerbs of roads etc.) with semi-permanent marks such as wooden pegs, with a small nail in the top if greater precision is required. A pipe nail can be driven into tarmac or asphalt, whilst it may be necessary to cut a cross with a cold chisel on stone or concrete or drive in a Hilti nail. Attaching bright fluorescent tape to buried marks greatly assists finding them again.

Much wasted labour can be avoided by establishing permanent reference points (by a peg surrounded with concrete or by driving a short length of small diameter tube), adjacent to the works but secure from damage by excavating machinery or construction traffic, from which the marks can be re-established quickly and easily if they are lost. Generally EDM and radial survey enables marks to be placed away from works.

General procedures

The general procedure in setting out is to establish a main control framework (a single base line will often suffice), usually by triangulation or traversing, from which the detail can be set out by means of off-setting, tie lines, radiation, intersection, etc. In the case of 'route' works (railways, roads, waterways, pipe-lines, etc.) the control framework is usually a traverse and it is convenient to utilize the main intersecting straights for this purpose, leaving only the curves to be established by other means. Resection is often used on site provided suitable software is available in the field.

For small sites, or when a high precision is not required, line and offset methods will often be adequate; the 3:4:5 triangle for setting out right angles, and the principles of equality of diagonals for checking the squareness of rectangles can be employed. Where greater precision is required, instrumental methods are to be preferred, and of course will be essential if use is to made of radiation and intersection.

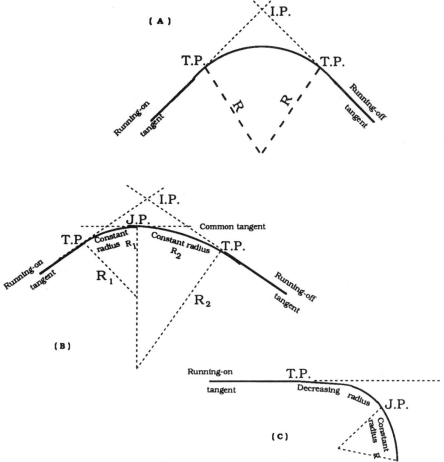

Figures 11. 1 A,B and C

In setting out, a system of coordinates is obviously called for, e.g. rectangular coordinates for off-setting methods, and polar coordinates for radiation methods.

11.2 Horizontal curves

For setting out horizontal curves a third system, using chords and deflection angles, is sometimes employed. Since in this system the location of each point is dependent on the location of the preceding point, the method is known as curve 'ranging'.

Setting out elements consisting of straight lines and curves are based on rectangular coordinates converted to radials from any suitable station. Many instruments now have a speech modulation system, or a digital display, to enable the operator to communicate with the prism holder. But on a noisy site a visual communication procedure still has some place.

In setting out road and railway works it is usual to establish pegs at 10 or 20 m intervals along the actual centre line of the route, these pegs being consecutively numbered from the commencement of the route. Thus any point on the centre line may be identified by its *chainage*, i.e. its distance along the centre line from the commencement.

In setting out curves it is convenient that some of the pegs *preserve running chainage*. In all examples used here the chainage is assumed to be increasing from left to right.

Proceeding in the direction of increasing chainage the first tangent to a curve is known as the *running-on tangent*, and the second as the *running-off* tangent. The junction of a straight and a curve is known as a *tangent point* (TP), the junction of two portions of a compound or combined curve as a *junction point* (JP), the point at which the running-on and running-off tangents meet as the *intersection point* and the external angle between the tangents as the *intersection angle*. (The internal angle, known as the *apex angle*, is rarely used in curve calculations.)

These terms are illustrated in Figures 11.1.

Types of horizontal curve

The types of horizontal curve with which the surveyor normally has to deal in connection with building and civil engineering works are as follows (see Figures 11.1 A to E):

(A) *Simple curves* - circular curves of constant radius.

(B) *Compound curves* - two or more consecutive simple curves of different radii.

(C) *Transition curves* - curves with a gradually varying radius (often referred to as 'spirals').

(D) *Reverse curves* - two or more consecutive simple curves of the same or different radii with their centres on opposite sides of the common tangent.

(E) *Combined curves* - consisting of consecutive transition and simple circular curves. This is the usual manner in which transition curves are used in road and railway practice, to link a straight and a circular curve, or two branches of a compound or reverse curve.

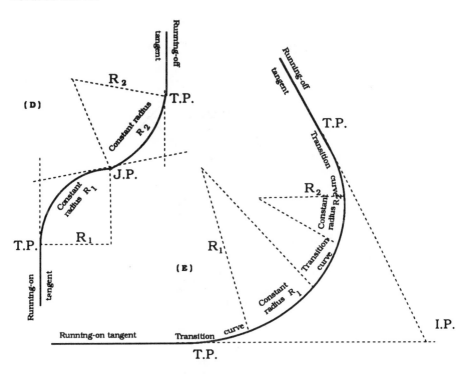

Figures 11.1 D and E

The problems connected with horizontal curves are twofold, firstly the design and fitting of curves of suitable radius, or rate of change of radius to suit site conditions, vehicle speeds, etc., and secondly the setting out of the designed curves in the field.

The first of these problems is generally the concern of the design engineer, the computation of setting out data and the field setting out only falling to the surveyor. For this reason, these last two portions of the work will be dealt with first, but design and fitting will also be considered briefly.

Circular curve geometry

Circular curves can be defined by 'radius' which is self-explanatory, or by 'degree'. The 'degree' D of a circular curve is defined as being the angle in degrees, subtended at the centre by a *chord* 100 m long, so that in Figure 11.2 we have

$$D^{\circ} = 2 \sin^{-1} 50/R \quad (R \text{ in m})$$

$$\text{or } R = 50/\sin \tfrac{1}{2} D^{\circ}$$

(If R is large, $D^{\circ} = 180/\pi = 5729.6/R$)

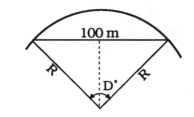

Figure 11.2

From Figure 11.3 we have :

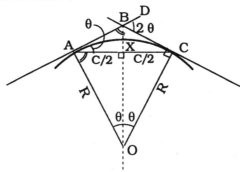

Figure 11.3

(a) Since $A = C = 90^{\circ}$, ABCO is a cyclic quadrilateral

$$CBD = AOC = 2\,\theta$$

(b) In the right angled triangles, ABX and OBA, angle ABO is common, therefore the triangles are similar, and hence,

$$AOB = BAX = \theta$$

(NB for 100 m chord, $\theta = \frac{1}{2}D'$)

(c) From triangle AOX

$$\sin \theta = c/2R$$

From Figure 11.4 we have

$$ABO = CBO = (90 - \theta)$$

therefore

$$\phi = 180 - (180 - 2\theta) = 2\theta$$

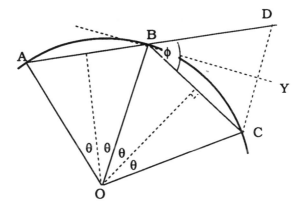

Figure 11.4

The above relationships are much used in curve surveying.

11.3 Transition curves

A transition curve is one in which the curvature varies uniformly with respect to arc, in order to allow a gradual change from one radius to another (a straight being merely a circular curve of infinite radius) and to permit a gradual change in the superelevation. It must, of course, have the same radius of curvature at its ends as the circular curves that it links.

By definition, the transition curve must have a constant rate of change of curvature with respect to arc, i.e. if ϕ is the tangential angle and s is the arc,

$$\frac{d^2\phi}{ds^2} = K, \text{ where K is a constant}$$

Consider, for the time being, the case of a transition curve linking a straight and a circular curve of (constant) radius R (see Figure 11.5).

$$\frac{d^2\phi}{ds^2} = K$$

whence

$$\frac{d\phi}{ds} = \int K\,ds = Ks + K_1$$

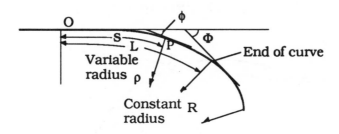

Figure 11.5

But, since the curvature at the origin O is zero,

$$\frac{d\phi}{ds} = 0 \text{ when } s = 0$$

and so $K_1 = 0$ and $\dfrac{d\phi}{ds} = Ks$

Integrating again,

$$\phi = \int K s\,ds = \tfrac{1}{2} K s^2 + K_2$$

but, again, since $\phi = 0$ when $s = 0$, $K_2 = 0$, the fundamental equation of the transition curve becomes that of the mathematical curve, the *clothoid*

$$\phi = \tfrac{1}{2} K s^2$$

This may also be written,

$$s = C \sqrt{\phi}$$

$$\text{where } C = \sqrt{\left(\frac{2}{K}\right)}$$

If L is the total length of the curve, then when $s = L$, $d\phi/ds = 1/R = KL$, therefore $K = 1/LR$, and the equation of the curve becomes

$$\phi = \frac{s^2}{2LR}$$

or

$$s = \sqrt{2LR\phi}$$

Note: since $d\phi/ds = Ks = s/LR = 1/\rho$, where ρ is the radius of curvature corresponding to the arc s, then

$$s\rho = LR = \text{constant}$$

This is an important and very useful property of the clothoid.

When $s = L$ (i.e. at the end of the curve) the total tangential, or ' spiral ' angle for the curve Φ is given by

$$\Phi = \frac{L}{2R}$$

This is half the tangential angle turned through by a circular curve of the same length, and of radius R.

The above equations do not lend themselves readily to the computation of curve components or to the field setting out of the curves and it is necessary to arrange them in a more convenient form.

Three systems of representing the curve are commonly used for these purposes, namely

(a) A system of rectangular coordinates
(b) A system of polar coordinates
(c) A system utilizing deflection angles and chords.

All these systems involve using approximations, the extent of the approximation being dependent on the degree of accuracy finally required. In this connection it must be borne in mind that even the maximum curvature of a practical transition curve is comparatively small and that the curvature shown in the diagrams is grossly exaggerated for the sake of clarity.

Rectangular coordinates
It is customary to use the tangent point as the origin, the continuation of the tangent as the X axis, and offsets from this continuation as the Y axis.

Considering a small element of arc ds at a distance s from the origin, it will be clear from Figure 11.6 that

$$x = \int \cos \phi \, ds$$

$$y = \int \sin \phi \, ds$$

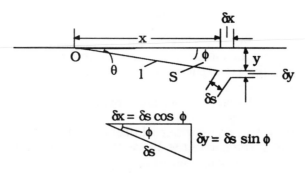

Figure 11.6

In order to perform these integrals it will be necessary to substitute for ϕ in terms of s or vice versa. The choice of the most convenient substitution depends on the relationship we require to establish. One illustration should suffice.

$$\phi = \frac{s^2}{2LR}$$

$$x = \int \cos \left(s^2/2LR\right) ds \qquad y = \int \sin \left(s^2/2LR\right) ds$$

whence, since $x = 0$ when $s = 0$

$$x = s - \frac{s^5}{40(LR)^2} + \frac{s^9}{3456(LR)^4} \quad \text{etc.}$$

whence, since $y = 0$ when $s = 0$

$$y = \frac{s^3}{6LR} - \frac{s^7}{336(LR)^3} + \frac{s^{11}}{42240(LR)^5} \quad \text{etc.}$$

By substituting $s = \sqrt{2LR\,\phi}$ expressions can be found for x and y in terms of ϕ. The power series for x and y, since s is less than L, approximate to

$$x = s$$

and

$$y = \frac{s^3}{6LR}$$

whence

$$y \approx \frac{x^3}{6LR}$$

This is the *cubic parabola* which is commonly used, without correction, as a transition curve where the final curvature is small. It should be noted however that it has a very important limitation. When the ratio y/x exceeds about 0.15 (equivalent to a deflection angle of 8° 34') the curvature ceases to increase as the distance round the curve increases, and begins to decrease again, so that the cubic parabola is useless as a transition curve outside this range.

The coordinate system actually used for the setting out may be based on any suitable point to which the curve based coordinates (x, y) are transformed by the methods of Chapter 2. This procedure gives great flexibility if field software is available.

Polar coordinates
It is convenient to take the tangent point as the origin, and the continuation of the tangent as the initial line. The amplitude θ is then referred to as the 'deflection' angle, and the radius vector l as the 'long chord'.

It will then be seen from Figure 11.6 that

$$\tan \theta = \frac{y}{x} \qquad l = x \sec \theta$$

In practical problems θ and l can be calculated from x and y where these have already been found.

Chords and deflection angles
As the system of coordinates utilizes the distance s round the curve from the origin and the deflection angle θ, which is the same as that found in the previous section, no further calculations are required at this stage.

Computation of curve components
The initial requisites for setting out any curve are

(a) The location of the straights and their intersection points.
(b) The determination of the intersection angles.

This information may be supplied by the planning engineer, determined in the field by direct measurement, or determined indirectly from field measurements (e.g. a traverse).

It will be necessary for some of the curve components (e.g. radius or tangent length, etc.) to be fixed, and these again will normally be supplied by the engineer or must be determined from traffic considerations or by site controls (e.g. property boundaries, etc.).

The present section assumes that the above data has been obtained, and will deal with the determination of the remaining components and the location of the tangent points, junction points and, where necessary, subsidiary intersection points.

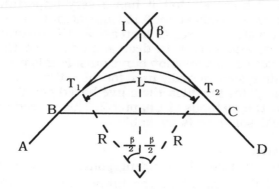

Figure 11.7

Simple circular curves

If the intersection angle β is fixed, the only 'free' components are the radius R, the length of the curve L, and the two equal tangent lengths $T_1\,I$ and $I\,T_2$. It will be seen at once from Figure 11.7 that $L = R\,\beta$ and that

$$T_1\,I \;=\; I\,T_2 = R \tan \tfrac{1}{2}\,\beta$$

The following examples should suffice to illustrate the necessary computations.

Example:

Given $\beta = 75°$, R = 1000 m, chainage I = 28 + 63.2 (Figure 11.7). Determine tangent length, curve length, and chainage of tangent points.

$$T_1\,I = I\,T_2 = 767.3 \qquad L = R\,\beta = 1309.0$$

Chainage

$$I = 28 + 63.2 \qquad\qquad T_1\,I = \; 7 + 67.3$$

Chainage $T_1 \;= \; 20 + 95.9$ $L = 13 + 9.0$

Chainage $T_2 \;= 34 + 49.0$

Example

Given (Figure 11.7) $\beta = 75°$, chainage I = 28 + 63.2 m and chainage $T_1 = 24 + 36.6$ m. Determine R, the length of curve, and the chainage of T_2.

(Chainage I = 28 + 63.2) - (Chainage T_1 = 24 + 36.6) = T_1 I = 4 + 26.6

$$R = 426.6/\tan 37°.5 = 556.0 \text{ m}$$

$$L = R \beta = 727.8 \text{ m}$$

Chainage T_2 = 31 + 64.4

Example

Given (Figure 11.7) : ABC = 149° 44' 40", BCD = 137° 43' 40", BC = 1023.3 m. A circular curve of radius 850 m is to be inserted, tangential to AB and CD. If the chainage of B is 30 + 15.6 m, find the chainages of the initial and final tangent points.

$$IBC = 180° - 149° \, 44' \, 40" = 30° \, 15' \, 20"$$
$$ICB = 180° - 137° \, 43' \, 40" = 42° \, 16' \, 20$$
$$\beta = 72° \, 31' \, 40"$$

$$BI = BC \sin ICB / \sin \beta = 721.6 \text{ m}$$
$$IC = BC \sin IBC / \sin \beta = 540.5 \text{ m}$$

$$L = R \beta = 1076.0 \text{ m}$$
$$IT = 623.6 \text{ m}$$

Chainage T_1 = 31 + 13.6
Chainage T_2 = 41 + 89.6

11.4 Compound circular curves

As will be seen from Figure 11.8, the centres O_1 and O_2 lie on a straight line O_2 J which is perpendicular to the common tangent PJQ. Note that in general O_2 J does not pass through I.

Assuming that the main intersection angle β is fixed, there are eight remaining components, R_1, R_2, T_1I, T_2 I, β_1, β_2, L_1 and L_2. Any three of these must be fixed for the curve to be defined uniquely. Five equations must therefore be established in order to obtain the remaining components. We shall now determine these five equations. Three are obtained simply from

$$\beta = \beta_1 + \beta_2$$
$$L_1 = R_1 \beta_1$$
$$L_2 = R_2 \beta_2$$

Consider the polygon $O_1 O_2 T_2 I T_1$, in which the algebraic sum of the projections of the sides onto any one side must be zero.

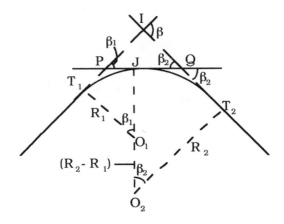

Figure 11.8

Consider first the projection onto $O_2 T_2$. We have

$$R_2 + 0 - I T_1 \sin \beta - R_1 \cos \beta - (R_2 - R_1) \cos \beta_2 = 0$$

whence

$$I T_1 \sin \beta = R_2 - R_1 \cos \beta - (R_2 - R_1) \cos \beta_2$$

Similarly, projecting onto $O_1 T_1$,

$$I T_2 \sin \beta = R_1 - R_2 \cos \beta + (R_2 - R_1) \cos \beta_1$$

Example

Given $\beta = 75°$, $\beta_1 = 30°$, $R_1 = 800$ m, $R_2 = 1000$ m, determine β_2, L_1, L_2, $I T_1$ and $I T_2$.

$\beta_2 = 45°$, $L_1 = 418.9$, $L_2 = 785.4$, $I T_1 = 674.5$, $I T_2 = 739.6$

Reverse circular curves

From traffic considerations it is desirable to introduce a section of straight and transition curves between the branches of a reverse curve, so that the pure reverse curve is rarely used in practice. It must however, be considered, the treatment being the same in principle as that applied to compound curves.

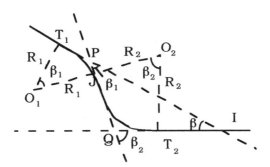

Figure 11.9

It will be seen from Figure 11.9 that the centres O_1 and O_2 lie on the straight line $O_1J\,O_2$ which is perpendicular to the common tangent PJQ. $O_1J\,O_2$ does not in general, pass through the main intersection point I of the running-on and running-off tangents $T_1\,P$ and $Q\,T_2$. Note that I may be at some considerable distance (infinite if $T_1\,P$ and $Q\,T_2$ are parallel) from the curves, and will commonly be inaccessible. The value of the main intersection angle however will be known, being the difference in bearing between the lines $P\,T_1$ and $Q\,T_2$.

As for compound curves there are eight remaining components, R_1, R_2,T_1I, T_2 I, β_1, β_2, L_1 and L_2, of which three must be fixed to define a unique curve system, and again, five equations must be established to obtain the remaining components. From triangle IPQ (Figure 11.9),

$$\beta_2 = \beta_1 + \beta$$

and in all cases β will be the difference (+ve) between β_1,and β_2.

As before,

$$L_1 = R_1\,\beta_1$$
$$L_2 = R_2\,\beta_2$$

As with compound curves, the sum of the projections of the sides of the polygon $O_1 O_2 T_2 I T_1$ onto any one side will be zero, leading, by an exactly similar process as that used before for the compound curve, to the equations

$$I T_1 \sin \beta = R_2 + R_1 \cos \beta - (R_2 + R_1) \cos \beta_2$$
$$I T_2 \sin \beta = R_1 + R_2 \cos \beta - (R_2 + R_1) \cos \beta_1$$

One particular case of reverse curves, when the running-on and running-off tangents are parallel, requires special treatment because the standard equations are not applicable.

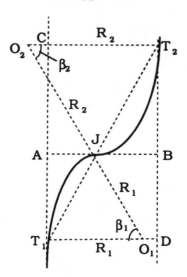

Figure 11.10

From Figure 11.10 all the following relationships will be perfectly clear.

$$\beta_1 = \beta_2 = \beta \qquad AJ = R_1 (1 - \cos \beta) \qquad JB = R_2 (1 - \cos \beta)$$
$$AB = AJ + JB \qquad T_1 A = R_1 \sin \beta$$
$$BT_2 = R_2 \sin \beta$$
$$T_1 C = T_2 D = (R_1 + R_2) \sin \beta$$
$$T_1 J = 2R_1 \sin \tfrac{1}{2} \beta \qquad J T_2 = 2R_2 \sin \tfrac{1}{2} \beta$$
$$T_1 T_2 = T_1 J + J T_2$$

11.5 Transition and combined curves

Once the rate of change of curvature of a transition curve (hereafter referred to as a 'spiral') has been fixed (by the methods discussed later), the only component remaining to be found to define the curve uniquely, is its length. Consequently the components of combined curves will be investigated in this section.

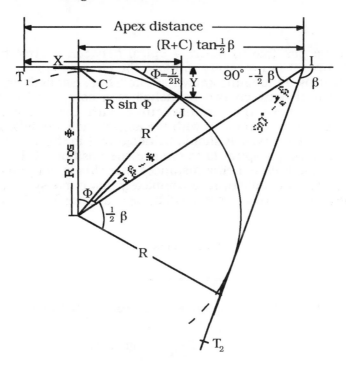

Figure 11.11

In the discussion which follows, for which it will be necessary to refer to Figure 11.11, the following notation will be used:

TJ = L is the total length of each spiral;
R the radius of the central circular arc (and of course the radius of curvature of the spiral at its junction with the circular arc);
c the shift;
T denotes springing point of spiral from straight;
J denotes junction point between spiral and circular arc ;
s, ϕ, x, y, θ, and l all have the meanings already assigned to them;
Φ, X, Y, and Θ refer to the particular values of ϕ, x, y, θ at the point J (i.e. when s = L).

To avoid confusion with the tangent lengths for purely circular curves, the distances IT_1, IT_2 etc. will be referred to as 'apex distances'.

In this section, the 'spiral' used will be the true transition curve, the clothoid, which, if the correction terms are ignored, becomes the cubic parabola. Some of the relationships developed below hold good whatever curve is used, and where this is so it will be indicated in the text.

Shift

For it to be possible to insert a spiral between a straight and a circular curve it is necessary to move the circular curve away from the straight by an amount known as the 'shift'. Similarly in order to insert a spiral between two branches of a compound curve it is necessary to move the circular curve with the smaller radius inwards, or the circular curve with the largest radius outwards, i.e. to separate the two curves. It should be noted that in this latter case the effect of inserting a spiral is to change the position of the common tangent. In the unusual circumstances that this is not practicable it will be necessary to substitute a completely new curve system.

For the single circular curve it will be seen at once from Figure 11.11 that

$$R + c = R \cos \Phi + Y$$

therefore

$$c = Y - R(1 - \cos \Phi)$$

This holds good whatever spiral is used. For the clothoid

$$y = \frac{s^3}{6LR} + \text{higher terms}$$

so that when $s = L$

$$Y = \frac{L^2}{6R} \quad \text{etc.}$$

But

$$\phi = \frac{s^2}{2LR}$$

therefore

$$\Phi \approx \frac{L}{2R}$$

But

$$1 - \cos \Phi \approx \frac{\Phi^2}{2} \approx \frac{L^2}{8R^2}$$

Thus we get

$$c \approx \frac{L^2}{24R}$$

Notice when

$$s = \frac{L}{2} \qquad y = \frac{L^2}{48R} = \frac{c}{2}$$

Thus the shift and the curve bisect each other approximately; a useful check when setting out.

Apex distance

For the circular curve it will be seen at once from Figure 11.11 that

$$T_1 I = X + (R + c) \tan \tfrac{1}{2} \beta - R \sin \Phi$$

This holds good whatever spiral is used. For the clothoid

$$x = s + \text{terms} \quad \text{thus } X = L + \text{terms} \quad \text{and } \Phi = \frac{L}{2R}$$

giving

$$X - R \sin \Phi = \frac{L}{2} + \text{terms}$$

$$T_1 I = T_2 I = L/2 + (R + c) \tan \tfrac{1}{2} \beta + \text{terms}$$

It can be shown (eg in the first edition of this book page 500) that the first small term is a generally negligible:

$$-\frac{L^3}{240 R^2}$$

Length of combined curve

It has been shown that the angle consumed by one (clothoid) spiral $\Phi = L/2R$ so that the angle consumed by two identical spirals $= 2\Phi = L/R$, leaving the angle available for the central circular curve as $\beta - L/R$. The length of the central circular arc is therefore

$$R(\beta - L/R) = R\beta - L$$

and the total length of the combined curve is

$$R\beta + L$$

Example

Referring to Figure 11.11 and using the notation of the previous sections, assume $\beta = 75°$, $R = 500$ m, $L = 400$ m, $(LR = 200\ 000)$ and chainage $I = (28 + 63.20)$ m.

Required to find the position and chainage of the tangent points, and the chainage of the junction points.

Shift:

$$c = \frac{L^2}{24R} \approx 13.333$$

The omitted correction is – 0.076.

Apex distance:

$$T_1 I = T_2 I = \frac{L}{2} + (R + c) \tan \frac{1}{2} \beta - \frac{L^3}{240\ R^2} = 200 + 393.84 - 1.07 = 592.77$$

Length of circular arc:

$$R\beta + L = 654.50 - 400 = 254.50$$

Chainages:

$$
\begin{aligned}
T_1 &= 28 + 63.20 - (5 + 92.77) = 22 + 70.43 \\
J_1 &= 22 + 70.43 + (4 + \quad 0\ ) = 26 + 70.43 \\
J_2 &= 26 + 70.43 + (2 + 54.50) = 29 + 24.93 \\
T_2 &= 29 + 24.93 + (4 + \quad 0\ ) = 33 + 24.93
\end{aligned}
$$

Wide roads and dual carriageways

For roads having a large formation width, and in particular for dual carriageways, it will be necessary to set out the individual kerbs, embankment edges etc. independently, the method of off-setting from the centre-line only being applicable to narrow roads. This makes no difference to the setting out of the curves themselves, but does affect the calculations of the curve components. Two cases arise:

(1) where the carriageway retains constant width throughout;

(2) where the carriageway is widened gradually along the spiral but maintains constant (increased) width round the circular portion. This is to meet traffic engineering considerations.

In both cases, the circular arcs remain truly concentric. (It should be noted that these have the effect that spirals used for each kerb utilize different speed (LR) values; methods which use the same speed (LR) values are possible but do not satisfy the other conditions set out above.)

These matters are mentioned here to illustrate that the design of motorway interchanges is sophisticated and the provision of the setting out data is something for which computer processing is necessary to calculate coordinates for use by the setting out surveyor.

11.6 Field setting out

As has been explained at the beginning of the chapter, a system of coordinates is required for all setting out, the three most convenient being rectangular coordinates, polar coordinates, and chords and deflection angles. Each of these will now be considered briefly.

Although in practice 'route' curves are nearly always 'combined' curves, in the following sections the setting out procedure for circular curves and spirals will be kept separate as they are set out separately. The procedure at the junctions between them will be dealt with in the section relating to spirals.

Circular curves by rectangular coordinates

These methods are convenient for simple equipment using tapes and optical squares for short curves, and are particularly useful for such situations as urban road intersections where the centres are inaccessible. They suffer from the disadvantage, when tapes are used, that the chords are not equal in length.

With total stations rectangular coordinates are converted to local polar coordinates referred to any suitable reference target.

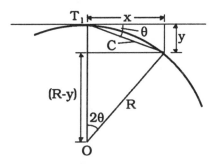

Figure 11.12

Taped offsets from the tangent

From Figure 11.12 using the tangent point as the origin, the distance along the tangent as the x coordinate and the rectangular off-set as the y coordinate.

$$R^2 - x^2 = (R-y)^2 = R^2 + y^2 - 2Ry$$

$$y \approx \frac{x^2}{2R}$$

When R is large compared with chord length C, $x \approx C$ and

$$y \approx \frac{C^2}{2R}$$

and the correction for x is

$$x \approx C - \frac{y^2}{2C}$$

If R = 100 m, and C = 20 m, then y = 2, and x = 19.90. The approximation gives 1.98, an error of 0.02 m. This is a suitable method for small tasks or infilling detail already set out.

Taped off-sets from the chord
This method is particularly useful for infilling additional chord points on any circular curve once the main chord points have been established. Refer to Figure 11.13. The midpoint of the chord is taken as origin, distances along the chord as the X coordinates, and rectangular off-sets as the Y coordinates.

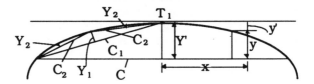

Figure 11.13

From the previous method

$$y' = \frac{x^2}{2R}$$

Because $y' = Y'$ when $x = C/2$

$$Y' = \frac{C^2}{8R}$$

and any value of y

$$y = Y' - y' = \frac{C^2 - x^2}{4R}$$

Then since chord $C_1 \approx C/2$

$$Y_1 \approx \frac{Y'}{4}$$

and so on for $Y_2 = \dfrac{Y_1}{4}$, giving the general rule of 'halving the chord and quartering the offsets'.

Circular curve ranging by theodolite

The first step in setting out the curve must be to establish the tangent points, which will involve measuring or deriving the intersection angle, calculating the tangent distances and pegging the tangent points, and determining their chainage, as has already been described.

Ranging the curve

The basic geometry shown in Figure 11.14 applies to three methods of setting out a circular curve:

(a) From one theodolite situated at a tangent point such as T_1 and taping chords C along the curve.
(b) From two theodolites situated at each tangent point.
(c) By radial survey from a total station at either tangent point.

In fact the last two methods can be used from any points whatsoever, as has been pointed out before. The traditional methods of working close to the curve will be described, as they are still useful, and also serve to describe the general procedure.

We will concentrate the description on the first method, by one theodolite and taped chords. It will be assumed that running chainage is to be maintained, and that the theodolite is to be set up at one or other of the two tangent points. Not shown in Figure 11.14 are $C_1 = T_1 P_1$ and $C_2 = T_2 P_2$, the initial and final subchords. All the remaining chords are equal to C. The deflection angles δ are referred to the running-on tangent T_1 I, and P_1, P_2 etc. are the chord points. Only δ_3 is shown in the figure.

The defection angles have to be calculated so that the theodolite circle may be set on I and each angle turned as the chords are taped from a previously fixed point. This chain process is bound to accumulate some slight error, but if a mistake is made, a clear indication of it will be given. Thus the work is quality assured and controlled. (If I cannot be seen, a reference point at a calculated bearing from I is used instead.)

The fact that the sum of deflection angles is equal to half the intersection angle is useful as a check on the arithmetic of the calculations.

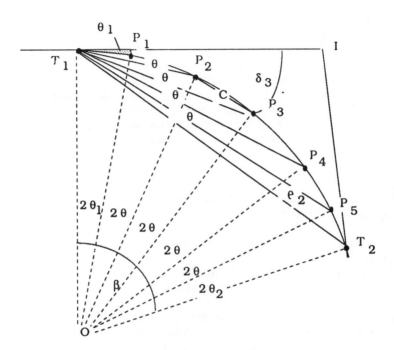

Figure 11.14

If the circular curve is defined by its 'degree' instead of its radius, and C = 100 m (which is common) then as previously shown θ = 0.5 D and the calculation of the deflection angles is considerably simplified. This is the main advantage of the 'degree' method. From Figure 11.14

$$T_1\ O\,P_1 = \theta_1 = \text{arcsin}\,(C_1\,/\,2R)$$
$$P_1\ O\,P_2 = \theta\ = \text{arcsin}\,(C\ /\,2R)\ \text{etc}$$
$$P_5\ O\,T_2 = \theta_2 = \text{arcsin}\,(C_2\,/\,2R)$$

Then the deflection angles δ are

$$\delta_1 = \theta_1$$
$$\delta_2 = \theta_1 + \theta$$
$$\delta_3 = \theta_1 + 2\theta$$

and so on until all sum to $\frac{1}{2}\beta$

From the above equations a table of deflection angles for setting out the curve can be prepared as in the following example.

Example

Two straights making a deflection angle of 75° intersect at I (chainage 28+6.32 m). A circular curve of 100 m radius, deflecting left, is to be set out in 10 m chords. Tabulate the setting out data assuming that the instrument is set up at T_1 and that the initial reading of the horizontal circle when the instrument is bisecting I is zero.

$$\text{Tangent length} = 60.88$$
$$\text{Arc length} = 130.90$$

Chainages:

$$T_1 = 22 + 54.40 \qquad T_2 = 35 + 6.34$$

$$\theta = 2° 51' 58''$$

$$C_1 = 10 - 5.44 = 4.56 \qquad\qquad C_2 = 6.34$$

$$\theta_1 = \theta C_1 / C = 1° 18' 23''$$
$$\theta_2 = \theta C_2 / C = 1° 48' 59''$$

The curve can be set out from the data of Table 11.1.

Table 11.1

Chainage 10 m chords	Deflection angles
22 + 5.44	
23	1° 18' 23"
24	4° 10' 16"
25	7° 02' 09"
26	9° 54' 02"
27	12° 45' 55"
28	15° 37' 48"
29	18° 29' 41"
30	21° 21' 34"
31	24° 13' 27"
32	27° 05' 20"
33	29° 57' 13"
34	32° 49' 06"
35	35° 40' 59"
35 + 6.34	37° 39' 00"

Once the instrument has been set up over the tangent point and the intersection point bisected, it is advisable to bisect the other tangent point, and measure the total deflection angle to check that it is in fact equal to half the intersection angle before proceeding any further.

This having been done, lay off the first deflection angle, stretch the tape a distance C_1 from T_1, line in the end of the tape and drive in a peg at P_1. Lay off the second deflection angle, stretch the tape a full chord's length C from P_1, line in the end of the tape and drive in a peg at P_2. Proceed in this manner until the last peg has been driven, and then measure the final subchord C_2. The amount by which this differs from the calculated value will be a measure of the accuracy with which the curve has been set out.

$$T_1 = 22 + 54.40 \qquad T_2 = 35 + 6.34$$
$$\theta = 2° 51' 58''$$
$$C_1 = 10 - 5.44 = 4.56 \quad C_2 = 6.34$$
$$\theta_1 = \theta C_1 / C = 1° 18' 23''$$
$$\theta_2 = \theta C_2 / C = 1° 48' 59''$$

The accuracy of the method is obviously dependent on the care which is taken in taping the chords and in lining in the pegs, and by the angle at which the two locating position lines meet. During the actual setting out the surveyor has an immediate idea of accuracy. If he is unhappy about the cut, a remedy has to be used such as moving the theodolite to another station.

Setting out by theodolite intersection
If the ground is unsuitable for taping it may be preferable to use two theodolites, one at each tangent point, lay off the deflection angles from both tangent points simultaneously, and locate the chord points by intersection. This of course presupposes that the chord points are all visible from both tangent points. Where the ground is undulating and accurate lining is difficult (whether using one theodolite or two), it will be preferable to change the position of the instrument as in the procedure for obstructed curves. This procedure is also necessary when the intersection angle between the line of sight and the tape becomes small. When the theodolite is set up at T_2 a fresh tabulation of deflection angles must be made, using, of course, the same values of θ_1, θ_2 and θ.

Setting out by polar coordinates
Circular curves can of course be set out using polar coordinates, with the tangent point as the origin, the tangent as the reference direction, the deflection angles as calculated in the last section as the amplitude, and the long chord (calculated from $l = 2R \sin \delta$) as the radius vector.

This is obviously the most convenient method when using a total station. It must be remembered that the design *horizontal* chords l have to be obtained from the *slant* EDM distances S, i.e. from S' = l sec α where α is the slope angle. Many instruments incorporate suitable computer algorithms for this calculation directly in the field, and some permit the prism holder to read the distance at his end of the line. Increasingly GPS differential fixes are being used to assist conventional methods with total stations in the setting out of planimetric data such as curves. It is most important to tie into to a base station surveyed on the same datum as the general design, and transform the GPS coordinates in sympathy.

Setting out transition and other curves
Although it is feasible to set out complex curves by deflection angles in a similar manner to that described for the circular curve, modern practice is to use a total station and a system of polar coordinates from almost any convenient position close to the site but free from the actual works themselves. These polar coordinates are computed from the cartesian values of the curves. It should be remembered that mistakes may occur so checks on key points from a second position are desirable and eyeballing the curve should be common practice.

Where numerous obstructions exist, or where earthworks or construction plant prevent the use of any permanent ground marks on the curve itself, it is more expedient to set out the curve from stations remote from the curve. In such cases, once the calculations have been made and tabulated, the curve can be quickly re-established whenever required.

The rectangular coordinates of all the chord points on the curve must be calculated, and the coordinates of the permanent stations on the same axes obtained, usually by traversing or by resection.The curve can then be set out either by intersection, or by radiation.

11.7 Design and fitting of horizontal curves

It has already been pointed out that the design of road and railway layouts is the province of the specialist engineer, and space only permits general principles to be discussed here. It has also been shown that both circular curves and clothoid spirals are each defined uniquely if any two of their properties are fixed. For circular curves these two will usually be selected from radius (or degree), length of arc, intersection angle, or tangent length. For spirals they will be selected from minimum radius, length of curve, maximum spiral angle, and shift. The combined curve, consisting of two identical spirals and a central circular arc, is the case most commonly met in practice, and as purely circular curves are very much simpler to deal with, such a combined curve only will be assumed in general in this section.

On severely restricted sites, curves may have to be designed from purely geometric considerations, but in the main, design is based on traffic requirements for safety and comfort, possibly modified by aesthetic considerations or site restrictions.

In most practical cases, the overall intersection angle is predetermined by the general layout, and the problem finally resolves into determining a suitable radius for the circular arc, and length for the spiral, from the traffic viewpoint.

It will be obvious that from traffic considerations the largest possible radius and longest possible spiral are desirable, but that some restriction will always arise from site conditions or cost. The first problems to tackle, therefore, are the determination of suitable minima for the central radius and the length of the spiral for given traffic speeds.

As the speed of vehicles using a particular road or railway is a variable quantity, a 'design speed' must be defined, which is normally taken as the 85 percentile speed, i.e. the speed that will not be exceeded by 85% of the vehicles using, or are expected to use, the road.

Minimum safe radius for circular curves

A vehicle travelling round a curve will be subject to a centrifugal acceleration, which must be combated by the superelevation of the track, combined with the reaction of the rails on the wheel flanges, in the case of a railway, or by the frictional force between the tyres and the road surface.

As in general, the track will have to accommodate vehicles travelling at a wide range of speeds, the superelevation (tan θ) cannot be so large that it will be uncomfortable for slow travelling or stationary vehicles. For railways, where the vehicles have a low centre of gravity, a *maximum* value of 0.1 is often adopted, and for roads, 1/14.5 (= 0.069).

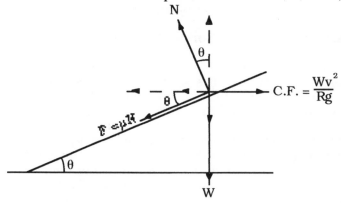

Figure 11.15

In Figure 11.15, we resolve the normal force N and the tangential force F horizontally and vertically to give

$$\frac{Wv^2}{gR} = N \sin \theta + F \cos \theta$$

$$W = N \cos \theta - F \sin \theta$$

Replacing F by $\mu N = N \tan \lambda$ (where μ is the coefficient of friction and λ the angle of friction) and dividing gives

$$\frac{v^2}{gR} = \frac{\sin \theta + \cos \theta \tan \lambda}{\cos \theta - \sin \theta \tan \lambda}$$

and dividing top and bottom of the right hand side by $\cos \theta$,

$$\frac{v^2}{gR} = \tan (\theta + \lambda)$$

An ideal situation arises if, for a given speed, there is no tangential force, i.e. $F = 0$, i.e. $\mu = 0$, and $\lambda = 0$, where

$$\tan \theta = \frac{v^2}{gR}$$

In practice, some assistance from the frictional force can be expected, and use is made of the empirical formula

$$\tan \theta = \frac{v^2}{gR} - 0.15$$

For design speed 64 km/hr, and maximum superelevation 1/14.5, a minimum safe radius R = 147 m.

Minimum length of spiral
Various theories have been put forward for finding a suitable length of spiral, but the most widely accepted is one that limits the rate of change (with respect to time) of centrifugal acceleration to a value lying between 0.15 m/ sec ^{3} and 0.6 m/sec ^{3}, depending on the degree of comfort required. A value of 0.3 m/sec ^{3} is most commonly used.

For the clothoid spiral the rate of change of curvature $(1/\rho)$ is constant, so that for constant speed, the rate of change of centrifugal acceleration

$$a = \frac{v^2}{\rho}$$

will be constant and equal to the change in acceleration divided by the time in which the change takes place, then

$$a = (v^2/R - 0)/t$$

where R is the radius at the end of the curve, and t is the time taken by a vehicle travelling round the curve at constant speed v. But $t = L/v$, so that

$$a = v^3/LR$$

$$a = (\frac{V}{3.6})^3/LR$$

where V is in km/hour.

It has already been shown that the product of arc and radius LR is constant for the spiral, and fixes the shape of the curve. It is known as the 'speed value' of the curve.

Once the central radius R has been fixed the minimum length of the curve can be found from the minimum LR value (or speed value). Assuming a = 0.3 m/sec ^{3} for v = 64 km/hr, LR =18 728 m = 18 800 say to a round figure.

If radius of central circular curve is 300 m, the minimum length of spiral = 62.7 m. If radius of central circular curve is the minimum for 64 km/ hr =147 m, the minimum length of spiral

$$= \frac{18\,800}{147} = 127.9\,\text{m}$$

Provided site conditions allow, it will always be preferable for both L and R to exceed the minima.

Other design considerations

The intersection angle being assumed fixed, and the minimum values of R and (LR) having been found, the problem remaining is to find a combination of L and R to suit the particular case. Two factors will influence the final choice of L and R.

If there are no site restrictions, the ideal combined curve is one in which the central circular arc is approximately equal to the length of each of the spirals. In this case we have, the angle consumed by one spiral is L/ 2R and by the circular arc is L / R so that the intersection angle β is given by

$$\beta = \frac{L}{2R} + \frac{L}{R} + \frac{L}{2R} = \frac{2L}{R}$$

If there are site restrictions, there may not be space for a central circular arc, in which case, the angle consumed by the spirals will equal the intersection angle, i.e.

$$\beta = \frac{L}{R}$$

11.8 Vertical curves

Whenever two gradients intersect on a road or railway, it is obviously necessary, from the traffic point of view, that they should be connected by a vertical curve, so that the vehicles can pass smoothly from one gradient to the next.

Although in practice road and rail gradients are comparatively flat and it is therefore relatively unimportant what type of curve (circular, parabolic, sinusoidal) is used, it is usually assumed that a curve having a constant rate of change of gradient (i.e. a parabola) is the most satisfactory, and it so happens that this is the easiest to calculate.

For modern high speed roads, some engineers prefer a curve in which the rate of change of gradient increases or decreases uniformly along the curve, but as roads designed to these standards have also very flat gradients, this is usually regarded as an unnecessary refinement. The simple parabola only will be considered.

Gradients are expressed as percentages, for example a +4% gradient is one in which the level rises 4 units vertically in 100 units horizontally, a positive gradient indicating a rising gradient, and a negative gradient indicating a falling gradient.

This method of representing gradients not only has the advantage of simplifying the calculation of rises and falls in level (e.g. a +4% gradient 260 m long rises 4 x 2.6 = 10.4), but also, apart from a factor of 1/100, such gradients are identical with those used in coordinate geometry or calculus, e.g. if the X axis denotes the horizontal, and the Y axis the vertical, then a 4% gradient is a mathematical gradient of 0.04 or dy/dx = 0.04.

Because of the well-known simple geometry of the parabola, many different methods are put forward for calculating the properties of vertical curves, but the approach here is selected because it is felt that it is the simplest to understand and because it will readily yield a solution to any problem that arises. In some cases a quicker solution can be found to a particular problem by utilizing simple geometry, but the same methods applied to another problem may become very involved.

Types of vertical curve
Vertical curves can take various different forms depending on the sign and magnitude of the intersecting gradients, passing from left to right over the curves (Figure 11.16).

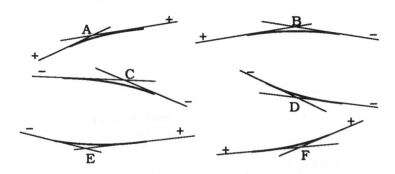

Figure 11.16

It will be observed that when there is a decrease in gradient, a curve convex upwards is formed, and when there is an increase in gradient a curve concave upwards is formed. The former are known as 'summit' curves, and the latter as 'valley' curves. It should be noted, however, that a true summit (highest point), or a true valley (lowest point), can only occur when there is a change of sign between the two gradients. It will be obvious that the gradient at the highest or lowest point is zero.

Simple parabola
Assume rectangular coordinates, with X axis horizontal and Y axis vertical in Figure 11.17.

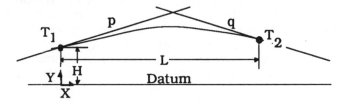

Figure 11.17

The basic requirement for the curve is that the rate of change of gradient (with respect to distance) shall be constant. This can be expressed in two different ways.

(a) $\dfrac{d^2y}{dx^2} = K$

(b) $\dfrac{q-p}{L} = K$

To explain (b) since the rate of change of gradient is constant, it is given by

$$\frac{\text{the change in gradient}}{\text{the distance in which it takes place}}$$

that is the initial and final gradients dy/dx are denoted by p and q respectively, and the horizontal length of the curve by L.

The gradient at any point on the curve can be found from

$$\frac{dy}{dx} = \int \frac{d^2y}{dx^2} \, dx = \int K \, dx = Kx + A$$

If point T_1 is taken as the origin for the X axis, then at T_1, x = 0, and dy/dx = p, whence A = p, and equation of the curve becomes

$$\frac{dy}{dx} = Kx + p$$

or

$$\frac{dy}{dx} = \frac{(q-p)x}{L} + p$$

Note that when

$$x = L, \quad q = KL + p, \quad \text{i.e. } K = \frac{(q-p)}{L}$$

The level of any point on the curve can be found from

$$y = \int \frac{dy}{dx} \, dx = \int (Kx + p) \, dx = \frac{1}{2} Kx^2 + px + B$$

If T_1 is taken as the origin for the Y axis, then at T_1, x = 0, y = 0, and so B = 0. It is usually more convenient to take the level datum as the origin for the Y axis, so that when x = 0, y = B = reduced level of T_1 = H say.

In this case, the equation becomes

$$y = \frac{(q-p)x^2}{2L} + px + H$$

This is an equation of the form $y = ax^2 + bx + c$ (a parabola with its axis vertical) where a is the $\frac{1}{2}$ (rate of change of gradient), b the initial gradient at T_1, and c the reduced level of T_1.

In evaluating these coefficients, care must be taken with their signs. The sign of b and c will be perfectly obvious in a practical problem, but the following will be of use in ensuring the correct sign for a.

The equation $y = bx + c$ ($= px + H$) is the equation of the straight line passing through the point T_1 and having a gradient p, i.e. the equation of the initial gradient tangent. If the curve is a summit curve it will fall away from this tangent and so the sign of a will be negative. If the curve is a valley curve, it will rise above the initial tangent and so the sign of a will be positive.

This will be demonstrated by numerical examples for each case illustrated in Figures 11.16.

(A) If $p = + 4\%$, $q = + 3\%$,   then $K = \dfrac{3-4}{100L}$ and

$$y = -\frac{x^2}{200\,L} + \frac{4\,x}{100} + H$$

(B) If $p = + 4\%$, $q = -3\%$,   then $K = \dfrac{-3-4}{100L}$ and

$$y = -\frac{7\,x^2}{200\,L} + \frac{4\,x}{100} + H$$

(C) If $p = -3\%$, $q = -4\,\%$,   then $K = \dfrac{-4+3}{100L}$ and

$$y = -\frac{x^2}{200\,L} - \frac{3\,x}{100} + H$$

(D) If $p = -4\%$, $q = -3\,\%$,   then $K = \dfrac{-3+4}{100L}$ and

$$y = \frac{x^2}{200\,L} - \frac{4\,x}{100} + H$$

(E) If $p = -4\%$, $q = +3\,\%$,   then $K = \dfrac{3+4}{100L}$ and

$$y = \frac{7x^2}{200\,L} - \frac{4\,x}{100} + H$$

(F) If $p = + 3\%$, $q = +4\,\%$,   then $K = \dfrac{4-3}{100L}$ and

$$y = \frac{x^2}{200\,L} + \frac{3\,x}{100} + H$$

Setting out data for vertical curves
It will be clear from the above derivation of equations, that when the gradients have been fixed, only one other factor, either the horizontal length of the curve or the rate of change of gradient, is required to define the curve uniquely. The choice of these factors is dealt with later and for the remainder of the present section it will be assumed that the curve has been defined, and that all that remains is to determine the setting out data.

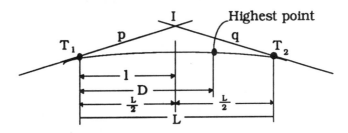

Figure 11.18

Two properties of the parabolic vertical curve must be established to assist in calculating the data. In Figure 11.18 let the horizontal distance from T_1 to the intersection point I be l, and the equation of the curve be

$$y = (q - p)\frac{x^2}{2L} + px + H$$

Substituting L for x, the reduced level of point T_2 becomes

$$\tfrac{1}{2}(p + q)L + H$$

But the reduced level of T_2 is also given by

$$H + pL + q(l - l)$$

Equating these two expressions for the reduced level of T_2 gives

$$l = \tfrac{1}{2}L$$

This very important relationship, that the horizontal distances from the tangent points to the intersection point are equal, is of considerable use in solving vertical curve problems, and yet is often overlooked.

For the two cases where there is either a true summit or valley, i.e. where the gradients are zero, we derive another useful property. In Figure 11.18, D denotes the distance to the highest point of the curve (or the lowest if the curve is a valley). (It should be noted that this will not coincide with I unless $p = -q$).

Remembering that the gradient at the highest (or lowest) point is zero, and making use of the definition, that the rate of change of gradient (which is constant) is given by the change of gradient, divided by the distance in which the change takes place,

$$\frac{0-p}{D} = \frac{q-0}{L-D} = \frac{q-p}{L}$$

from which

$$D = \frac{-p}{q-p}L \quad \text{and} \quad L-D = \frac{q}{q-p}L$$

It will have been seen (Figure 11.16) that a true summit (highest point) or true valley (lowest point) only occurs when there is a change of sign between p and q, so that the term $q - p$ in the above equations will always be given by the numerical sum of the two gradients, and that the signs in the two equations are consistent.

If at a summit, $p = +4\%$ and $q = -3\%$, then $D = \frac{4}{7}L$ and $L-D = \frac{3}{7}L$

and at a valley, $p = -4\%$ and $q = +3\%$, then $D = \frac{4}{7}L$ and $L-D = \frac{3}{7}L$

Location of tangent points
This is best illustrated by numerical examples. Refer to Figure 11.19.

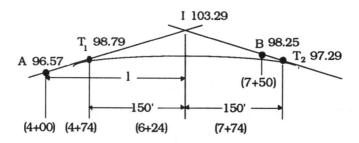

Figure 11.19

Example
A rising gradient of +3% and a subsequent falling gradient of – 4 % are to be connected by a parabolic vertical curve of horizontal length 300 m. The reduced level of a point A at chainage 4 + 00 on the first

gradient is 96.570 m, and the reduced level of a point B at chainage 7 + 50 on the second gradient is 98.250 m. Find the chainage and reduced levels of the tangent points.

Let the horizontal distance from A to the intersection point I be s. Then since the distance AB is 750 – 400 = 350 m, the reduced level of I is given by

$$96.570 + 3\,s\,/\,100 = 98.250 + 4\,(350 - s)/100$$
$$s = 224.0 \text{ and } 350 - s = 126.0$$

$$\text{Chainage of } I = 4 + 00 + 224.0 = 6 + 224.0$$
$$\text{Chainage of } T_1 = 6 + 224.0 - \tfrac{1}{2}s = 4 + 74.0$$
$$\text{Chainage of } T_2 = 6 + 224.0 + \tfrac{1}{2}s = 7 + 74.0$$

$$\text{Reduced level of } I = 96.570 + 3 \times 224.0/\,100 = 103.290$$
$$\text{Check from } T_2 = 98.250 + 4 \times 126.0\,/\,100 = 103.290$$

$$\text{Reduced level of } T_1 = 103.290 - 3 \times 150\,/100 = 98.790$$

$$\text{Reduced level of } T_2 = 103.290 - 4 \times 150\,/100 = 97.290$$

Example
Points A and B, and C and D, data for which are given below, lie on two intersecting gradients which are to be connected by a parabolic vertical curve of horizontal length 500 m. See Figure 11.20.

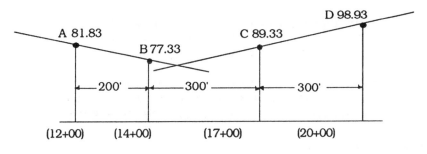

Figure 11.20

Determine the chainage and reduced level of the tangent points.

Horizontal distances are AB = 200 m, BC = 300 m and CD = 300 m, thus the gradients are AB = – 2.25 % and CD = + 3.2 %.

Let the distance from B to I be s, then the reduced level of I from two points gives

Point	Chainage	Reduced level
A	12 + 00	81.830
B	14 + 00	77.330
C	17 + 00	89.330
D	20 + 00	98.930

$$77.330 - 2.25 \; s/100 = 89.330 - 3.2 \; (300-s)/100$$

and

$$s = -44.0 \text{ and } 300 - s = 344.0$$

Since s turns out to be negative the point I lies to the left of B as in Figure 11.21.

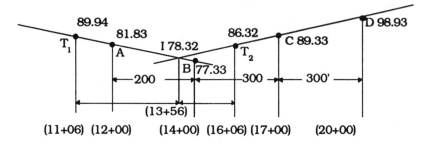

Figure 11.21

The chainage of I = 14 + 00 − 44.0 = 13 + 56.0
Chainage of T_1 = 13 + 56.0 − 250.0 = 11 + 06.0
Chainage of T_2 = 13 + 56.0 + 250.0 = 16 + 06.0

Reduced level of I   = 77.330 + 2.25 x 44.0 / 100 = 78.320
Reduced level of T_1 = 78.320 + 2.25 x 250.0 / 100 = 83.940
Reduced level of T_2 = 78.320 + 3.2 x 250.0 / 100 = 86.320

Reduced levels on curve
Once the chainage and reduced level of the tangent points have been determined, it remains to determine the values of the spot levels along the curve. This is normally done for a series of equidistant points, usually referred to as 'chord' points, although the horizontal distances between them are in fact the horizontal components of the chords.

The calculations are simplified if the curve is set out in equal 'chords', and the total horizontal length of the curve is an exact multiple of the 'chord' length. In general, both these conditions can not be satisfied if running chainage is to be preserved, but a convenient compromise can be obtained if the length of the curve is chosen to make the final 'chord' only a short chord,

It should be noted that, for a parabola, second differences are constant, and this provides an alternative method of calculating the curve levels. For practical purposes, however, it is usually suitable to calculate levels to the nearest 0.005 m but to achieve this by the difference method it is usually necessary to calculate the second differences to two further decimal places, so that very little time is saved, unless the second difference is exact to 0.005.

The calculation of the second differences acts as a valuable check.

Example

In this example the running chainage is preserved, and the curve length is not adjusted. The levels are calculated for the 50 m even chainage points for the curve used in the example of Figure 11.19 in which we have

$$p = +3\% \qquad q = -4\% \qquad L = 300$$

$$Ch\, T_1 = 4 + 74 \qquad RL\ of\, T_1 = 98.79$$
$$Ch\, I = 6 + 24 \qquad RL\ of\ I = 103.29$$
$$Ch\, T_2 = 7 + 74 \qquad RL\ of\ T_2 = 97.29$$

The equation of the curve is

$$y = ax^2 + bx + c$$

where

and

$$a = \frac{q - p}{2L} = \frac{-4 - 3}{200 \times 300} = 1.166 \times 10^{-4}$$

$$b = p = \frac{3}{100} \qquad c = H = 98.79$$

The levels of the curve are tabulated in Table 11.2 whose columns are as follows:

(1) The chainages of points along the curve.
(2) The x coordinates.
(3) ax^2.
(4) bx.
(5) The sum of columns (3) and (4).
(6) The reduced levels of points on the curve.
(7) The first differences between these levels.
(8) The second differences between the levels, which should be constant.

11.9 Choice of design constants for vertical curves

It has been shown that, once the value of the two intersecting gradients, and the position (vertically and horizontally) of their intersection point have been fixed, only one further property is required to define a vertical curve uniquely. For comparison purposes, the most convenient additional property to fix is the (horizontal) length.

Table 11.2

1	2	3	4	5	6	7	8
4 + 74	0	0	0	0	98.79		
5 + 00	26	– 0 08	0.78	0.70	99.49	0.91	
5 + 50	76	– 0.67	2.28	1.61	100.40	0.32	–0.59
6 + 00	126	– 1.85	3.78	1.93	100.72	– 0.26	–0.56
6 + 50	176	– 3.61	5.28	1.67	100.46	– 0.85	–0.59
7 + 00	226	– 5.96	6.78	0.82	99.61	– 1.43	–0.58
7 + 50	276	– 8.89	8.28	– 0.61	98.18		
7 + 74	300	– 10.50	0.00	–1.50	97.29		

The determination of a suitable length for a particular vertical curve is normally the responsibility of the road engineer, but it is felt that at least the principles influencing the choice have a proper place in this volume.

It will be appreciated that, where the length of the curve is derived from traffic considerations, these will be minimum lengths, and the actual curves will usually be longer, even if only to obtain a whole number of chords, or to preserve running chainage. In consequence, any calculations made to assist in the choice of length, need only be approximate.

Safety and free flow of traffic are the first considerations in traffic engineering. With this in view, comfort and prevention of stress on vehicle and driver are the main objectives for valley curves. For summits, whilst these are still of equal importance, visibility over the crest must also be taken into account. As in most (but not all) practical cases, a longer curve will be required, visibility will be the overriding criterion.

Design by limitation of vertical acceleration
A rational method of design, takes into account the design speed of the road. (The design speed is usually taken to be the 85 percentile speed, i.e. the speed that is not exceeded by 85% of the vehicles using

the road.) For safety and comfort the vertical acceleration must be limited to a reasonable value.

Design by vision distance

For summits, in addition to applying other criteria, the question of visibility over the brow of the hill must also be considered.

The *vision distance* D, is defined as being the length of the sight line between two points at driver's eye level h above the road (see Figure 11.22), h being normally assumed to be 1.143 m (2.75 ft). Assuming a circular arc for the vertical curve we have its radius R limited to a minimum value given by:

$$h \approx \frac{0.5 \, D^2}{2R}$$

therefore

$$\frac{p-q}{L} = \frac{1}{R} = \frac{8h}{D^2} \qquad \text{or} \qquad L = \frac{D^2 \, (p-q)}{8h}$$

L cannot be less than the vision distance D for an acceptable curve at a given design speed.

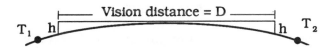

Figure 11.22

The required length of this sight line for two way roads (single carriageway) is based on the minimum distance required for overtaking, and for one-way roads (dual carriageway), on the minimum stopping distance, both of these being dependent on the design speed. Typical values are given in Table 11.3.

Table 11.3

| Design speed | | Vision distance | | | |
| | | Two way | | One way | |
mph	kph	ft	m	ft	m
40	65	950	290	300	91
50	80	1200	366	425	129
60	96	1400	427	650	198
70	113	1650	503	950	290

Curve controlled by fixed level

In certain cases none of the above criteria will apply, and the curve may have to be designed to pass above or below some point of fixed level, for example under or over a bridge. A numerical example should demonstrate this problem. See Figure 11.23.

Example

A falling gradient of 1.8% is followed by a rising gradient of 2.5%, their intersection occurring at chainage 8 + 73.2 and RL 72.56. In order to provide sufficient clearance under an existing bridge, the RL at chainage 7 + 95.8 on the curve must not be higher than 74.20.

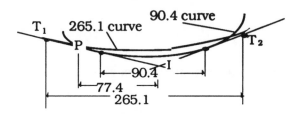

Figure 11.23

Find the greatest length of simple parabolic vertical curve that can be used. Let P be the control point under the bridge. Distance PI is given by

$$PI = (8 + 73.2) - (7 + 95.8) = 77.4$$

$$RL \text{ of } T_1 = 72.56 + \frac{1.8\,L}{200} = H$$

Equation of curve is

$$y = H - \frac{1.8\,x}{100} + \frac{(2.5 + 1.8)\,x^2}{200L}$$

RL of P = 74.20

$$= 72.56 + 0.009L - 0.018\,(0.5\,L - 77.4) + 0.0215\frac{(0.5\,L - 77.4)^2}{L}$$

whence

$$L^2 - 355.5\,L + 23\,963 = 0$$

from which

$$L = 265.1 \text{ or } 90.4$$

Since 0.5 x 90.4 is less than 77.4, the second solution is impracticable.

In problems of this kind a quadratic equation will always occur producing two solutions as shown in Figure 11.23. Common sense, aided where necessary by a rough sketch, will indicate which solution is applicable.

Excavations for setting out curves

Obviously, curves to be setout to a design do not follow the existing ground, which has to be excavated of filled as part of the process. The detail of formations and their establishment on the ground are based on the shapes of horizontal and vertical curves. These matters are considered in Chapter 12.

12 Volumes of earthworks

Chapter Summary

The calculation of the volumes of earthworks is of major importance in civil engineering work. The topic is dealt with by various commercial computer packages which model the problem with as little intervention from the user as possible.

However, the irregularities encountered in civil engineering operations, and the complex nature of the ground surface, make necessary the imposition of some *a priori* constraints, such as the break lines marking the edge of an embankment or cutting.

Although the accuracy with which ground surfaces can be modelled is a matter of current investigation, there is no escaping the fact that it depends on the density of points used to described the surface. A very varied surface needs many points, but a few only are needed for smooth surfaces.

Accuracy is also related to costs and need. The surfacing of an airfield runway and its related volumetric-cost calculations demand much tighter geometrical accuracy than in ordinary roadwork, or in quarrying.

In this chapter, we concentrate upon traditional roadwork calculations which illustrate many of the basic problems involved, both in the calculation of volumes and in the setting out of formations.

12.1 Prismoidal and end-area formulae

Because land surfaces are invariably irregular, all earthwork calculations must of necessity be approximate, the precision obtained being entirely dependent on the density of the measured ground heights in relation to the nature of the surface.

Volumes are usually obtained by calculating the volume of the regular geometrical solid whose surfaces most nearly represent the actual ground surfaces. The geometrical solid most frequently used for this purpose is the prismoid, which is defined as a solid figure having plane parallel ends and plane sides (see Figure 12.1).

It will be clear that the area A of a cross-section, parallel to the ends, will be a function of the perpendicular distance x of the cross-section from one of the ends, and that if the cross-section area A is plotted against the distance x (see Figure 12.2), the volume will be equal to the area under the curve $V = \int A dx$.

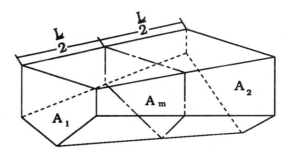

Figure 12.1

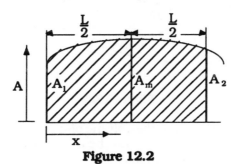

Figure 12.2

An approximation to the area under the graph can be found, using Simpson's rule, by taking three ordinates, A_1, A_m, and A_2, equally spaced at distance L /2 apart, i.e.

$$V = (L/6)(A_1 + 4A_m + A_2)$$

(Note : A_m is the cross-section area midway between A_1 and A_2 and not the mean of A_1 and A_2).

For a true prismoid, the cross-sectional area is a quadratic function of x, the distance from one end, so that the above relationship is exact, and the equation is known as the *prismoidal rule*.

When this rule is applied to the portion of a normal road or railway cutting or embankment, between two parallel cross-sections, as shown in Figure 12.3, where three of the side faces are planes, if the ground surface is also a plane, the figure is a true prismoid.

Since the rule is also exact when the cross-sectional area is a cubic function of x, the rule will also give the true volume, if either

(a) the slope of the ground at right angles to the centre-line is uniform (i.e. PQ is a straight line), and the central height h is a quadratic function of x (i.e. the longitudinal profile on the centre-line is parabolic), or

(b) the transverse profile PQ is parabolic, and h is directly proportional to x, (i.e. the longitudinal profile is a straight line).

Thus it will be seen that, provided the ground surface curves smoothly, the rule will give a very good approximation to the true volume for this type of solid.

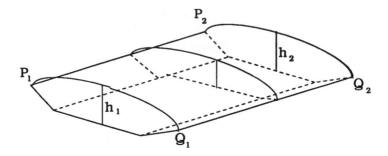

Figure 12.3

A less accurate estimate of the volume of a prismoid can be obtained by using the trapezoidal rule to find the approximate value of the area under the curve in Figure 12.2, i.e.

$$V = \frac{1}{2} L (A_1 + A_2) \text{ approx}$$

which, when applied to volumes, is known as the *end-area rule*. The best accuracy will be obtained from this rule when A_1 is approximately equal to A_2 and the accuracy will decrease as the difference between A_1 and A_2 increases. As an extreme example, for a pyramid or cone, where one end-area is zero, and the true or prismoidal volume is base x height/3, the rule gives base x height/2, an error of 50%.

12.2 Determination of volumes from contours

This method, which depends largely for its accuracy on the contour vertical interval, and on the accuracy with which the contours have been determined and plotted, is extremely valuable where very large volumes are involved, e.g. reservoirs, land reclamation schemes, open-cast mining etc., and is also very useful in the preliminary stages of route projects (roads, railways, canals, etc.) for making initial estimates of cost, comparison of alternative routes, selection of best profile, and so on.

The method consists of splitting the solid, along the contour planes, into a series of horizontal slabs, each slab being then regarded as a prismoid whose length is the contour vertical interval, and whose end-areas are the areas enclosed by the contour lines at the height in question, the areas being taken off the map or plan by planimeter or by calculation from coordinates.

If the prismoidal rule is to be applied to each slab individually, it will be necessary to interpolate the contour lines midway between those already plotted, in order to obtain the mid-area of each slab. It is usually adequate, however, to take the slabs in pairs and find the volume by Simpson's rule. Any portions of the solid which are not embraced by two contour planes will have to be treated separately and the volume of the nearest appropriate geometrical solid (usually a pyramid or wedge) found.

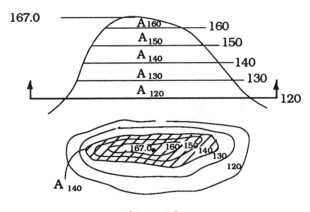

Figure 12.4

Referring to Figure 12.4 (vertical interval = 10 m) by end-area rule,

$$V = \frac{10}{2} [A_{120} + 2(A_{130} + A_{140} + A_{150}) + A_{160}] + \frac{7}{3} A_{160}$$

By prismoidal (or Simpson's) rule,

$$V = \frac{10}{2} [A_{120} + 4 A_{130} + 2 A_{140} + 4 A_{150} + A_{160}] + \frac{7}{3} A_{160}$$

When the new profile (i.e. the ground profile after the works have been carried out) is anything other than a horizontal plane, the new contour lines for the finished work will have to be superimposed over the existing ground contours, and the end-areas of the prismoids will then be the areas on the plan enclosed between the new and the old contour lines for the height in question.

In Figure 12.5 the new contours for the benching shown in the cross-section have been superimposed onto the contours for a natural hillside, and the prismoidal end-areas at heights 20 m and 25 m above datum indicated by hatching.

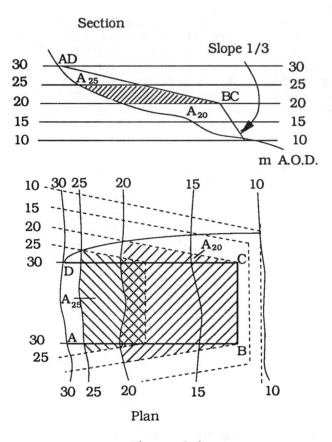

Figure 12.5

In all but the simplest earthwork projects, the plotting of the contours for the finished work may require some thought, and Figure 12.5 also illustrates some of the principles involved.

It will be assumed that the embankment surrounding the benching on sides AB, BC and CD has a slope of 1/3 (i.e. 1 vertical in 3 horizontal), and that the vertical interval is, as shown, 5 m. Then if at any point of known height on the plan we move away along the line of greatest slope for a distance of 3 x 5 = 15 m, the level will have dropped 5 m and a point on the next contour line will have been found. For example, if from points B and C (both at 20 m above datum) we move out 15 m in a direction normal to BC, two points on the 15 m contour will have been found; if we move out 30 m, two points on the 10 m contour will have been found, and so on. As for a plane, the contours will be a series of equidistant parallel lines, and the contours for the embankment falling away from the line BC can now be plotted.

Similarly, if we move away 15 m from C in a direction normal to CD, and 45 m away from D in the same direction, two further points on the 15 m contour for the embankment will have been found. Since the normal to CD at D, in this case, lies along the 30 m contour, the point we have just located on the 15 m contour for the embankment lies below existing ground level, and the 15 m contour for the finished work will only extend part of the way along the line between the two points by which its position has been established. Clearly at the point where the new and the old 15 m contours intersect the new and existing ground levels are the same, and the actual new contour line will not extend beyond this point.

By linking up all such points where new and existing contours of the same value intersect, the position of the toe of the embankment can be established. This is also shown in the figure.

Figure 12.6 shows the same principle applied to a section of road running in cutting and on embankment, with again some of the prismoid end-areas marked to indicate the areas that will have to be calculated or taken off with a planimeter. It should be noted that, in both these cases, the cross-sectional view is not required for the purposes of the volume calculations, and is only included to illustrate the meaning of the construction lines on the plan. A very small part of a project, drawn to a large scale, has been used for the sake of clarity, but it must be realized that in the ordinary way, due to the difficulty of obtaining sufficiently accurate contours, the accuracy of the volumes obtained under these conditions would be considerably inferior to those obtained by the cross-section method described below.

However, if the ground has been described in terms of triangular facets of a sufficiently small size to describe the ground adequately, and volumes are calculated from three-dimensional coordinates, accurate results can be obtained. This method is described in section 12.7.

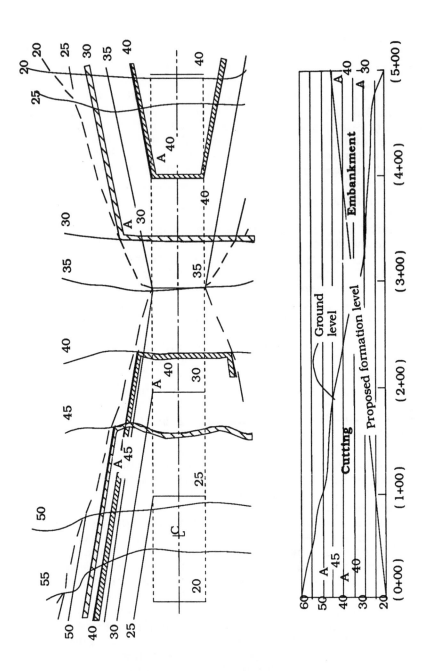

In the road example illustrated in Figure 12.6, the road gradient is uniform, and the plan centre-line straight. If the vertical profile of the road is laid to a vertical curve, the new contours crossing the road will no longer be equidistant, and the side-slope contours will cease to be equidistant parallel lines, and become instead equidistant parallel curves. Unless the change in gradient is very acute, the curvature of the side-slope contours is very slight.

The case where the road centre-line is curved on plan is different, as will be seen from Figure 12.7. The line of greatest slope of the side-slope is usually normal to the edge of the road, so that the same technique of stepping off intervals of n times the vertical interval (where the side-slope is 1/n) from points of known height on the edge of the road, still applies, as shown in Figure 12.7.

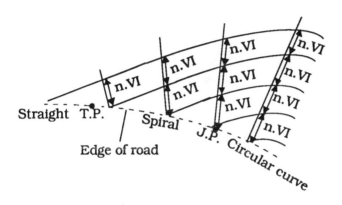

Figure 12.7

Figure 12.8 demonstrates one further application of contour geometry to the solution of earthwork volume problems. Here a tunnel 16 m in diameter, and with its axis horizontal, is to be driven into a hillside whose contours are shown.

In order to be able to plot, on the plan view, the position of the interpenetration curve between the tunnel and the hillside, and the shape of the end-areas of the horizontal slabs, it is necessary in this case to draw the front elevation of the mouth of the tunnel, and from this view obtain the true-length (such as ab) of the horizontal distances from the centre-lines to the lines of intersection of the contour planes with the walls of the tunnel. In the figure this has been achieved by drawing the auxiliary view of the tunnel mouth, so that these intersection lines can be obtained by simple projection. The positions of points on the plan view of the tunnel mouth denoted by the intersection of the old and new contour lines at 58 m above datum, and the shape of the slab end-area at the same height, are marked on the diagram. The embankment carrying the road up to the tunnel mouth has been omitted, as the construction for this would be similar to that shown in Figure 12.6.

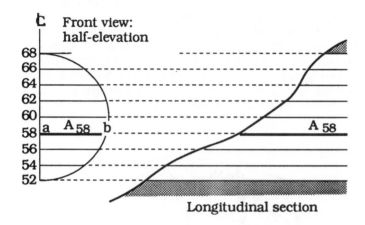

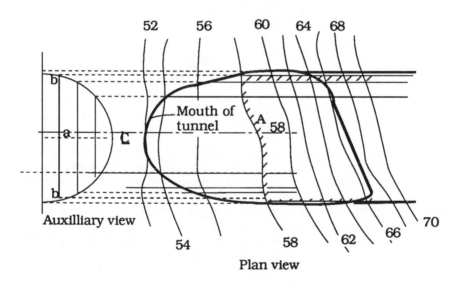

Figure 12.8

12.3 Determination of volume from spot heights

This method, which depends for its accuracy only on the density of the levels taken, is usually used for large open excavations such as reservoirs, for ground levelling operations such as parks or playing fields, or for building sites. The use of photogrammetric methods of heighting from air photographs, coupled with the use of electronic computers for data processing, has opened up a field in which, by enormously increasing the number of spot heights that can be used, the method can be extended to almost any earthwork project for which photogrammetrical heights are sufficiently accurate.

The coordinates and heights of points may also be obtained rapidly by total station, usually from carefully selected points at changes of slope or bends in plan. Efficient gathering of data by field methods is still a skilled task. Because such points appear at irregular intervals over the ground it has become common to describe them as *random points* . This terminology is most unfortunate because they are *systematically* selected to model the ground and its break lines and are certainly not random. The technique of gathering height data from a regular grid is *statistically random.*

In normal ground methods, the site is 'gridded' by a series of lines forming squares (or occasionally rectangles), and the ground levels are determined at the intersections of the grid lines, together with

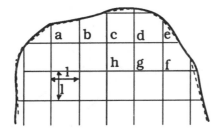

Figure 12.9

such additional points (referenced to the grid by offsets or tie lines) as may be necessary to pick up break lines: special boundaries or exceptional ground irregularities or discontinuities. The spacing of the grid lines will depend on the nature of the ground, and should be sufficiently close for the ground surfaces between the line to be reasonably regarded as planes.

There are two reasons for using grids

(a) They are simple to use in the field, and
(b) Subsequent computer calculations are regularized.

The proposed formation levels at the corners of the grid squares are obtained from the designer's drawings, and the volume within each square is taken as being the plan area of the square, multiplied by the average of the depths of excavation (or fill) at the four corners of the square. The volumes of the portions lying between the outermost grid lines and the random boundaries of the site are taken as the plan areas of the nearest equivalent trapezia or triangles, each multiplied by the average of their corner depths (see Figure 12.9).

Because all the depths at the internal intersections of the grid lines are used in the calculation of the volume of more than one square, a formula on the lines of the following can be evolved

$$V = \frac{l^2}{4} \left(\sum h_1 + 2 \sum h_2 + 3 \sum h_3 + 4 \sum h_4 + \dots \right) + \sum R$$

where l is the length of side of square, h_1 the depths such as at a and e which are used once, h_2 the depths such as at b, c, and d, which are used twice, h_3 the depth such as f which is used three times, h_4 the depth such as g, h etc. which are used four times, and R the volume of extra peripheral trapezia and triangles, which must, of course, be calculated individually.

Certain difficulties will arise if there is a changeover from cut to fill, or vice versa, within one grid square (see Figure 12.10).

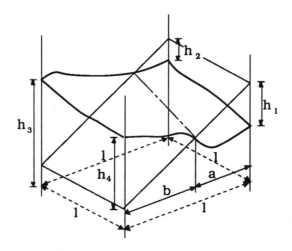

Figure 12.10

If $h_1 = -h_4$ and $h_2 = -h_3$, the calculation, $V = l^2 (h_1 + h_2 + h_3 + h_4)/4$ will produce zero volumes for the square, which is patently incorrect, and whatever the values of the depths the calculation will only produce the net volume cut or fill, whichever is the larger.

If the grid lines are fairly close together, the formulae

$$V_{cut} = l^2 (h_1 + h_2 + 0 + 0) / 4$$
$$V_{fill} = l^2 (0 + 0 + h_3 + h_4) / 4$$

will not produce excessive inaccuracy.

To avoid mistakes and to assist in the tabulation of the figures, it is often convenient to extract the ground levels from the field book, and the formation levels from the drawings, and to record them in coloured inks on a gridded plan of the site in the manner indicated in Figure 12.11. (The colours indicated are not used in this book.)

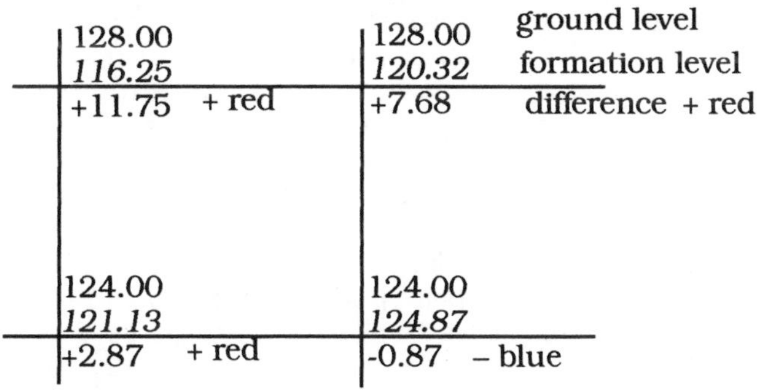

128.00		128.00	ground level
116.25		*120.32*	formation level
+11.75	+ red	+7.68	difference + red
124.00		124.00	
121.13		*124.87*	
+2.87	+ red	-0.87	– blue

Figure 12.11

The first figure (sepia) is the ground level, the second below it (green) is the proposed formation level, and the difference where positive (red) indicates cut, and where negative (blue) indicates fill.

Volume from triangular network: Delauney tesselation

Where points have been carefully selected at breaks of slope or changes of direction in plan, more efficient estimates of volume can be obtained by forming a system of triangles connecting these points together and calculating the volume for each triangular column from the formula

$$V = \frac{A}{3} \, (h_1 + h_2 + h_3)$$

where A is the area of the triangle. The volumes of all triangular columns are added to give required cuts and fills.

The selection of triangles has been automated by an algorithm due to Delauney, which selects a network of the most nearly equilateral triangles from any given set of points, according to the rule that the circumcircle of any triangle cannot contain another point. The system produces a unique set of triangles except where there are four concyclic points. In this case an abitrary rule makes the automatic choice of the two possibilities. The algorithm starts at any outside point, proceeds to find the nearest two neighbours by examining distances, then moves to one of them to proceed to finding nearest neighbours. The selection of triangles is easy to do manually, but takes a little thought, and much calculation, when programmed for a computer. See section 12.7.

12.4 Determination of volume from cross-sections

This method is widely used for engineering projects of all types, and lends itself particularly well to route projects, roads, railways, waterways, etc. It illustrates general principles and has applications in small tasks. Much work formerly undertaken in the field, such as cross-sectioning after the centre line has been set out, can now be achieved from the computer model provided it is sufficiently accurately formed. Work practices vary according to the technology available.

Cross-sections are taken on lines at right angles to the main job centre-line and the volume between adjacent cross-sections is found for a prismoid with the areas of the cross-sections as its end and, if necessary, its mid-areas. For route surveys it is usually convenient to take the cross-sections at the round number chainage points, with additional sections at points of unusual irregularity or discontinuity.

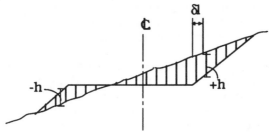

Figure 12.12

When the ground is very irregular it is best to plot the profiles of the existing ground and the proposed formation to a convenient scale and obtain the areas of the cross-sections by planimeter or counting squares etc. In this connection it is important to note that it is customary, for clarity, to plot cross-sections to an exaggerated vertical scale, and suitable allowance must be made for this if the areas of the cross-sections are to be found by any of the above methods.

When the ground surface is reasonably regular, and the ground profiles at the cross-sections can be satisfactorily represented by uniform slopes, it is more expeditious to make use of suitable formulae for calculating the cross-sectional areas.

As referred to in the previous section, photogrammetrical heighting, coupled with computer processing, can greatly assist with this method. In such cases, the area of the cross-sections can be obtained from a series of equidistant spot heights. As will be seen from Figure 12.12, if the depths are determined at sufficiently close intervals dl, the areas will be given by

$$A_{cut} = dl \sum (+ h) \qquad A_{fill} = dl \sum (- h)$$

and because depths taken beyond the end of the section will be zero, there is no need to determine the overall width of the cross-section independently.

When normal ground methods of survey are employed and the ground slopes can be regarded as uniform, the following calculation methods will be found useful.

It will be noted that the general case considered here is the *three-level* cross-section, in which the ground slopes are assumed to be uniform from A to E, and from E to D, in Figure 12.13.

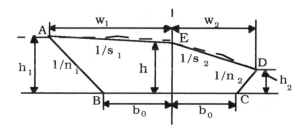

Figure 12.13

Although such a situation, with a sudden change of slope exactly on the route centre-line, will rarely, if ever, occur in practice, this assumption gives a much closer approximation to the true ground profile when this is a smooth curve, as shown in the figure, than would be given by the assumption of a straight line from A to D. Where the ground surface approximates to a plane, however, the *two-level* cross-section is quite justifiable.

In Figure 12.13 slopes are expressed as tangents, i.e. 1 vertical in n or s horizontal, the rest of the symbols being self-explanatory.

Calculation of half breadths
Unless the field method of *slope staking* has been used, the half breadths w_1 and w_2 and the corresponding h_1 and h_2 will not be known, and they will have to be determined from the other data. The ground slopes can be obtained by direct measurement with a clinometer or gradienter, or by calculation from two spot heights at

known positions on the slope. The formation breadth and level, and the values of the side slopes will have been supplied by the designer.

$$h_1 = h + \frac{w_1}{s_1}$$

therefore

$$w_1 \left(\frac{1}{n_1} - \frac{1}{s_1} \right) = h + \frac{b_0}{n_1}$$

giving

$$w_1 = \frac{\left(h + \dfrac{b_0}{n_1} \right) s_1 n_1}{s_1 - n_1}$$

applying when the ground rises from the centre-line, and similarly

$$w_2 = \frac{\left(h + \dfrac{b_0}{n_2} \right) s_2 n_2}{s_2 + n_2}$$

applying when the ground falls away from the centre-line.

It will be realized that, if a slope falling away from the centre-line is regarded as being negative, one equation will suffice; e.g. for slopes as shown in the figure, if the formation breadth is 40 m, the central depth is 7 m, both the side slopes are 1/5, and the ground slopes are both 1/20, we have

$$w_1 = \frac{\left(7 + \dfrac{20}{5} \right) 20 \times 5}{20 - 5} = 73.3$$

$$w_2 = \frac{\left(7 + \dfrac{20}{5} \right) 20 \times 5}{20 + 5} = 44.0$$

Calculation of cross-sectional areas

Once the half breadths w_1 and w_2, and the side heights h_1, h_2 have been found, the area A of the cross-section can be found by inspection, or from coordinates taking the origin at any point such as F, giving

$$2A = h(w_1 + w_2) + b_0(h_1 + h_2)$$

It should be clear that by turning the diagrams upside down, all the above equations apply equally to embankments having the same characteristics.

In addition to the above, one further type of cross-section, that shown in Figure 12.14, must be considered.

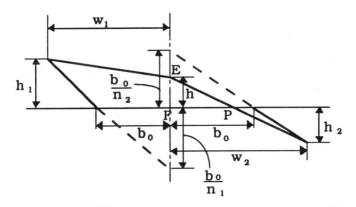

Figure 12.14

It should be noted that in this case it is more common for n_1 and n_2 to be unequal, the slope for the fill usually being flatter than that for the cut. Referring to Figure 12.14, for the left-hand side, we have the same equation as before, i.e.

$$w_1 = \frac{(h + \frac{b_0}{n_1})\, s_1 n_1}{s_1 - n_1}$$

and for the right-hand side

$$w_2 = \frac{(-h + \frac{b_0}{n_2})\, s_2 n_2}{s_2 - n_2}$$

Clearly, if P lies to the left of the centre-line, the above equations will be interchanged.

Since FP = $h\, s_2$, the coordinates of all points with respect to an origin at F are known, and therefore the areas of the cut and of the fill can be evaluated to give

$$2A_{cut} = -b_0 h_1 - w_1 h - h^2 s_2$$

$$2A_{fill} = -b_0 h_2 + h s_2 h_2$$

It should be clear, by turning the diagram in Figure 12.14 upside down, that if P lies to the left of the centre-line, all these equations for cut and fill must be interchanged.

12.5 Slope staking

Slope staking is a trial-and-error process for locating and staking, in the field, the points at which the proposed side slopes of cuttings or embankments will intersect the existing ground surface. As it will be seen that the depths of cut and fill at these points, as well as the cross-section half-breadths, are calculated as part of the procedure, not only are the limits of the earthworks defined on the ground but data for the calculation of the earthwork volumes is provided.

It will be obvious that the proposed formation levels, the formation breadth, and the values of the side slopes must have been settled before any work can be done.

The alternative to slope staking is to set out points from coordinates, itself an iterative process, made easier by the ability of the prism holder to read the setting out measurements and data at his end of the line. Since slope staking can be carried out with simple equipment, it may be a fall-back position, if other equipment fails.

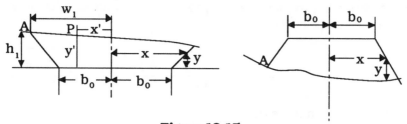

Figure 12.15

Principle

In Figure 12.15, any point on the side slope will be defined by the equation

$$x = b_0 + n y$$

Consequently, at A the values of $x = w_1$ and $y = h_1$ will also satisfy this equation.

The field procedure consists of placing the staff at a series of points on the cross-section such as P, determining the values of x' and y' at P, (by taping and levelling) and finding out whether they satisfy the equation. When they do, P is at the point A, and w_1 and h_1 are found. A is marked by a peg or batter rail.(See later.)

This procedure may sound laborious, but in fact, after a few hours' experience, very few trials are required to locate the correct point. It must be appreciated that even on fairly smooth ground, a correspondence of levels of 0.01 m is all that can be expected, which at a ground slope 1/10 is equivalent to a variation in x of 0.1 m. By evolving a suitable system for the fieldwork, calculation and booking, the work can be made to proceed quite rapidly.

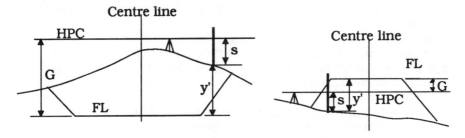

Figure 12.16

Calculation

Once the level has been set up (Figure 12.16), and the height of the horizontal plane of collimation HPC has been found, then G, known as the *grade staff reading*, can be calculated from the formation level FL as follows.

(a) For cuttings, $G = HPC - FL$
(b) For embankments, $G = FL - HPC$

This is a constant for one cross-section and one instrument position. From this, the values of y ' can be found corresponding to the various values of S, the staff reading.

(a) For cuttings $y' = G - S$
(b) For embankments $y' = G + S$

If the ground is horizontal, the half-breadth will be given by

$$w = b_0 + nh$$

(where h is the depth at central-line), and it is convenient to calculate this for each cross-section, as a starting point for estimating w_1 which will be greater or less than this value according to whether the ground slopes upwards or downwards away from the centre-line (vice versa for embankments).

Field work and batter rails
The cross-sections are usually taken at equidistant intervals along
the centre-line of the route, often at the round-number chainage
points, and the work will be easier if the ground levels on the centre-
line have been determined first.

Since slope staking cannot be carried out until the formation levels
have been decided, the centre-line ground levels will almost certainly
have been found at some previous time. A plentiful supply of
temporary bench marks is also desirable, and the chainage pegs can
be conveniently used for this purpose.

If the values of w are all expected to be less than 30 m, the level can be
set up at a convenient point for reading the staff on several cross-
sections, and the distance from the centre-line measured with a tape,
the staff man calling out the distance x ' for each point where he sets
the staff.

If the values of w are greater than 30 m, it will often be more
convenient to measure x ' tacheometrically, using the stadia hairs, or
by EDM, although this, of course, requires the levelling instrument to
be set up on the centre-line for each cross-section.

Clearly if pegs are left in the ground at the toe of the slope, they are
immediately lost at the beginning of excavation. Therefore *repere
marks* in the form of *batter rails* are erected for each cross-section
following the slope staking or other form of setting out. For a
discussion of these and other reperes see Chapter 10.

12.6 Curvature correction

When the centre-line of a road or railway etc. is curved on plan, the
portion between two adjacent cross-sections is no longer a true
prismoid, and a correction is required.

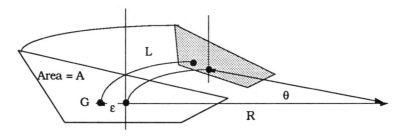

Figure 12.17

Consider first a portion of cutting or embankment of constant, but
unsymmetrical, cross-section area A, having an arcual length L, and
whose centre-line has a constant horizontal radius R, as illustrated
in Figure 12.17.

If the cross-section is unsymmetrical, its centroid G will be situated at a horizontal distance e from the centre-line, where e (referred to as the eccentricity) is regarded as being positive, if G lies on the opposite side of the centre-line from the centre of curvature.

If θ is the angle subtended at the vertical axis at the centre of curvature, by the two vertical planes containing the identical cross-sections at the ends of the solid, then the volume of the solid will be that generated by rotating the cross-section A through an angle θ.

By the theorem of Pappus, the volume of this solid will be the product of the area A and the length of the path of the centroid, i.e.

$$V = A(R + e)\theta$$

$$= AL + \frac{AeL}{R}$$

In this expression, AL is the volume of a prismoid of length L, and the curvature correction is A e L /R, being positive or negative according to whether ε is positive or negative.

If the cross-section is symmetrical, G lies in the centre-line and the correction is zero.

In the general case, the shape of the cross-section will not be constant, so that neither A nor e will be constant, and if the horizontal curve is a spiral, the radius R will not be constant.

As, in practice, the ratio e/R will always be comparatively small, it is usually sufficient to calculate the correction for each cross-section and then use either Simpson's rule, the prismoidal rule, or the end-area rule as appropriate, to determine the volume in the normal manner.

In calculating the value of the correction, it is useful to remember that for the clothoid spiral the radius of curvature is inversely proportional to the distance round the spiral from the origin, and that the product of these two quantities is equal to the LR value for the curve.

Calculation of eccentricity
Because the correction term to be applied to the cross-sectional area A is Ae/R, equations for Ae are derived in this section. If the actual eccentricity itself is required for any purpose, it can always be obtained numerically by dividing Ae by A, which will already have been found.

Three cases only will be considered: the three-level cross-section with unequal side-slopes, the same with equal side-slopes, and the part cut-part fill cross-section. All other cross-sections can be very simply derived from these.

In each of these cases the expressions will be found in terms of the half-breadths, which themselves can be found as before.

It will be remembered that the perpendicular distance of the centroid of a triangle from its base is 1/3 of its height, and that the area is half the base times the height. Refer to Figure 12.18.

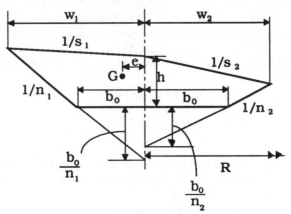

Figure 12.18

Assuming e to be positive when G lies to the right of the centre-line, taking moments about the centre-line we have

$$Ae = (h - \frac{b_o}{n_1}) \frac{w_1^2}{6} - \frac{b_o^3}{6 n_1} + (h - \frac{b_o}{n_2}) \frac{w_2^2}{6} - \frac{b_o^3}{6 n_2}$$

If $n_1 = n_2 = n$ this simplifies to

$$Ae = \frac{1}{6} (h - \frac{b_o}{n}) (w_1 + w_2) (w_1 - w_2)$$

For the part cut-part fill section in Figure 12.19, by the same process, we have

$$Ae_{cut} = (h - \frac{b_o}{n_1}) \frac{w_1^2}{6} - \frac{b_o^3}{6 n_1} + s_2^2 \frac{h^3}{6}$$

$$Ae_{fill} = (h - \frac{b_o}{n_2}) \frac{w_2^2}{6} + \frac{b_o^3}{6 n_2} - s_2^2 \frac{h^3}{6}$$

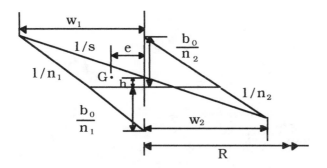

Figure 12.19

12.7 Earthworks from computer models

When the surface of the ground has been modelled by computer algorithms the volumes of earthworks are calculated in a different manner. The surface model, described by a series of triangles, is intersected by the geometrically primitive surfaces of the formation, usually consisting of planes and lines. A typical problem is illustrated in Figure 12.20. The line ST is a part of the series of many straight lines marking where the embankment meets the existing ground.

Figure 12.20 shows ABC, a typical triangular facet of the ground surface, intersected, in the line ST by the side slope plane of the embankment. This side slope is part of the plane PQR, which is defined by the road design. The plane ABC is given by its coordinated points. The task has three objectives.

(a) To calculate the coordinates of S and T.
(b) To find the equation of the line ST, for plotting by computer, and setting-out in the field from coordinates.
(c) To calculate the volume of the solid RABTS which is a portion of the embankment.

The complete task consists of many similar calculations. All data recorded by coordinates in the computer files can then be used to create cross-sections and three-dimensional visualizations etc. as required. These visualizations may be stereoscopic perspective views constructed as indicated in Figure 1.1 of Chapter 1. The whole process is converted to a computer algorithm based on the transformation of the coordinates of the solid. Equations are written for the projecting lines and the coordinated points on the plane of the drawing found as indicated in Chapter 13.

Altenatively, isometric views may be drawn using skew coordinate axes as in Figure 12.20.

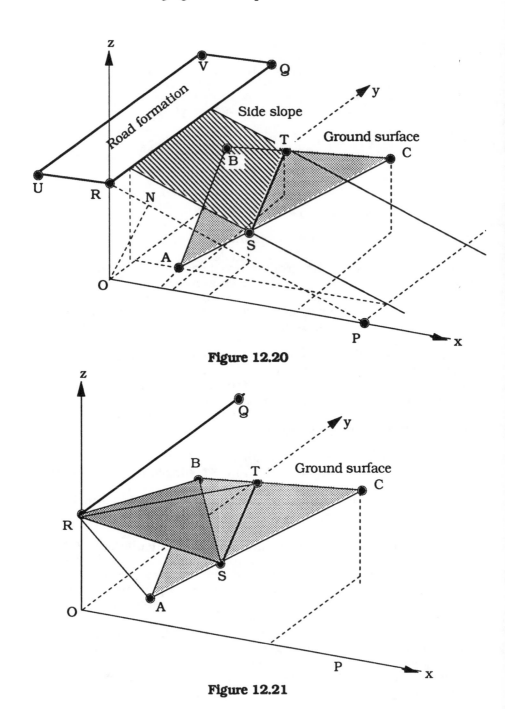

Figure 12.20

Figure 12.21

Example
The procedure is best explained by an example using the data of Table 12.1 in which the coordinates of the ground surface triangle ABC and of the formation plane PQR are listed. See Figures 12.20 and 12.21.

Table 12.1

Point	x	y	z
P	50	0	0
Q	0	100	10
R	0	0	10
A	10	10	0
B	0	50	5
C	70	30	10

Equation of the plane PQR
The first stage is to find the equation of the plane PQR from the coordinates of these points. This equation is written

$$Ax + By + Cz = D$$

where the coefficients A, B, C and D are given by the determinants

$$A = \begin{vmatrix} y_P & z_P & 1 \\ y_Q & z_Q & 1 \\ y_R & z_R & 1 \end{vmatrix} \qquad B = \begin{vmatrix} z_P & x_P & 1 \\ z_Q & x_Q & 1 \\ z_R & x_R & 1 \end{vmatrix}$$

$$C = \begin{vmatrix} x_P & y_P & 1 \\ x_Q & y_Q & 1 \\ x_R & y_R & 1 \end{vmatrix} \qquad D = \begin{vmatrix} x_P & y_P & z_P \\ x_Q & y_Q & z_Q \\ x_R & y_R & z_R \end{vmatrix}$$

The numerical values are:

$$\begin{vmatrix} 0 & 0 & 1 \\ 100 & 10 & 1 \\ 0 & 10 & 1 \end{vmatrix} \qquad \begin{vmatrix} 0 & 50 & 1 \\ 10 & 0 & 1 \\ 10 & 0 & 1 \end{vmatrix}$$

$$\begin{vmatrix} 50 & 0 & 1 \\ 0 & 100 & 1 \\ 0 & 0 & 1 \end{vmatrix} \qquad \begin{vmatrix} 50 & 0 & 0 \\ 0 & 100 & 10 \\ 0 & 0 & 10 \end{vmatrix}$$

From these numerical values we obtain the equation of the plane PQR as

$$1000\,x + 0\,y + 5000\,z = 50000$$

Note that, although this equation can be simplified to

$$x + 5z = 50$$

and normalised to

$$0.1961\,x + 0.9806\,z = 9.8058$$

it has been retained in the form taken within the computer software.

The reader may like to verify, from the section ORP, that the perpendicular distance ON from the origin of coordinates O to the plane, is 9.8058, and that its direction cosines are 0.1961 and 0.9806.

Table 12.2

POINT T	Equations					Soln T
x	1	0	0	-70	0	18.42
y	0	1	0	20	50	44.74
z	0	0	1	-5	5	6.32
plane PQR	1000	0	5000	0	50000	$t = 0.2632$

POINT S	Equations					Soln S
x	1	0	0	-60	10	31.82
y	0	1	0	-20	10	17.27
z	0	0	1	-10	0	3.64
plane PQR	1000	0	5000	0	50000	$t = 0.3636$

Coordinates of T and S

The equations of the line BC are

$$x + (x_B - x_C)t = x_B$$

$$y + (y_B - y_C)t = y_B$$

$$z + (z_B - z_C)t = z_B$$

where t is the ratio BT/BC. Since this line also intersects the plane PQR at T, the coordinates of T also satisfy the equation

$$1000x + 0y + 5000z = 50000$$

The numerical values of the coefficients and their solution are given in Table 12.2 for the two points T and S.

Calculation of volumes

The volume of fill is calculated from the addition of a series of tetrahedra such as RABS, RBTS etc. Taking the example for RABS, the volume V is evaluated from the determinant

$$6V = \begin{vmatrix} x_R & y_R & z_R & 1 \\ x_A & y_A & z_A & 1 \\ x_B & y_B & z_B & 1 \\ x_S & y_S & z_S & 1 \end{vmatrix}$$

The numerical values are given in Table 12.3, giving the V = 2000 cubic units.

Table 12.3

Volume		x	y	z		Vol
RABS	R	0	0	10	1	2000
	A	10	10	0	1	
	B	0	50	5	1	
	S	31.82	17.27	3.64	1	

Equations of line ST

To plot points on the toe of the new embankment, or set them out in the field, we use the equations of the line ST.

$$x + (x_S - x_T)t = x_S$$

$$y + (y_S - y_T)t = y_S$$

$$z \quad + \quad (z_S - z_T)\,t \quad = z_S$$

$$
\begin{aligned}
x \quad &+\ 13.40\,t \ = \ 31.82 \\
y \quad &-\ 27.47\,t \ = \ 17.27 \\
z \quad &-\ 2.68\,t \ = \ 3.64
\end{aligned}
$$

Thus the coordinates of a point on a cross-section perpendicular to the centre-line at $y = 20$ can be found from

$$t = \frac{-2.73}{-27.47} = 0.099\,38$$

giving

$$x = 30.49 \qquad\qquad z = 3.91$$

From these coordinates the bearing and distances from an instrument set-up can be calculated for setting out.

Normally all these computations are carried by menu-driven software mounted on portable computers for use on site.

12.8 Mass-haul diagrams

Mass-haul diagrams are used for route earthwork projects to assist in designing the best profile and in organizing the actual work in the most economical manner. Such diagrams relate only to the longitudinal movement of earth along the route, and take no account of transverse movement of material.

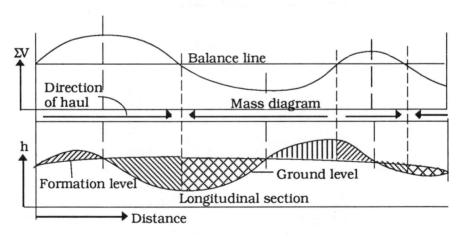

Figure 12.22

The mass-haul diagram is a graph showing the cumulative volume of excavation (as ordinate) plotted against the centre-line chainage (as abscissa), and is usually (although not necessarily) plotted to the same horizontal scale as, and projected up from, the longitudinal section (see Figure 12.22).

The ordinate (aggregate volume) can be regarded as a measure of the volume of earth contained in the bowl of a hypothetical scraper of infinite capacity, as it moves in the direction of increasing chainage, along the route centre-line.

When this imaginary machine is cutting, the volume of earth in the bowl increases, and the greater the depth and/or width of the cut, the greater will be the rate of increase of volume of material in the bowl, and the steeper will be the gradient of the plotted curve. When the machine stops cutting, and starts spreading, the ordinate will cease to increase, and commence to decrease, and a local maximum (zero gradient) will occur on the curve. Similarly, when the machine changes over from filling to cutting, a local minimum will occur on the curve.

From the above, it will be apparent that a positive gradient to the curve indicates cut, and a negative gradient indicates fill, and in fact, since

$$V = \int A dx$$

the gradient of the mass-haul curve at any point will be equal in sign and magnitude (subject to a possible scale factor) to the cross-sectional area of the cutting or embankment.

The shape of the curve for a particular project, i.e. a horizontal line when there is no cut or fill, an inclined straight line when the cross-sectional area is constant, and a curve with a changing gradient when the cross-sectional area is varying, will soon become apparent from the table of cross-sectional areas from which the volumes, and hence the cumulative volume, are calculated.

Although certain indications, i.e. whether it is cut or fill, and hence the sign of the mass-haul curve gradient, the position of changes from cut to fill, and hence the position of local maxima on the curve, etc., may be obtained from the longitudinal section, care must be taken in trying to relate gradients on the curve with the depths of cut and fill on the section. If cuttings and embankments with side slopes are used, the areas of the cross-sections will not be directly proportional to the depths of cut and fill, and if the formation width on the value of the side slopes varies, there will be even less correspondence between the gradients of the mass-haul diagram and the ordinates of the longitudinal section.

Since the ordinates on the mass-haul diagram represent cumulative volumes, it will be further apparent that for any two points on the curve having the same ordinate, the volume in the bowl of the hypothetical machine will be the same, and therefore, the volume that has been dug, and the volume that has been spread between those two chainages will be equal.

Two points on a continuous curve that have the same ordinate must, of course, embrace at least one local maximum or minimum, and the vertical distance between the points and this highest or lowest point will represent this volume which has been dug and deposited again, i.e. is equal to the volume of the embankment between the two chainages. Two points on the curve having the same ordinate will, of course, be indicated by the points at which a horizontal line drawn on the diagram cuts the curve. Any such horizontal line is referred to as a balance line since there is a balance of cut and fill between the chainages of all the points where it cuts the curve (see Figure 12.22).

Since a positive gradient indicates cut, and a negative gradient indicates fill, it should be clear that when the curve lies above the balance line the direction of the movement of the earth is forwards (i.e. increasing chainage) and when the curve lies below the balance line, the direction of movement is backwards. Thus for a continuous balance line, as in Figure 12.22, the points where it cuts the curve also indicate the chainages at which the direction of haul changes.

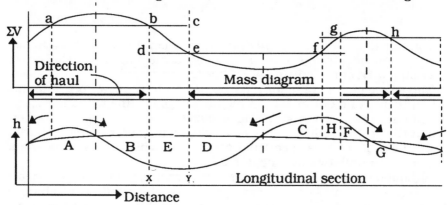

Figure 12.23

If two, or more, balance lines are drawn on the diagram, as shown in Figure 12.23, the earth from the cutting A is moved forwards and just fills the embankment B. Similarly the spoil from cutting C is moved backwards and just fills embankment D. If this scheme is adopted, there is no material available to fill the embankment E between chainages X and Y, and filling will have to be obtained from a borrow pit alongside the route in the vicinity of E, or otherwise imported on to the job. Similarly, as the spoil from cutting F, when run forwards, will be sufficient to fill G, the earth that is excavated from cutting H will have to be dumped alongside the route adjacent to H, or otherwise removed from the job.

Borrowing and wasting
It is customary to refer to these procedures as *borrowing*, when material is brought on to the job, and *wasting*, when material is removed from the job. It will be seen that when the balance line jumps down (as from ab to ef) borrowing must take place, and when it jumps up (as from ef to gh) wasting must occur.

If the borrowed material is brought into the embankment E, transversely, all along the route between the chainages X and Y, no balance line need be drawn for this portion. If, however, the material is imported at one point, the appropriate lines can be drawn. The continuation of the balance line ab to c, and then jumping down to e, would indicate that a volume equal to the ordinate ec was being imported at Y, and since the curve between X and Y lies below the balance line ac the direction of haul would be backwards.

If, however, the line ab jumps down to d, and then continues to e and f, the material is being imported at X and since the curve now lies above the balance line, is being moved forwards. The material could, of course, be brought in at any single point between X and Y, which would be represented by a balance line somewhere between ac and df.

It has been seen that drawing more than one balance line on the diagram introduces the necessity to waste or borrow, which, apart from that required at one end of the job to accommodate any overall imbalance in the cut and fill, will involve an additional amount of excavation and filling, and before this can be justified some further factors will have to be investigated.

The area under the mass-haul curve, the product of volume and distance hauled, is known as the haul H

$$H = \int V \, dx$$

The units of H are *cubic metre-metres* (m^4).

Examples
Two simple examples will demonstrate how the mass-haul curve can be used to find out whether intermediate borrowing or wasting is desirable, and if so, the best positions for it to be carried out.

Referring to the material that has to be imported to fill embankment E in Figure 12.23 , the total volume required is given by the ordinates db or ec. If it is brought in at x and run forwards, the haul will be given by the area bde, enclosed between the curve and the balance line de. If it is brought in at y and run backwards, the haul will be given by the area bce, enclosed between the curve and the balance line bc. As the second of these two areas (due to the shape of the curve) is smaller, importing at y and running backwards is the more economical procedure. It should be noted that the average haul distance for this work can be found (if required) by dividing the haul by the volume, i.e. dividing the area bce by the ordinate ec.

The whole job is represented by the curve in Figure 12.24, which has an overall excess volume of cut over fill equal to the ordinate qs, which will have to be wasted somewhere. If the single balance line pq is used, the direction of haul will be forwards, the whole of the excess will be wasted at the end of the job, and the haul will be given by the area enclosed between the curve and the balance line pq. If, however, the balance line rs is used, the directions of haul will be both backwards and forwards from x, the whole of the excess volume will be wasted at the beginning of the job, and the haul will be given by the sum of the two areas enclosed between the curve and the balance line rs, which is considerably less expensive than if the balance line pq

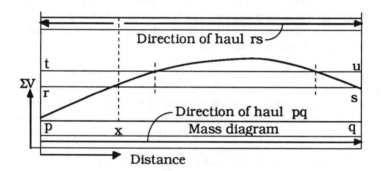

Figure 12.24

were used. It will be realized that the area of the rectangle prsq represents the haul involved if the whole of the excess volume were moved from one end of the job to the other.

Costing
In this project the most economical scheme, from the point of view of haul, is one employing the balance line tu, in which the sum of the areas enclosed between the curve and the balance line is the minimum possible. This however involves increasing the volume to be wasted by the ordinate rt, and introducing an equal volume (su) to be borrowed at the end of the job. To decide whether this will in fact produce a cheaper scheme than that of employing the balance line rs, will require a knowledge of the actual costs of borrowing, wasting, and hauling so that the cost of an increase in borrow and waste can be compared with the saving arising from a decrease in haul.

In looking for the most economical scheme a further factor must be taken into account.

Free haul and overhaul

When the excavated material is loaded into lorries, dumpers etc. for transport, the cost is not directly proportional to the haul (i.e. product, volume x distance), since the cost of loosening, getting out, loading, dumping, spreading and consolidating will be constant per cubic metre regardless of the distance run by the vehicles. When the earth has to be moved short distances, it may not be loaded into vehicles at all, or only into short-run vehicles.

For these reasons it is a common practice to divide the material into two categories.

(a) Material that is to be moved through distances less than an agreed amount known as the free-haul distance (fixed by the type of plant envisaged for the job), such material being charged at a unit price per cubic metre, regardless of the distance moved (provided of course it is less than the free-haul distance). This volume of material is known as the *free-haul volume.*

(b) Material that has to be moved through distances greater than the free-haul distance, and which is charged for at the same rate per cubic metre as the free-haul volume, plus an extra charge at a unit rate per m^4 of haul, for the distances it is moved in excess of the free-haul distance. This volume of material is known as the *overhaul volume* and the product of the volume and the excess distance through which it is moved as the *overhaul.*

Example

Refer to Figure 12.25, on which an additional balance line equal in length to the free-haul distance (say 100 m) has been drawn. Between chainages X and Y the total volume of material to to be moved is given by the ordinate ac (= 1000 m^3) at P pounds sterling per m^3. Of this volume bc (= 300 m^3) is the free-haul volume and ab (= 700 m^3) is the overhaul volume.

The total haul between X and Y is given by the whole area between the curve and the balance line pq. Of this, the area A (= 21 000 m^4) represents the free-haul for the free-haul volume bc, all of which is moved a distance less than 100 m, the average haul distance for this material being 21 000/300 = 70 m.

The area B represents the free-haul for the overhaul volume ab, being the haul for moving it the first 100 m of the journey, and so included for in the rate of P pounds per m^3. The areas C and D (= together, 43 000 m^4) represent the haul for the overhaul volume ab, being moved through a distance beyond the first hundred m, and so the overhaul, which is charged at £Q per m³. The total cost of this part of the job is therefore made up as follows:

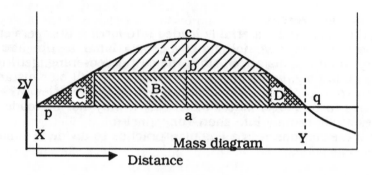

Figure 12.25

Free-haul volume + overhaul volume

$$= 1000 \text{ at } £P \text{ per m}^3$$
$$= 1000 \times £P$$

Overhaul

$$= 43\,000 \; £ \text{ per m}^4 \text{ at } £Q \text{ per m}^4$$
$$= £43\,000 \; Q$$

Total cost

$$= £(1000\,P + 43\,000\,Q)$$

If required, the average distance moved by the overhaul volume is

$$(43\,000 + 70\,000)/700 = 161 \text{ m.}$$

The average overhaul distance
$$= 43\,000/700 = 61 \text{ m, i.e.}$$

$$161 \text{ m minus FHD} = 161 - 100 = 61 \text{ m}$$

and the average distance moved by all the material is

$$(43\,000 + 70\,000 + 21\,000)/1000 = 134 \text{ m}$$

Bulking and consolidation
In practical earthwork projects, one further factor has to be taken into account, which however does not in any way affect the foregoing arguments.

If 1 m³ of solid rock is excavated, and subsequently used as fill, however carefully the consolidation is carried out it will occupy a larger volume, perhaps 1.3 m³ in its new position. This is known as bulking, and the material is said to have a *bulking factor* of 1.3.

If, on the other hand, 1 m³ of clay is excavated and is consolidated carefully at its optimum moisture content, when used as filling, it may only occupy say 0.9 m³, in which case it is said to have a *consolidation factor* of 0.9.

All material, of course, when loosely loaded into a vehicle for transport, will occupy a greater volume than it did on the ground, and this is also referred to as bulking, the bulking factor in this case being somewhat larger than that referred to above. This bulking of material in transit, however, although it affects the constructor in deciding his haulage costs, does not affect the mass-haul curve, for which only volumes in the ground are considered.

The bulking and consolidation factors first referred to, do, of course, affect the mass-haul diagram, which aims at balancing cut and fill. All that is necessary to allow for this phenomenon is, prior to calculating the cumulative volumes from which the curve is plotted, to convert all volumes of fill into *equivalent cut*. This is most conveniently done by dividing all the cross-sectional areas for fill by the bulking or consolidation factor, since to fill a void of 1 m³ with rock fill having a bulking factor of 1.3 will only necessitate 1/1.3 = 0.77 m³ of virgin rock being excavated. Similarly to fill a void of 1 m³ with earth having a consolidating factor of 0.9, it will be necessary to excavate 1/0.9 = 1.11 m³ of virgin soil.

Conclusion
From the foregoing discussions, it can be seen that the mass-haul diagram can be used in two ways.

(a) To assist the designer in determining the most suitable profile and horizontal layout for the job. In this case a diagram is drawn for a trial profile and layout, the curve is analysed, trial balance lines, etc. are drawn and modifications made to the profile and/or lay-out in order (as far as controlling heights, ruling gradients etc. allow) to obtain balances of cut and fill, keep the total volume of earth moved to a minimum and, in particular, keep the overhaul as small as possible. It must be realized that any modification of the profile or lay-out will necessitate the recalculation of the individual and cumulative volumes, and replotting the curve.

(b) To assist the constructor in planning and organizing the actual work. In this case the profile has been settled, and the volumes calculated and plotted. Trial balance lines and the resultant calculation of haul volumes, hauls, haul distances, directions of haul, etc. are then carried out in order to find the most economical sequences of construction, determine the requirements for plant and transport, locate the best positions for borrow pits, spoil heaps, and so on.

The treatment of all complex issues of design, setting-out, and costing is available to the designer and surveyor as suites of computer software to carry out the tasks described in this chapter.

Computer software is also available to create three dimensional models which may be viewed from any vantage point for the illustration of projects to prospective clients or to support arguments in public enquiries. Animated videos of sequential graphics are also used to advantage. Even more realistic effects are achieved using stereoscopy, as illustrated in Chapter 1, Figure 1.1.

13 Industrial and engineering surveying

Chapter Summary

This chapter is concerned with the surveying work which arises in the dimensioning and construction of industrial and engineering projects. Some attention is given to large-scale metrology, theodolite triangulation methods and the processing of data.

Photogrammetric and other vision methods are excluded.

13.1 The objectives of industrial and engineering surveying

The fabrication and assembly of complex engineering structures such as oil rigs and ships, often from parts made elsewhere and brought together for assembly, poses severe problems of dimension control. Tolerances vary from a few micrometres in industrial work to a few millimetres in civil engineering structures.

Drive shafts have to be carefully aligned if engine life is to be adequate; magnets of large particle research accelerometers have to be positioned to provide a sweet path of say 1 mm over distances in excess of 20 km; the antennae of radio telecopes require to be constructed to sub-millimetre tolerances; nuclear reactor boilers, weighing hundreds of tonnes, have to be placed in position to within centimetres, to name but a few applications.

Often this demanding work has to be carried out in a hostile and difficult environment, or to a tight time-scale.

To avoid extra costs, speed may also be required, so that corrective measures may be taken with the minimum delay. This demand for immediate results sometimes rules out some systems of measurement, such as photogrammetry, although it and television systems are now being used to effect. The increasing use of computers on-line has increased the scope for theodolite systems.

A system of measurement may often combine old and new techniques in order to give optimum efficiency and corroborative alternative results.

Many standard surveying techniques are applicable to industrial and engineering surveying. Thus, although all chapters of this book have some relevance, this is especially true of Chapters 11 and 12. Engineering surveying deals with the measurement and construction of typical rectangular and curved features, such as buildings, tanks, tunnels and roads.

It is common practice to base many dimension problems on a numerical coordinate system developed and processed by computer software, and to present results in digital form for future use. Methods of working have become very versatile as a result. It is not possible to describe particular systems here. We deal only with the essential principles which enable the surveyor to apply common sense to a working medium which is dominated by computers and a very special close-knit relationship with the construction industry.

13.2 Basic principles and introduction

Most civil engineering and building construction projects lie at the lower end of the accuracy scale where tolerances of about 5 mm are sufficient. By contrast, industrial or large-scale metrology projects, such as engine shaft alignment, require accuracies of 10 micrometres or better.

Although many principles are common to both industrial and engineering surveying, apart from the tolerances demanded, the working environments differ, as do the instruments required and the terminology used. The *site* and *setting-out* of civil engineering, become the *workpiece* and *building* of large-scale metrology.

For various reasons, metrology requires immediate results, which can only be obtained with direct measurements. These require to be almost free of systematic error, and must give *real-time* results without the need for complex numerical processing. The tooling telescope typifies this principle, in contrast with the standard theodolite. This situation is changing considerably with the advent of on-line processing.

Measurement and positioning
Problems generally take one of two forms:

(1) Measuring or dimensioning existing components.
(2) Positioning components to a designed schedule.

In engineering surveying, structures are set out with reference to fixed points such as bench marks and pegs. The setting out is made by trial and error using surveying equipment of all kinds according to need: from EDM to tapes and off-set squares and levelling.

In large-scale metrology, the key reference points are usually some holes in a *jig* or *fixture* to accept a *tool*, from which the assembly of components is controlled. The *tools* are specially designed fittings within which dimensioned parts must fit. For example a cube may be set at some attitude as defined by the four points A, B, C and D of Figure 13.1, which gives a schematic idea of the layout.

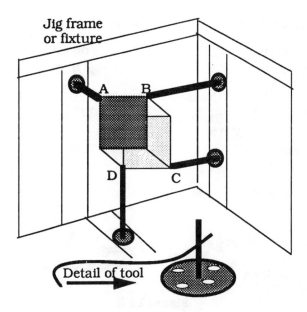

Figure 13.1

The jig is a large stable frame (not all of which is shown), which creates the reference system from which all components are assembled using the tools attached to the jig. These tools, custom made for the job in hand, are fitted to the jig by a unique arrangement of holes and dowels so that the required point, such as A, is correctly positioned eventually to accept the cube.

The component itself is not likely to be a cube, but some machinery with reference points on it. The cube can be thought of as a local cartesian coordinate system describing the component.

The process does not end there, because the "cube" shown may be used to fix further cubes within it. Thus the assembly can consist of placing one reference system within another until all components are in their correct positions within the machine. The process consists of a series a coordinate transformations based on physical reference points on each "cube".

Building tools
To effect these transformations, which in practice may amount to moving large heavy objects through very small distances, special *building tools* are constructed. These tools also need to be located in pre-arranged places, with their movement axes also aligned correctly. A schematic presentation of some building tools is shown in Figure 13.2.

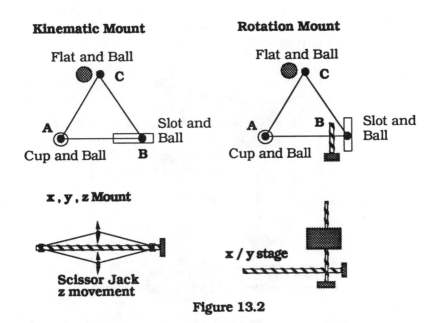

Figure 13.2

Figure 13.2 illustrates some basic concepts about building tools.

The kinematic mount for three balls, mounted below the tool, gives very stable location: at A the cup-and-ball establishes position: the slot and ball at B prevents any rotation about a z axis through A perpendicular to the plane ABC; and the flat-and-ball at C allows rotation about the axis through AB by a fine screw pressing downwards on the base plate.

The rotation version of the mount also allows small rotations about the z axis through A.

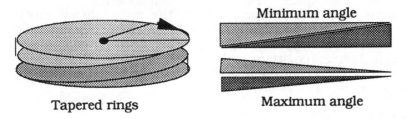

Figure 13.3

The cartesian or (x, y, z) mount consists of a scissor jack for z movement, mounted on a pair of perpendicular arms for (x, y,) movement, all moved by fine screws. Sometimes it is better to mount the (x, y) movement on the z movement.

Apart from the standard three foot-screw system common to most theodolites and other fixtures, a tapered wedge system is commonly used since it can accept great weights.

A practical method of introducing a tilt at a required azimuth is by two tapered rings first set to a required angle, which are rotated together to a calculated orientation to produce the required tilt and final orientation. Calculation of the correct settings is achieved by spherical trigonometry.

In practice, various combinations of these basic tools are used according to need. In some cases the whole construction process has to be programmed by computer to optimize efficiency, and records kept of all assembly stages, for future dismantling.

In civil engineering construction, often a skilled blow from a hammer, or the use of wedges and jacks, is sufficient to move a component into position.

13.3 Techniques and methods

Traditional metrology methods, using speedy uncomplicated optical and mechanical techniques, are still efficient and accurate. They too can be linked to computer-driven digital data systems.

However, surveyors have been able to enter the field of *Large Scale Metrology*, and have developed measurement systems which incorporate electronic theodolites and total stations linked on-line to computers. Lately video cameras have been added to these instruments.

Current practice consists of a series of complex measurement systems working either in "real time" or off–line to control survey work and the construction of components, using all manner of hybrid measurement systems, including photogrammetry.

Although many triangulation systems adopt the basic surveying principles of intersection, resection and polar fixation, described elsewhere in this book, the high tolerances demanded require much greater attention to detail and great care over instrumentation and computation.

13.4 Alignment and positioning

At the outset it is most important to stress that *alignment* should be considered separately from *positioning*.

Often accurate position is not required whereas alignment may be vital. For example the rollers of a paper mill or a conveyor belt need careful relative alignment to reduce unnecessary wear on bearings, but their absolute positions are not so stringently required.

Alignment

Alignment is achieved via the establishment of straight lines, either mechanically or optically. It is practicable to establish a straight line mechanically to a tolerance of 0.01 mm by straight edge or tooling bar up to a length of two or three metres. Beyond this length, nylon fishing line (0.25 mm diam) stretched nearly to breaking point has been use profitably. Offsets from these lines are measured in various ways.

The advantage of mechanical systems is that they do not suffer from the effects of refraction, due to the presence of temperature gradients. At CERN in Geneva many survey control networks have been strengthened by alignments from nylon lines. The use of such lines can be a useful device for carrying bearings from one room to another in the surveys of interiors.

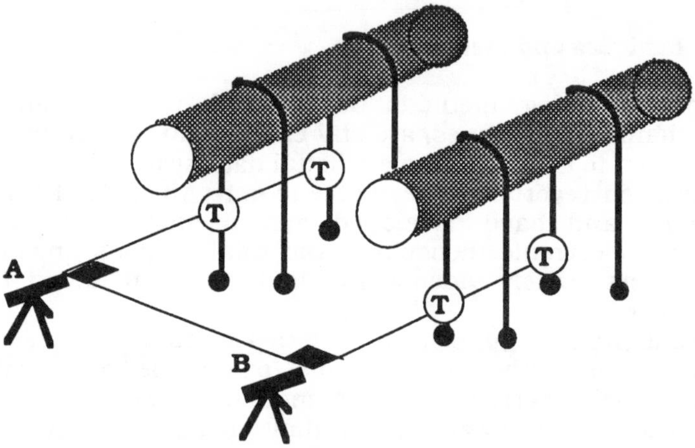

Figure 13.4

Optical alignment uses either normal (incoherent) light viewed in a telecope, or laser (coherent) light viewed indirectly on a screen or sensor. Optical systems suffer from the effects of shimmer and refraction mainly due to temperature gradients close to heating systems. For greatest accuracy a three component system is used involving a light source, a lens or zone plate, and a target screen.

An important stage in any form of works is the transference of the alignment of a component to the line of sight of a telescope. This line of sight may then transfer direction to others as required, either by right angles created by pentagonal prisms, or by any angle set out by theodolite. The first stage may be achieved by co-planing, or the Weisbach triangle, or by autocollimation or autoflexion.

Co-planing or bucking in

The techniques of co-planing or "bucking in" are used frequently for applications such as the alignment of rollers and engine shafts as shown in Figure 13.4.

Two tooling or theodolite telescopes are co-planed at A and B on the targets T or on the plumb lines direct. This means that the directions of the roller axes are transferred to the telescopes.

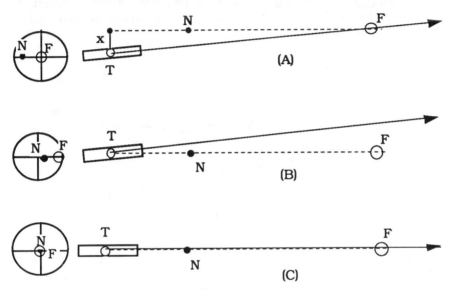

Figure 13.5

The right angles between telescopes are established either by pentagonal prisms or angle measurement. If many rollers have to be aligned, the datum direction AB is established as a semi-permanent feature. This is usually by a collimator or by a plane mirror or planizer or by datum marks or wires. See below for further explanations.

It is essential to change face in theodolite work, because of the alteration to collimation with focus. If telescopes are mutually pointed, the unique pointing routine must be adopted. See below.

Another method of transferring the axis of rotation of a roller is by autocollimation or autoreflection from a mirror attached to the end face of the roller. See later comment.

Technique

The technique of co-planing requires that the telescope can be separately translated sideways, and also rotated about its primary axis.

Refer to Figure 13.5 showing the plan view of operations.

The telescope T has to be pointed so that its optical axis lies in the plane of the wires N (near) and F (far) as shown in Figure 13.5 (C). An initial position is adopted, as shown in Figure 13.5 (A). Both targets will not be in sharp focus at the same time. The line of sight is pointed at F with N to its left side.

The whole instrument is racked through a distance x until target N is a little to the right of the cross-hairs. Rotation of the telescope to point on F should bring N nearly on line. Iteration is required to establish perfect alignment.

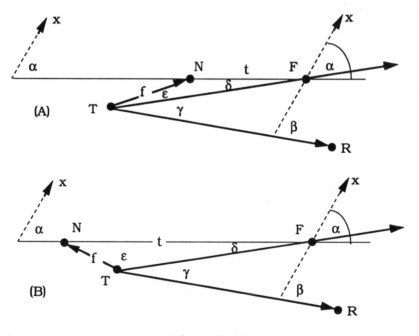

Figure 13.6

It should be remembered that alignment is made on the far target and shift on the near target. Although the amount of the overshoot to be set when sighting N is in proportion to TN/TF, this can only be estimated because both targets cannot be seen simultaneously.

It should also be clear that, if collimation alters with focus, perfect alignment cannot be achieved unless the procedure is reversed. In theodolite work the whole operation of transferring a direction to another line must be repeated on the other face, and a mean taken. In contrast with most theodolite telescopes, tooling telescopes exhibit very little collimation change with focus (2").

The Weisbach triangle

An alternative to co-planing is the Weisbach triangle. Refer to Figure 13.6. In this method, the theodolite T is place slightly off the line NF either, as in case (A), outside NF, or, as in case (B), within it. In the latter case, the points N and F are often reference targets on walls inside a building from or for which a bearing is required.

Assume the direction of the line NF, in Figure 13.6 (A) is known, and its bearing with respect to the x axis is α. We require the bearing β of the line TR, where T is the theodolite and R the reference object.

From the figure

$$\beta = \alpha - \delta + \gamma$$

The technique is to measure the angle γ and calculate the angle δ from other measurements. No special equipment is needed.

The angle δ is obtained from

$$\sin \delta = \frac{f \sin \varepsilon}{t} \tag{1}$$

The angle ε is observed at the same time as γ, and the distances f and t are taped. Provided the angles ε and δ are kept small ($< 1°$ or $179°$) no great accuracy is required for the distances. An error analysis shows this clearly.

Differentiating (1) partially we have

$$\cot \delta \, d \, \delta = \frac{df}{f} - \frac{dt}{t} + \cot \varepsilon \, d\varepsilon$$

Since the angles are small, for error analysis, it is sufficient to use

$$\frac{d \, \delta}{\delta} = \frac{df}{f} - \frac{dt}{t} + \frac{d\varepsilon}{\varepsilon} \tag{2}$$

Suppose $f = 1.500$ m, $t = 3.500$ m and $\varepsilon = 00° \, 40' \, 00'' = 2400''$, then

$$\delta = 00° \, 17' \, 09'' = 1029''$$

Suppose the limit of $d \, \delta$ is $10''$, then we have

$$\frac{1}{102.9} = \frac{df}{1.5} - \frac{dt}{3.5} + \frac{d\varepsilon}{2400}$$

Considering each measurement independently we obtain the error bounds of

$$df = 14.5\,mm \quad dt = 34\,mm \quad d\varepsilon = 23''$$

The angle γ must be observed to the 10" tolerance required of the angle δ. The bearing transfer is the combination of three angles obtained to within 10" tolerance, giving the final error bound as 10 $\sqrt{3}$". It must be remembered that the angle γ may have to be observed several times to meet the required specification, and that the bearing should be carried through the longer of the two directions TF. The inter-target distance NF will depend on the limitations of the site. In tunnel work, this is the diameter of the shaft down which the wires are hung.

DATUM ALIGNMENT

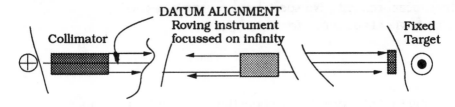

DATUM POSITIONING

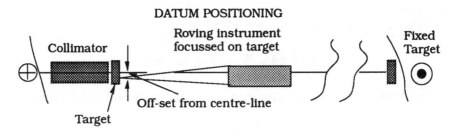

Figure 13.7

13.5 Collimators

A *collimator* is a telescope, focussed at infinity, to bring parallel light to a focus at its cross hairs; or, conversely, a light source placed at the cross hairs will project a parallel beam from the telescope. A collimator set to point at a reference direction may be used throughout an alignment task, as depicted in Figure 13.7.

To check that the collimator remains undisturbed, a distant fixed target is established on the datum line. Because the roving instrument is also focussed at infinity, it can view the collimator cross hairs from zero distance. Notice this arrangement produces alignment only.

Alignment problems do not require the roving instrument to be collinear with the collimator axis, only parallel to it. This can be achieved anywhere within the aperture size of the telescopes.

If, on the other hand, the roving telescope has to be collinear with the reference axis, the lateral off-set can be seen by focussing on a graticule or target marked with concentric dimensioned rings. The telescope may be then be moved laterally to centre. The alignment has to be checked looking at the collimator with infinity focus.

Distance along the line is not always required to any great accuracy, if at all.

Auto-reflection

When we look at ourselves in a mirror we are auto-reflecting. To avoid complexities and double images, a front-silvered plane mirror is used in technical work. Figure 13.8 shows the optical arrangement of auto-reflection.

The top diagram shows the the telescope with its line of sight normal to the surface of the plane mirror M. With the telescope focussed on a target T mounted on the front of the telescope, an image appears to be at T', a distance r behind the mirror. It is central to the cross-hairs.

If the mirror is tilted through an angle I the reflected ray moves through 2I, and is no longer appears centrally, as shown in the middle diagram.

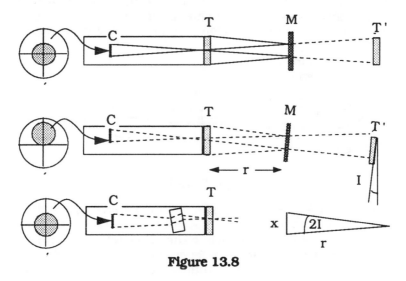

Figure 13.8

If the telescope is fitted with a parallel plate micrometer, the apparent lateral shift x can be corrected and measured. Thus the tilt I, in radians, is given by

$$I = \frac{x}{2r}$$

Since the parallel plate micrometer of a tooling telescope may operate in two orthogonal directions, orthogonal components of tilt can be measured separately. Thus the mirror can be aligned, or its misalignment allowed for in calculations.

In ordinary theodolite work, no front target is needed. The mid-image position is obtained from the mean of four tangential observations of the aperture image.

The process of auto-reflection only ensures that the collimator axis is *parallel* to the mirror normal. If position is also required, the mirror is fitted with scales or dimensioned rings or a grid, so that the *position* of the line of sight can be seen from scale readings when focussing directly on the mirror face.

One important application of this technique is in the placing of an industrial component with the aid of a collimator, mirror and tape. If the collimater is set at the desired attitude, the component can be put in position by trial and error until autoreflection is achieved at the correct distance from the collimator.

Auto-collimation

A similar system involves viewing the *projected* image of the telescope cross-hairs in the mirror, instead of a front target. Greater accuracy is possible at short distances which can be zero because the telescope is focussed on infinity. A typical layout is shown in Figure 13.9.

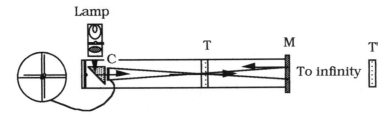

Figure 13.9

Light from the lamp projects an image of the cross-hairs C to the mirror M and back. The observer sees two sets of cross-hairs in the field of view. One set is steady, while the other reflected set moves as a result of vibration and air turbulence. When the telescope is auto-collimated the two sets of cross-hairs coincide. To distinguish the reflected image from the direct image, the hand can be passed over the aperture to obscure the reflected image momentarily.

The application of auto-collimation is similar to auto-reflection, except that it cannot be used with a parallel plate micrometer to measure small offsets or mirror tilts. The reason for this limitation is that a ray of light passing through the plate is also deviated back to its original position on returning.

Reference mirrors or planizer

A reference line may be established by a plane mirror from which a roving telescope auto-reflects or auto-collimates. The mirror may also carry scales for offset measurements. To give a wider scope for viewing, use is made of two orthogonal mirrors, acting like a Porro prism, which may rotate about a horizontal axis. Such devices, shown in the Figure 13.10, are obtainable commercially, or can be constructed in-house.

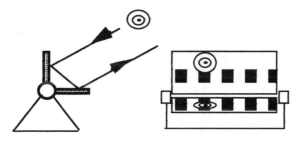

Figure 13.10

The circular graticule or auto-reflected target is viewed after the two reflections which return the incident ray of light parallel to the original direction, provided they all lie in a plane perpendicular to the mirror axis.

The device establishes a series of reference vertical planes whose distances apart can be measured on the scale printed on the mirrors. Again a distant fixed reference target is used to ensure that the mirrors are not disturbed during the measurements.

A plane mirror mounted on the front of a theodolite telescope has many uses, which arise from the fact that the telescope can be pre-set to any required direction in space. Auto-viewing the mirrors from a laser is also convenient.

13.6 Mutual pointing of telescopes

When two theodolite telescopes are pointed at each other, a good practical solution is obtained by sighting a centrally placed target in front of each objective lens in turn. Transparent targets make mutual pointing even quicker.

In metrology, however, greater accuracy at close range is obtained if the fine cross hairs of each instrument graticule are used as targets. A sharp image can be obtained in each of three circumstances:

(1) when both telescopes are focussed on a common intermediate plane between them;
(2) when both are focussed at infinity, i.e. collimated;
(3) when both are focussed on a common plane behind one of them.

The optics of the three cases are illustrated in Figure 13.11 which shows a plan view of the two theodolites whose primary axes of rotation are situated at B and C marked in the middle diagram.

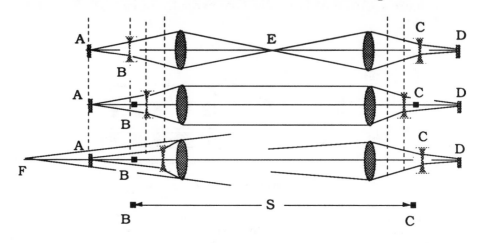

Figure 13.11

In each diagram, points A and D are the cross-hairs of the telescopes whose fixed objective and sliding internal lenses are shown diagrammatically.

To assist in establishing the first case, the beginner should view a piece of bright card at some intermediate point E, not necessarily midway. This means that the separation b between the instruments must be greater than twice their minimum focussing distance, i.e. usually about 3 m. Some slight alteration of the focus of one instrument is needed to obtain a clear view of the other's cross-hairs, illuminated by a bright card held outside the eyepiece. After some practice, the preliminary card at E is not required.

The second case is easy to obtain once the position of infinity focus for each instrument has been noted. This is usually about two turns of the focussing screw back from its maximum position.

The third case may seem somewhat surprising. In most theodolites the internal lens can go beyond the position needed for infinity focus. One telescope is set to this position. The other needs to be focussed on the point F, which is easy to do if the backgound is bright.

If one of the telescopes is kept fixed as a reference collimator, the infinity case is used. But if both telescopes are rotated as in theodolite intersection, the original position cannot be recovered because it is not unique. The mutual pointing merely establishes the lines of sight parallel to each other.

Pointing procedure
Refer to Figure 13.12. A good initial pointing is made by iterative pointing at the centre of each telescope. Consider the usual case with a front focus point.

By front focus point
Consider the case when an intermediate point is used as shown in diagrams (a) and (b). After the lines of sight are made parallel, they are equally inclined to the line BC by an angle β as shown in diagram (a).

Both telescopes are brought to focus on real images of the crosshairs at E_C and E_B. Telescope B is rotated through β to bring it into the required direction BC. (Diagram (b). Both telescopes are then set to infinity focus, and keeping telescope B fixed, telescope C is rotated into direction CB. Usually the process has to be repeated because the initial rotation of B is incorrect.

Notice that the angle γ of diagram (b) can be observed, and if the distances BC = b and BE_C = d are measured, the angle β can be calculated from

$$\beta = \frac{d}{b}\gamma$$

Since a mid point is usually chosen for E_C, $\beta = \frac{1}{2}\gamma$, and no calculation is normally required. The procedure can be very fast with two observers accustomed to the work, and no actual plane is placed at E.

By back focus point
When the telescopes are too close together to focus on an intermediate point, the second case of diagram (c) has to be adopted. In this case, a virtual image of cross-hairs A appears at E_B which lies outside the line BC.

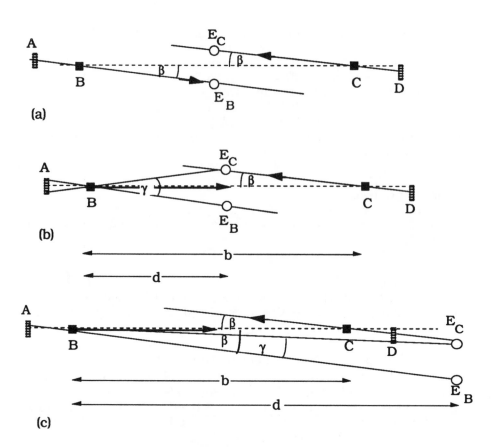

Figure 13.12

In this case a trial and error procedure is more difficult, because the telescope B has to overshoot the image E_C. ($\beta > \gamma$).

Often both theodolites are so close together that one observer can manipulate them both more easily.

13.7 Laser alignment

Similar to the optical telescope which uses incoherent light, the laser beam of monochromatic, or coherent, light is much used for alignment when the atmosphere is stable, and refraction either predictable or absent. Work in such places as tunnels can give poor results because asymmetric temperature gradients cause serious bending of the beam over long distances. Over shorter distances, of say less than 50 m, good work can be done.

As with collimators and mirrors, the alignment of the laser line should always be controlled by a reference target. Best results are obtained from the three point system, consisting of laser source, lens or zone plate, and target. See Figure 13.13.

The roving lens or zone plate is shown in Figure 13.13 in a common application aligning a shaft or bearings at a power station. The long vertical stand is kept vertical by a precise level.

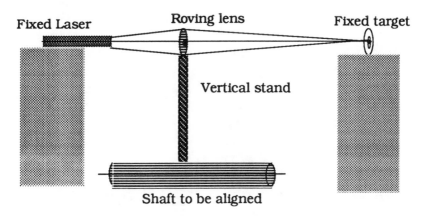

Figure 13.13

13.8 Mechanical alignment by piano-wire or nylon line
When optical conditions are unsuitable or inconvenient, alignment in the horizontal plane may also be done by nylon fishing line under

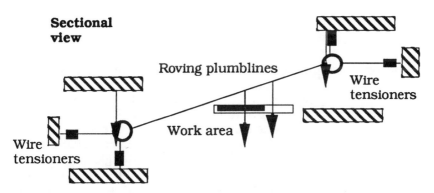

Figure 13.14

tension to near breaking point. Offsets from the line are measured by a travelling microscope or by special electro-optical sensor which uses the line as a reflector.

Piano wires are also much used to align lift shafts and escalators in a typical arrangement such as Figure 13.14. The much stronger piano wire is much more difficult and dangerous to work with than nylon fishing line.

Alignment by inverted plumb line: Rylance method

An extension of the inverted pendulum (see later) to permit the transfer of alignment down a shaft was developed by the late K. Rylance in the UK. The arrangement is as in Figure 13.15.

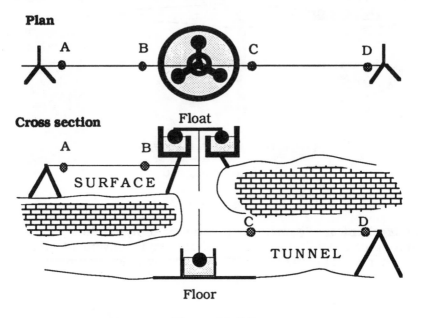

Figure 13.15

Unlike the inverted plumbline, in this arrangement the foot of the downward line is free to move laterally, but so too is the top. Thus the "vertical" vector can move its plan position within the range of the hole in the water container. As the two lines AB and CD pull against each other with a slight tension, all three lines become coplanar after a short while. The surface bearing is transferred underground using the targets A, B, C and D by two Weisbach triangles, one on the surface, the other in the tunnel, not necessarily simultaneously. The Rylance method does not require the wide shaft of the two-wire method, but only a small pipe sufficient to carry the slightly inclined downward line lying in the common vertical plane of the system. Vertical drops of up to 50 m give a transfer accuracy of of 3". The method is cheap and very positive.

13.9 Bearing transfer by gyro-theodolite

If a north seeking gyroscope is attached to a theodolite, the direction of astronomical north can be determined to an accuracy of about 3" from a series of careful observations.

One version of the gyro-theodolite incorporates a rapidly spinning wheel W suspended by tape T above a theodolite as indicated in Figure 13.16. A pointer attached to the cage carrying the wheel may be read against a scale to which theodolite horizontal circle readings are related when the two units are clamped together.

The rapid spin of the wheel and the spin of the Earth cause the gyro to oscillate about north. This oscillation is observed on the scale. The turning points P of the oscillation are recorded on the scale, as are the corresponding theodolite horizontal circle readings C_P.

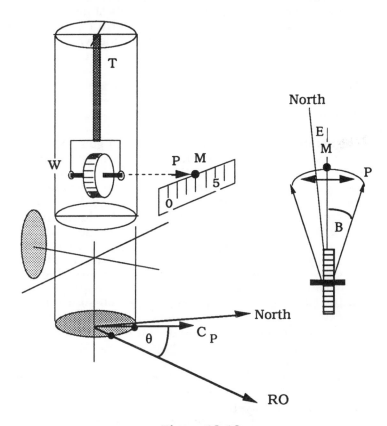

Figure 13.16

From a series of oscillations, with M approximately first to the left and then to the right of north, the estimated values of M are converted to angular corrections E to yield a circle reading on north. This reading is connected to the RO by angle θ, giving the azimuth of the required reference line.

The instrument has to be calibrated for *tape-zero* at every set-up, and for *index* by comparison against a known azimuth, determined astronomically as in Chapter 6.

A complete observation, including calibration, takes about two hours if the best accuracy (3") is required.

If a projection bearing is required, the projection convergence is applied. See Chapter 4.

Although the system has greatest use in mining and tunnelling work, it is occasionally required for industrial applications when wire and laser methods are unsuitable or impractical, for example to calibrate weapon guidance systems in the bowels of a ship.

Because the gyroscope needs a stable platform for operation, the technique does not function well on oil platforms where it would be most useful. Readers are referred to reference (17) for more details.

13.10 Vertical alignment

Mechanical and optical (visual and laser) methods are employed in vertical alignment. It has the disadvantage that a distant reference point cannot usually be incorporated into the on-going measurement process, such as the erection of a tall building. Dependence has to be place on the level sensor. Therefore this should be checked against a level line using a pentagonal prism to deviate the beam horizontally.

The *vertical* or horizontal datum is established by bubble tube, pendulous compensator, liquid surface or simple plumb line. Special clinometers with electronic read-out giving accuracies of 0.05" are available: a mercury pool surface is level to better than 1". Refraction is a special problem with optical methods, while wind affects mechanical systems if not shielded. A long transparent plastic tube makes an efficient cheap screen for a plumb line.

The pentagonal prism is much used to transfer a horizontal line through a right angle to look vertically upwards or downwards. Telescopes can be fitted with a prism which rotates about the line of sight within acceptable tolerances. This can be checked like the optical plummet. See Chapter 3. An alterative arrangement is to mount two fixed telescopes above each other, one looking upwards and the other downwards as in the autoplumb.

To check a system, auto-reflection from a mercury pool can be used to advantage. An upward-looking system, which utilizes a corner cube reflector to view the mercury pool, is checked as in Figure 13.17 (A) and a downward-looking system by the arrangement of (B).

In both systems, the line of sight of the telescope, set horizontally with the aid of the precise levelling mechanism L, is directed vertically by the pentagonal prism P, and reflected back parallel to itself by the horizontal mercury surface M. Any departure from verticality, as in (B), is seen at once in the field of view. This arrangement is used to calibrate and test the system which is used without the mercury pool.

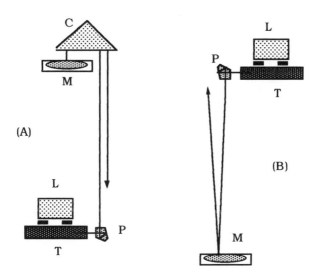

Figure 13.17

Theodolite methods

The theodolite telescope can point vertically to the near-zenith for upward alignment, provided the collimation circle is small. See Figure 13.18. Rotation of the alidade about the primary axis of theodolite T3 causes the line of sight to sweep out a collimation cone into which the target C can be centred. The vertical circle of the theodolite is used for levelling to about 0.5' accuracy.

Downward plumbing can be achieved with a pentagonal prism attachment as in Figure 13.17 (B). Some old instruments were capable of pointing vertically downwards through a hollow primary axis, others were fitted with an auxiliary side telescope for this purpose.

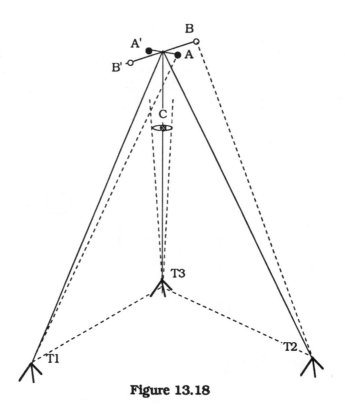

Figure 13.18

Alternatively, if the site permits, a vertical line can be set out by two theodolites, T1 and T2, set at right angles to each other, as indicated in Figure 13.18. The process is repeated on both faces to give positions A, A' and B, B' (greatly exaggerated here), the mean of which gives the correct solution. The instruments are very carefully levelled using the vertical circle or compensator, and the process is repeated to ensure that the result is within tolerance.

Laser systems
The laser has the big advantage that a projected dot can be thrown on a screen for easy detection. The laser suffers from all the errors of any other optical system (collimation, verticality and refraction), and must be calibrated at frequent intervals against the theodolite or mercury pool systems described above.

Modern systems using invisible lasers operate with special detection devices mounted on x-y rails. Systems using photographic or television recording of visible spot movements are also available to monitor building vibrations.

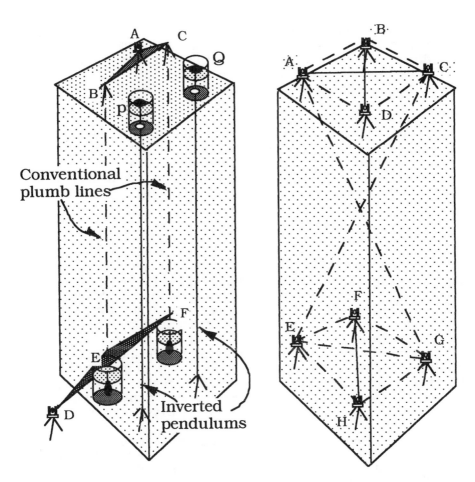

Figure 13.19

13.11 Alignment and positioning

In many applications, both alignment and position are required to be transferred vertically. Two methods are commonly used for vertical drops of up to 100 m. Drops greater than this are handled by a series of shorter repeated sections, or by gyrotheodolite for bearing transfer and plumbing for position.

Two-wire method
The method involves two wires suspended from the ground to a tunnel or from the roof of a building to the ground. The alignment of the wire plane relative to others is determined at top and bottom, by

simultaneous co-planing or Weisbach triangles. The oscillation of the wires is damped by immersion of the bobs in cans of water or oil. The wires need not be suspended down the same hole. Within reason, the further they are apart the greater is the accuracy of alignment. Figure 13.19 shows the general arrangement.

Triangulation method

A three-dimensional triangulation can produce satisfactory results provided no angles are less than 10° and the theodolites are very carefully levelled and several observations made on both faces. As shown in Figure 13.19, a complete transfer of height bearing and length can be made by this method. It is commonly used to check the other direct methods described in this chapter.

13.12 Length measurement

Length is determined directly by scale, or stick micrometer, or steel tape, or invar wires; indirectly by subtense measurement, or by electromagnetic waves. Standards often need to be traced back to a scientific laboratory, especially when components are assembled from various manufacturing sources. An accuracy of from 0.1 mm to 0.001 mm is normally required in large-scale metrology.

The method used depends on the lengths to be measured and the accuracy required. Stick micrometers and clock gauges are excellent for lengths up to two metres; tooling bars made from heavy steel girders may be used up to ten metres; and beyond this steel and invar tapes or wires come into their own. A variety of methods will be needed.

The most accurate method of length measurement is by laser interferometer, see Chapter 7, which can achieve results to 0.0001 mm, provided refraction is also known adequately. The need for a solid straight bar on which to measure distance by interferometer makes it unsuitable other than in a laboratory. A good compromise is to calibrate an invar wire against a laser interferometer for subsequent use in a practical environment, such as by the CERN *Distinvar* system.

Carbon fibre rods, although expensive, make very stable length standards, used as subtense bars with theodolite systems. The ends are registered by a typical cup-and-ball system of targets.

Conventional surveyors' EDM systems are not accurate enough as a rule, although developments will no doubt alter this situation very soon.

Tape standardization

As with EDM, all tapes must be calibrated against a known length standard, which is ultimately traceable to an international standard. This is especially necessary for industrial surveying in which components made in different factories and different countries have to conform in the final construction.

A common procedure is to set up a temporary base against which working tapes are calibrated. This base is measured with a calibrated invar tape used only for this purpose. The following example is typical of the process.

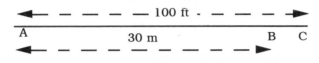

Figure 13. 20

Example
A 100 ft invar band is calibrated at a standards laboratory for use in catenary with 20 lb pull at 68° F (20°C). The accepted length is 100.000 ft. A site base has to be set up to calibrate 30 m steel tapes.

Three collinear marks A, B and C are placed on a solid concrete floor of a workshop, such that AB ≈ 30 m and AC ≈ 100 ft ≈ 30.48 m. The standard tape is used to measure AC on the flat under a tension of 20 lb at temperature 8.5°C giving

$$AC = 99.9660 \text{ ft}$$

The temperature correction is given by

$$(t - t_s) \, C \, l = (8.5 - 20) \times 9 \times 10^{-7} \times 100 = -0.0010 \text{ ft}$$

The coefficient of linear expansion $C = 9 \times 10^{-7}$ for invar.

Because the tape has been standardized in catenary and is used on the flat, a correction has to be applied amounting to

$$\frac{W^2 l^3}{24T^2} = 0.0319 \text{ ft}$$

The weight of the tape (found by weighing the reel with and without the tape) is 1.75 lb. The taped base length is therefore

$$99.9660 - 0.0010 + 0.0319 = 99.9969 \text{ ft} = 30.47906 \text{ m}$$

The short length BC, measured with a precise calibrated scale, is 0.4820 m, giving the site base AB as

$$29.9971 \text{ m}$$

A working tape gave a mean reading of 29.995 for the base. The calibration scale factor F is therefore

$$F = \frac{29.9971}{29.995} = 1.000\,0700$$

All lengths measured by this working tape have to be scaled by F. For example, a length of 246.993 has to be corrected to

$$246.993 \times F = 247.010\,\text{m}$$

13.13 Vertical distance measurement

Vertical distance measurement arises in the construction of buildings, towers, dams etc. Three methods, shown respectively in Figures 13.21 A, B and C, are available for this work: levelling, taping, EDM. Triangulation is also used as a backup to the others.

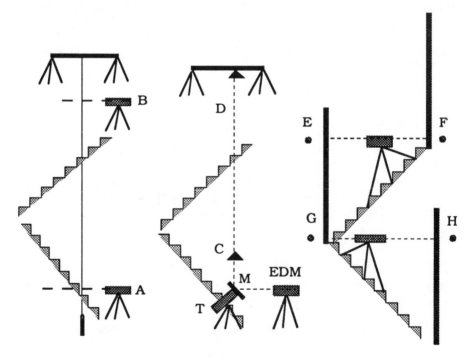

Figure 13.21

Taping
A steel tape is suspended vertically under the tension T at which it is standardized on the flat. One level is kept at A to read the tape

throughout the measurements, whilst the other, B, makes readings at various levels up or down through the building or excavation.

Small corrections may have to be applied for the calibration scale factor of the tape and for the added extension to the different lengths of tape involved at each level.

The calibration factor is proportional to the length s used for the reading the tape whose total length is L.

The extension factor for the stretch of the tape ds is given by

$$ds = s\frac{T - T_s}{AE} + (L - 0.5\ s)\frac{T}{AE}$$

where the field tension is T and the standard tension T_s. If these tensions are equal, the correction is

$$ds = (L - 0.5\ s)\frac{T}{AE}$$

EDM measurement

In the EDM method, the differential distance CD is found by subtracting the two distances folded by the mirror M. The reflector points C and D are tied into others by levelling. The method is particularly practical in large engineering structures and deep shafts. See Chapter 7 for more details.

Levelling

Levelling up a stair-well is a common way of transferring heights from floor to floor in building surveys and construction. The height of collimation method of booking is usually preferred so that other height data for floors and ceilings can be recorded. Often, temporary bench marks, such as E, F, G and H, are set out during the process for further construction use.

13.14 Positioning and targets

The main problem which arises in close range observations and very high accuracies of 10 μm is the establishment of the physical position of a telescope or the intersection of the three axes of a theodolite to this tolerance. Targets too have to be centred to this accuracy. The most satisfactory way to proceed is to adopt a cup-and-ball system. The Rank Taylor Hobson company has established world standard for this system with a sphere of 2.25 inches in diameter to which most tooling telescopes, fittings and mechanical mounts conform.

Tooling theodolites are designed with a two foot-screw levelling system which enables the instrument to pivot about a small ball resting in cup, so that the height of instrument is not altered when levelling. The older Brunson Jig Transit had four foot-screws for this purpose.

The centering of a target within a sphere is achieved by means of a shadow graph in which an image of the target is projected on a screen. When the sphere is rotated slightly the image remains fixed if the target is central. The same procedure can be adopted by viewing the target through a telescope and rotating the sphere in its cup.

Tooling balls are expensive; a cheaper alternative is the ordinary snooker ball, whose centre can be machined out quite easily. The differing colours are helpful too. The cup-and-ball system can also be adapted to accept instruments with a standard thread. The ball is eventually locked in place by a holding ring or cage not shown in Figure 13.22.

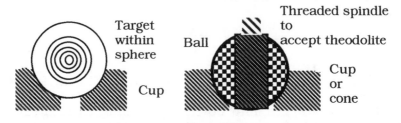

Figure 13.22

To enable small radial offsets from the line of sight to be measured directly, accurate circular rulings are indispensable, such as on the target shown in Figure 13.22, located inside a tooling ball mounted on a cup. The target can be illuminated by light guided into the ball by a rod of Perspex.

Another useful idea is to project a light from a telescope to a large steel ball bearing to view the small reflected light.

Small reflective corner cubes can also be inserted into the ball for use with EDM.

Instrument stands and fixtures

Industrial surveying requires such accuracies that conventional surveying tripods are unsuitable, because they lack rigidity, and inconvenient, because instruments need to be set up in unusual positions. Special stands and attachments are required.

A typical instrument stand has great rigidity, and a head capable of fine x, y, z movements. Stands may also need to be very high, or attached to structures, or hang from roofs. Instruments may also need to be mounted inside pipes and other awkward places. Targets too need special fixtures. For most purposes, the custom-designed tooling equipment made by Rank Taylor Hobson in the UK is happy in most situations. The compatibility and versatility of telescopes, ball and cone mounts, pentagonal prisms, targets, mirrors, stands etc is remarkable.

13.15 Example of alignment and distance in sports

A good example of the proper use of a scale to measure distance but not straightness, combined with a line of sight for alignment, is the arrangement shown in Figure 13.23 to judge the distance jumped by an athlete. The *alignment* is provided optically by the collimator viewed via the pentagonal prism whilst the observer views the footprint through the beam splitter by sliding the telescope along the bar which carries a scale on which the *distance* jumped is recorded. Thus the bar need not be very straight nor aligned well, but the collimator's direction has to be maintained.

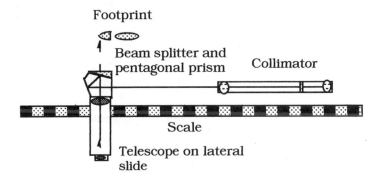

Figure 13.23

A similar arrangment is used in a *tooling bay*, where two identical systems assembled perpendicular to each other can read off, or set out, cartesian coordinates (x, y) directly, as with a map grid. Thus a pattern, such as an aircraft wing shape, can be moulded to shape under the control of the two telescopes. Height is supplied by levelling.

13.16 Theodolite systems

The development of electronic theodolites with on-line real-time computer processing has turned a conventional field surveying system into a valuable and practical addition to industrial surveying instrumentation. Either in its own right, or as a back-up to more conventional systems, the theodolite based provision of coordinated points can deal with the location of fiducial marks on fittings, in the same way as a jig does. In addition it is easier to quantify the accuracies attained. Theodolite methods however have limitations, and need to be used in conjunction with other methods. We shall now discuss this topic in some detail.

The principles of theodolite intersection systems

Figure 13.24 shows the minimum configuration with two theodolites and a scale bar. We will describe this elementary system before extending our ideas to deal with several theodolites, scale bars and field procedures needed to operate computer software.

The two theodolites are mutually pointed for relative orientation, and the terminals of the scale bar are observed. Thus a three-dimensional right handed coordinate system (x, y, z) is set up with one theodolite A as the origin, where the axes are aligned with z to the zenith, and the x axis is in a plane containing the other theodolite's primary axis, also set vertically at B.

It is sufficient to ignore the 10 µm effect of earth curvature up to distances of 10 m, and refraction effects are usually ignored. The scale bar need not be horizontal, nor aligned in any special way.

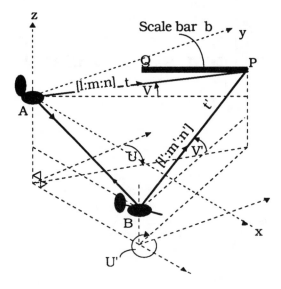

Figure 13.24

At each theodolite A and B record the respective bearings and vertical angles (U, V) and (U', V') to point P. From these calculate the respective direction cosines (l, m, n) for AP and (l', m', n') for BP from

$$
\begin{array}{llll}
l &=& \cos V \sin U \\
m &=& \cos V \cos U \\
n &=& \sin V
\end{array}
\quad \text{and} \quad
\begin{array}{llll}
l' &=& \cos V' \sin U' \\
m' &=& \cos V' \cos U' \\
n' &=& \sin V'
\end{array}
$$

We assume the length AB = 1, and from the slope angle α between A and B, compute the coordinates of B from

$$x_B = \cos \alpha \text{ and } \qquad z_B = \sin \alpha$$

Thus we now have the direction cosines of AP and BP and coordinates of A and B

$$A(0, 0, 0) \quad B(x_B, 0, z_B) \quad (1, m, n) \quad (1', m', n')$$

If we assume that the rays AP and BP intersect, which they will do subject to small errors of pointing, we solve for the coordinates of P as follows.

Let AP = t and BP = t', then the coordinates of P are given by

$$
\begin{array}{lll lll}
x_P & = & 0 + 1t & x_P & = & x_B + 1' t' \\
y_P & = & 0 + mt & y_P & = & 0 + m' t' \\
z_P & = & 0 + nt & z_P & = & z_B + n' t' \qquad (3)
\end{array}
$$

Therefore we have, by eliminating t from the above equations,

$$t' = \frac{m \, x_B}{1 \, m' - m \, 1'}$$

Substitution in these equations gives the required coordinates of P. Similarly we obtain the coordinates of Q, from observations to Q from A and B, and calculate the length of the scale bar based on the initial assumption that AB = 1. The coordinate system then has to be scaled to fit the correct length of the scale bar. Before giving a worked example we consider the matter of quality control on intersection.

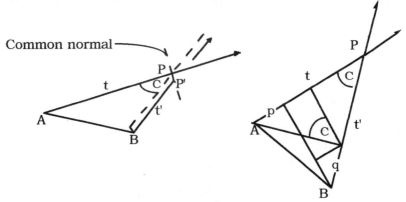

Figure 13.25

Quality control on intersection

The above solution assumes that the two observed rays from A and B intersect at P. In fact they will not do so by a small error of mispointing or by a large amount if there has been a misidentification of the target. To check for this mismatch, a more rigorous mathematical solution is needed as follows.

Table 13.1

Direction	Bearing U	Vertical V
AB	90°00'00"	– 00°08'32"
AP	9°25'41.4"	–5°06'33.0"
AQ	30°59'02.1"	– 4°27'15.0"
BA	270°00'00"	+00°08'30"
BP'	333°25'48.2"	– 4°32'59.5"
BQ'	356°30'33.1"	– 5°05'18.2"

We now consider that the rays do not intersect and that their common normal PP' has to be found (see Figure 13.25). The angle C between the rays AP and BP is given by

$$\cos C = l\,l' + m\,m' + n\,n'$$

Hence we may calculate sin C. The projection p of the ray AB on the ray AP is given by

$$p = l(x_B - x_A) + m(y_B - y_A) + n(z_B - z_A)$$

and the projection q of the ray BA on BP by

$$-q = l'(x_B - x_A) + m'(y_B - y_A) + n'(z_B - z_A)$$

Thus from Figure 13.25

$$AP = t = (p + q \cos C) / \sin^2 C$$

Hence from these equations we have the coordinates of P, and since

$$t' = q + t \cos C$$

from equations (3) we have the coordinates of P'. The length of this common normal PP' is a measure of the quality of the intersection, and may be used to alert the observer to a bad measurement which should be repeated. This is practicable because the theodolites are usually linked directly to a computer in the field.

Table 13.2

Line		Direction cosines					
AP	l	0.163 15986	C	35.870 144 90		
	m	0.982 57229	t	1.527 764 51		
	n	− 0.089 05370	t'	1.683 703 26		
BP'	l'	− 0.445 88030	x	0.249 269 88	x'	0.249 269 88
	m'	0.891 57055	y	1.501 139 07	y'	1.501 139 07
	n'	− 0.079 32660	z	− 0.136 053 00	z'	− 0.136 053 00
		Normal	$\Delta =$	1.3038 − 05		
AQ	l	0.513 24281	C	34.357 234 20		
	m	0.854 72246	t	1.768 714 19		
	n	− 0.077 66160	t'	1.520 564 25		
BQ'	l'	− 0.060 64860	x	0.907 779 85	x'	0.907 779 90
	m'	0.994 21095	y	1.511 759 75	y'	1.511 761 63
	n'	− 0.088 69240	z	− 0.137 361 20	z'	− 0.137 340 10
		Normal	$\Delta =$	2.11020 − 05		

Practical example

Table 13.1 gives the necessary data to set up a coordinate system as described in the above section. The invar scale bar, b, was found to be 2.0000 m long, when calibrated against a laser interferometer,

First compute the value of z_B from tan (− 00° 08' 31"), giving the unscaled coordinates of A and B as

Point	x	y	z
A	0	0	0
B	1.0	0	− 0.002 4776

Table 13.2 gives the salient stages in the calculations described in the above section.

In the calculations below it will be seen that the ray mismatches, Δ, are very small, of the order of a few micrometres. These figures are quite typical of theodolite intersections at this range, provided carefully selected targets are used, and the theodolites are placed on stable stands; not on conventional tripods.

Table 13.3

Mean Coords	P		Q	
	x	0.249 269 86	x	0.907 779 87
	y	1.501 139 66	y	1.511 760 69
	z	− 0.136 046 50	z	− 0.137 350 70
Distance		PQ	=	0.658 596 95
		Scale factor	=	3.036 758 67
Final scaled coordinates	P		Q	
	x	0.756 97	x	2.75671
	y	4.558 60	y	4.59085
	z	− 0.413 14	z	− 0.41710

To assist data handling by computer, the two theodolites are oriented with respect to the axes systems as a preliminary process to identify the four quadrants for the software. On-line information is supplied about quality of intersection and coordination from least squares processing, so that observations can be repeated if not to standard.

Resection and least squares processing
When more than two theodolites are employed, or when the two instruments are not intervisible, a system of space resection is used to link the instruments.

At least three points have to be observed from two theodolite stations, creating observation planes which intersect in the line joining the instruments to yield its direction cosines (two parameters). The orientation of one theodolite has also to be found, thus giving three parameters in all. The system has also to be scaled by observation to a scale bar as in the intersection problem. A figural consideration helps the observer to select satisfactory positions for the points to be observed. The planes must cut at good angles.

Normally at least five points are chosen to give redundancy for checking and statistical assessment of quality.

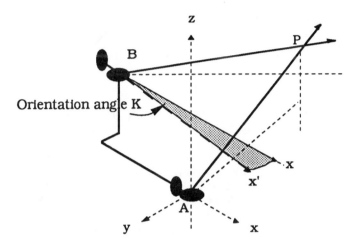

Figure 13.26

The points intersected need not be permanently marked, unless their coordinates are required as part of the survey. Indeed it is common pratice to use only one target mounted on a pole for the resection operation. The target, held at any suitable position, is intersected from both theodolites in turn, then moved to another four points in succession. These control positions are of no value other than to link one theodolite to the other in an arbitrary system of coordinates. The procedure will be explained by an example.

Example
Theodolite positions A and B have to be established relative to each other from observations to the six points shown in Figure 13.27. Points L and R mark the extremities of a 2 m scale bar.

Theodolite A is chosen as the origin of a right handed cartesian system (x,y,z) with an arbitrary orientation, which becomes the fixed reference of the system.

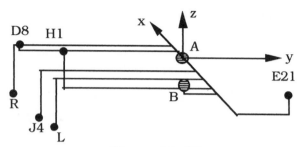

Figure 13.27

(Note, in selecting the axes, we have resorted to the normal convention of surveying so as to preserve the observation equations given elsewhere in this book.)

To simplify the selection of provisional coordinates and to establish approximate orientation at all stations, an obvious direction such as a wall is used as a guide. Each telescope is pointed at the wall before setting the horizontal circles to zero.

This initial direction at point A is taken as the x axis of the local system. The approximate direction at station B is assumed in error by the orientation error K which can be quite large. In our example it is about 2.5°.

A target is held in turn at the points D8 to R of Figure 13.27 and horizontal and zenith angles read. These are listed in Table 13.4. The targets L and R, marking the ends of a subtense bar, whose calibrated length is 1999.454 mm, are used to scale the system, after the relative orientation has been established.

Because the approximate length of line 1–2 is 3100 mm, initial coordinates (in mm) of B are assumed to be :

x	y	z
3100.0	0.0	0.0

Using these coordinates for B and the origin at A, provisional coordinates of the six unknown points are computed. The x and y values are computed by the intersection formulae of Chapter 8 equations (7) and (8).

From these coordinates the approximate distances from point A are calculated, which in turn are used to give the approximate values of the z coordinates at the new stations.

Zenith distances from these new points to station B are calculated.They differ from the observed values giving the O-C terms in the observation equations for zenith distances. The sizes of these terms indicate that the pairs of vectors from stations A and B to new points are not coplanar with the line AB.

All other O-C terms for all directions, and for the zenith distances from A are zero, because the observed values were used to obtain the new coordinates in the first place.

The constraint, keeping the length of AB fixed, is expressed as an observation equation with a high weight of 1000, and the O-C term zero.

The direction equations are expressed in the form of equation (5) Chapter 8, the length equation for AB in the form of equation (11) Chapter 8, and the vertical angle equations in the form expressed in section 2.3 of Chapter 2.

There are 22 parameters and 25 observation equations. There is no space to give the full solution here, which needed six iterations to converge because the geometry is poor. The original data are given in radians in Table 13.4 and the last two solutions for point B are given in Table 13.5.

Table 13.4

Point	U From1	ZD from 1	U From2	ZD from 2
D8	4.735 010 39	1.553 3238 4	5.602 996 56	1.561 294 75
E21	2.928 119 49	1.464 999 70	2.366 079 84	1.371 223 16
H1	4.005 390 04	1.316 870 12	4.904 055 22	1.159 296 56
J4	4.357 825 34	1.763 292 18	5.421 695 64	1.790 004 44
L	4.207 232 52	1.722 964 60	5.247 429 36	1.778 771 51
R	4.796 785 35	1.749 556 44	5.698 068 52	1.715 857 62

This method of theodolite resection, without the need for marked controls, is a powerful tool in industrial work, where flexibility and speed are essential. The data are usually processed on-line, so that the theodolite position can be used for the task in hand, which is usually to measure up components by the method of intersection. On-line processing also allows the data to be assessed statistically to reject bad observations and maintain the necessary quality.

In a very large task, such as a ship survey, several theodolite positions are needed. Checks on the work can be made by taking observations to targets mounted on stretched nylon lines which pass through the whole measurement space. These mechanical checks are preferred to optically aligned targets, or laser beams, which can suffer from the effects of refraction.

Very long scale bars are required in large tasks. Often these are established by spheres seated in sockets on carbon fibre bars measured in a workshop by laser interferometer. The differential method of EDM measurement can also yield sufficient accuracy, as can the Distinvar system using an invar band in catenary.

Table 13.5

	x	y	z	
D8	79.049	– 3493.834	61.068	
E21	– 4606.757	998.634	500.568	
H1	– 2654.918	– 3107.625	1060.767	
J4	– 1153.971	– 3117.081	– 647.845	
L	– 1733.182	– 3134.048	– 549.216	
R	266.125	– 3145.790	– 570.440	
				K
B	– 2993.280	– 747.267	20.543	– 0.049 2844
cn	0.010	– 0.032	0.005	– 1.851E– 05
B	– 2993.270	– 747.300	20.548	– 0.049 3029

13.17 Hidden points

It often happens that a point to be surveyed cannot be seen from either or one of the theodolites used for intersection. Either the theodolites have to be moved to positions where the point can be seen, or an alternative approach is necessary. Although in such cases some form of distance measurement can be used provided its accuracy is sufficient, three alternative theodolite methods are of general interest:

(a) by hidden points bar,
(b) by fitting a sphere,
(c) by reflection in a mirror.

Combinations of these can also be used to check the work. In all cases it is essential that the bars or mirrors are kept in a stable position during the measurement. Hydraulic articulated arms with clamp attachments, although expensive, provide a good solution. An old theodolite used with telescope mounted devices also makes a stable, versatile and controllable holding device.

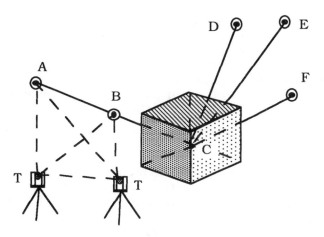

Figure 13.28

Hidden points bar

Refer to Figure 13.28. The points A and B, visible from the two theodolites, are intersected with C held in contact with the hidden point. From the coordinates of A and B and the known distance BC we compute the coordinates of C. Because the length of AB is also known. a partial check on the coordinates of A and B can be made. The formulae to be used are:

$$l = \frac{x_B - x_A}{AB} \qquad m = \frac{y_B - y_A}{AB} \qquad n = \frac{z_B - z_A}{AB}$$

$$x_C = x_B + l\,BC \qquad y_C = y_B + m\,BC \qquad z_C = z_B + n\,BC$$

In practice it is vital that the bar is kept still during the intersection. Various forms of clamping arms are available for this purpose.

If the hidden point is at B, with A and C intersected, greater accuracy is achieved. Sometimes the hidden points bars have to be shaped specially to enable them to be fed into inaccessible parts of machinery. The hidden end is located automatically in position by a cup and ball.

Hidden points by fitting a sphere

A simpler practical alternative is to use a bar, of known length r, with one target. The target is placed in at least three different attitudes with one end located at the hidden point C. From the calculated coordinates of the intersected points, such as D, E and F of Figure 13.28, a sphere of radius r is fitted to the points. The coordinates of its centre are the required coordinates of C. Usually at least five bar positions are used as a check.

The solution is by variation of coordinates and least squares if more than three positions of the bar are observed.

Approximate coordinates of C are estimated. Computer software may take the mean coordinates of the observed points, since this places the provisional centre on the correct side of the sphere. The computed radii to each observed point yield the absolute terms in observation equations, which are of the form:

$$-L\,dx_C - M\,dy_C - N\,dz_C = r - \overset{*}{r} + v$$

L, M, and N are the direction cosines of the typical ray such as CD. See Chapter 2 equation (1). There is no index parameter for r because its length is known. The solution and error estimation follow standard lines explained in Appendix 5.

Experience proves that this method is not only more practical, but also more accurate than the hidden-points bar method. Where there is limited access to the point C, and a suitably well conditioned sphere cannot be described, the other method is preferred.

Fixation from one theodolite
It is possible to fix the position of a hidden point from observations at only one theodolite made to three points on a bar. Normally this would be used as a last resort. However the principle of the method is interesting. Refer to Figure 13. 29.

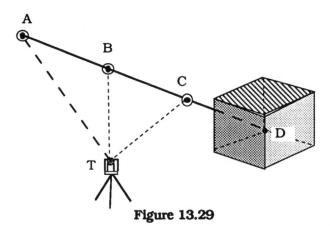

Figure 13.29

Bearings U and vertical angles V are taken to the three points A, B and C at known positions on the bar AD, whose end point D is located at the hidden point D.

From the observed angles at T, the angles at T in the slant plane ACT are calculated from the cosine formula of spherical trigonometry. Next, the base angle BCT is calculated by the resection method of

Chapter 8 equation (6). The sides TA, TB and TC are then calculated and the coordinates of points A, B and C obtained by direction and distance. The solution for D follows as before. Readers will notice that this procedure is a general application of the method of stadia tachymetry.

If possible, the position of the rod will be varied to obtain a check. One method of doing this is to alter the distances between the targets on the rod, or to use a rod with more than the minimum of three targets.

Mirror observations

Sometimes it is convenient to observe behind an object by means of a ray or rays reflected from a plane mirror whose position has been established by intersecting points on its surface. The technique works well provided the mirror is no great distance from the point to be fixed. In Figure 13.30, which illustrates the method of fixation of a point from one theodolite only, the distance MA is shown much too great, simply to make the principle clear. The method is practicable within a tooling bay where the mirror, or mirrors, are permanently in place.

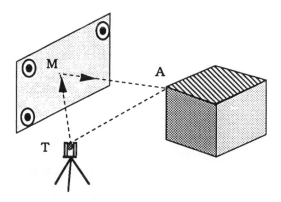

Figure 13.30

Targets on the mirror face are intersected to obtain the equation of the plane in the form:

$$Lx + My + Nz = p$$

where L, M and N are the direction cosines of the normal to the plane from the origin of coordinates, and p its length. The mirror point M is calculated from the intersection of the vector TM with this plane.

If the known direction cosines of the ray TM are l, m and n, the required direction cosines l', m' and n' of the ray MA are calculated from

$$\cos \theta = lL + mM + nN$$

where θ is the half angle between the rays MT and MA. Thus θ is known and we have:

$$\cos \theta = l'L + m'M + n'N$$
$$\cos 2\theta = ll' + mm' + nn'$$

Since the three vectors TM, MA and the mirror normal are coplanar we have a third equation:

$$\begin{vmatrix} l & m & n \\ l' & m' & n' \\ L & M & N \end{vmatrix} = 0$$

The last three equations are solved to give the direction cosines of the ray MA. Finally the point A is computed by intersection from the theodolite T and the reflected ray MA.

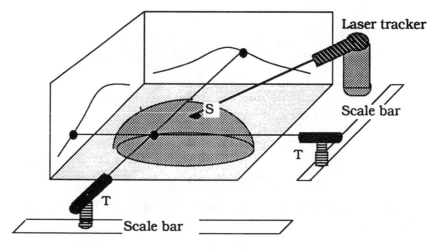

Figure 13.31

13.18 Setting out and building

On completion of the site plan for an engineering survey, the design work has to be completed. Eventually the design has to be implemented in the field. If possible the same control points are used to maintain consistency and design fidelity.

Before beginning the setting out, a field check is essential. In particular it is vital to see that new works, such as road alignments, match up with the existing terrain. For instance sometimes the alignment of an existing road has to be established. This is done by marking several centres of the road over a distance of about 100 m, then eye-balling a best mean fit line to the centre marks. Similar techniques are used for off-sets from walls, directions of power lines etc. These alignments are compared with the setting out data and plans for accuracy.

In the factory or workshop a reference system also has to set up to control construction. This often takes the form of a special tooling bay in which parts are constructed from full scale design drawings mounted on walls. In both industrial and engineering surveys reference markers or *reperes* are commonly used to control day to day work.

The tooling bay
Figure 13.31 shows the general layout of a typical tooling bay in which the sectional scale drawings, mounted on the tooling bay walls, are used to locate positions from the sighting telescopes T.

These are used in turn to control the pattern maker as he works the scale prototype, say of an aircraft wing section. The same technique is used to control jig fixtures. Computer driven telescopes are also used to create a large-scale three-dimensional coordinatograph.

Polar methods using very costly laser controlled angle and distance measurers, or laser trackers, capable of accuracies of a few micrometres over ranges of tens of metres, are also to be found. Distance measurement is by laser interferometer, equipped with servo-controlled pointing, reflecting light from special retrospheres.

Comment
The field of industrial and engineering surveying is subject to rapid change as research into sophisticated automated systems progresses. Much progress has been made towards the design and construction of a fully automated robotic system to carry out surveys of hostile environments such as nuclear reactors.

On-line computer processing of theodolite and EDM data, and the rapid processing of photogrammetric and television data, herald even more advances in this branch of surveying.

Appendix 1
Useful data

Fractions and multiples of units

power of 10	prefix	symbol
− 1	deci	d
− 2	centi	c
− 3	milli	m
− 6	micro	μ
− 9	nano	n
− 12	pico	p
− 15	femto	f
− 18	atto	a
+ 1	deka	da
+ 2	hecto	h
+ 3	kilo	k
+ 6	mega	M
+ 9	giga	G
+12	tera	T

Units of length

1 metre (trad) = length of a quadrant of the Earth x 10^{-7}
$\qquad\qquad\quad$ = 0.5 π R 10^{-7} where R is the radius of the Earth

Radius of the Earth R $\approx$ 6.4 x 10^6 m

1 metre = distance travelled by light in vacuo during a period of 1/ 299 792 458 th of a second

1 British foot = 0.3048 metre (m) exactly (one inch= 254 mm)

1 British statute mile $\quad$ = 5280 feet (ft)
$\qquad\qquad\qquad\qquad\quad$ = 1760 yards (yd)
$\qquad\qquad\qquad\qquad\quad$ = 1609.344 m
$\qquad\qquad\qquad\qquad\quad$ $\approx \frac{8}{5}$ km

1 nautical mile = average distance subtended by a sexagesimal minute of arc on the surface of the Earth
$\qquad\qquad\quad$ = 1.853 18 km

1 ft = 12 inches (in)
1 inch = 25.4 mm exactly
1 yard = 3 ft
1 chain = 22 yards = 66 feet

Area

1 are (a) = 10 square metres (m^2)
1 hectare (ha) = 10^2 ares = 10^4 m^2
$\qquad\qquad$ = the area of the average football pitch

1 acre = 10 square chains ≈ 0.405 ha
$\qquad$ = about half the area of a football pitch

Angle

1 radian (rad) = the angle subtended by the arc of a circle equal
$\qquad\qquad$ to its radius
1 radian = $\frac{180}{\pi}$ sexagesimal degrees (°)
$\qquad$ ≈ 57.3° ≈ 3438' ≈ 206265 "
$\qquad$ = $\frac{200}{\pi}$ centesimal degrees (gon) (g)
2 π radians = one cycle = 360 sexagesimal degrees (°)
$\qquad\qquad$ = 400 centesimal degrees (g)

60 sexagesimal minutes of arc (') = one sexagesimal degree
60 sexagesimal seconds of arc (") = one sexagesimal minute

100 centesimal minutes (c) = one centesimal degree (g)
100 centesimal seconds (cc) = one centesimal minute (c)

one sexagesimal minute ≈ 0.003 rad
one sexagesimal second ≈ 3 centesimal seconds (cc)
$\qquad\qquad$ ≈ 0.000 05 rad

Temperature conversion
Fahrenheit to Celsius

$$C = \frac{5}{9}(F - 32)$$

Equivalent pressure units

The conversion of barometer readings in inches (in) or millimetres (mm) of Mercury to millibars (mb) is given in the following table. In accurate EDM work the barometer has to be calibrated against a mercury (Fortin) or other accurate barometer.

Most altimeters reading in feet (ft) or metres (m) require the scale zero index of 1000 ft (or other) to be subtracted from readings before using the table.

in	mb	mm	mb	ft	mb	m	mb
19	643	490	653	– 1000	1050	– 250	1044
20	677	515	687	0	1013	0	1013
21	711	540	720	1000	976	250	982
22	745	565	753	2000	941	500	953
23	779	590	787	3000	907	750	925
24	813	615	820	4000	874	1000	879
25	847	640	853	5000	842	1250	871
26	880	665	873	6000	812	1500	845
27	914	690	920	7000	783	1750	820
28	948	715	953	8000	754	2000	795
29	982	740	987	9000	727	2500	749
30	1016	765	1020	10 000	701	3000	705
31	1050	790	1053	12 000	651	3500	664

The Greek alphabet

α	alpha	ν	nu
β	beta	ξ	xi
γ	gamma	o	omicron
δ	delta	π	pi
ε	epsilon	ρ	rho
ζ	zeta	σ	sigma
η	eta	τ	tau
θ	theta	υ	upsilon
ι	iota	φ	phi
κ	kappa	χ	chi
λ	lambda	φ	psi
μ	mu	ω	omega

Appendix 2
Spherical trigonometry

Three basic formulae

In this appendix we derive the three basic formulae of spherical trigonometry, and quote several others which may be derived from them. Napier's rules for the relationships between the circular parts of a right-angled spherical triangle are also given.

The formula for the spherical excess of such a triangle is derived, and the appendix concludes with a semi-graphic treatment of the partial differentiation of a spherical triangle.

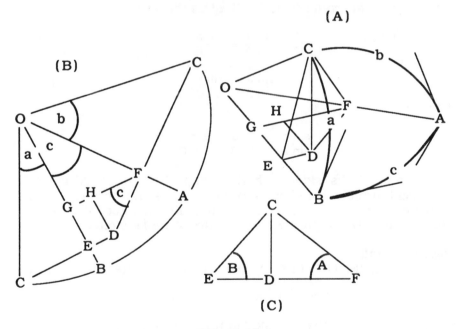

(A)

(B)

(C)

Figure A2.1

The spherical triangle

The triangle ABC formed by three great circles of a sphere, centre O and unit radius, is shown in Figure A2.1(A). If the sphere is *developed* along the radius OC it can be presented as the sector of a unit circle shown in Figure 1(B).The figure CEDF of Figure 1(A), in which CD is perpendicular to the plane OAB, is developed as Figure 1(C).

The *angles* of a spherical triangle are the angles between the tangents to the sphere at each apex point.

The *sides* are the angles subtended by the linear arcs at the centre of the sphere. This definition of sides means that the size of the sphere is of no importance. Thus we can treat a unit sphere as being quite general.

If the reader has difficulty in visualizing the three-dimensional nature of Figure A2.1(A), Figure A2.1 (B) should be redrawn at a larger scale, cut out and folded along OA and OB to create a three-dimensional model, with CD represented by an elastic band.

Sine formula

In triangles OCE, OCF, CED and CFD of Figure A2.1 (A), we have respectively

$$CE = \sin a : \quad CF = \sin b : \quad CD = \sin a \sin B : \quad CD = \sin b \sin A$$

Equating the two expressions for CD gives the sine rule

$$\frac{\sin a}{\sin A} = \frac{\sin b}{\sin B}$$

Similarly

$$\frac{\sin a}{\sin A} = \frac{\sin c}{\sin C}$$

therefore

$$\frac{\sin a}{\sin A} = \frac{\sin b}{\sin B} = \frac{\sin c}{\sin C} \tag{1}$$

This formula links *sides* with their opposite *angles*. Although a convenient formula, care has to be taken with signs, because the sine is positive in both the first and second quadrants. For this reason, a more apparently complicated alternative formula is often preferred.

Cosine formula

Consider triangles FDH, CDF and OCF of Figure A2.1(A) in turn to give

$$HD = DF \sin c = \sin b \cos A \sin c$$

From triangles OCE, OFG and OCF in turn we have also

$$HD = GE = OE - OG = \cos a - \cos b \cos c$$

Equating values of HD gives the cosine formula:

$$\cos a = \cos b \cos c + \sin b \sin c \cos A \tag{2}$$

This is the most useful formula of spherical trigonometry. It relates two sides and their included angle to the side opposite. There is no sign ambiguity between the first and second quadrants.

Cotangent formula

Consider triangles HFD, CDF and OCF of Figure A2.1 (A) to give

$$HF = DF \cos c = \sin b \cos A \cos c$$

From triangles OFG, OCF, CED, and OCE we have

$$HF = GF - GH = GF - ED = \cos b \sin c - \sin a \cos B$$

Equating values of HF gives

$$\sin b \cos A \cos c = \cos b \sin c - \sin a \cos B$$

Dividing by sin b gives

$$\cos A \cos c = \cot b \sin c - \frac{\sin a \cos B}{\sin b}$$

$$\cos A \cos c = \cot b \sin c - \sin A \cot B \qquad (3)$$

This cotangent formula relates four adjacent parts of the triangle.

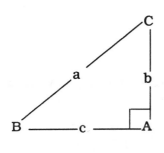
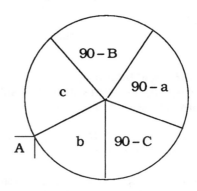

Figure A2.2

Napier's rules for a right-angled triangle

If one angle of a spherical triangle is 90° the various formula simplify considerably. For example from equation (1), if angle A is right , we have

$$\sin a \sin B = \sin b$$

The Scottish mathematician, Napier, showed that thirty formulae for a right angle spherical triangle can be derived by applying the simple rules which bear his name.

Consider the triangle ABC right-angled at A. A circle with five sectors is drawn as shown in Figure A2.2. The right angle A is marked outside the circle as indicated. Into the sectors are written the five remaining parts of the triangle, according to the following scheme.

Parts adjacent to the right angle are written as they stand, but the complements of the others are inserted in correct order as shown.

The required formulae are obtained by applying the rules

sIne of the mIddle part = product of the tAngents of **A**djacent parts
= product of the c**O**sines of the **O**pposite parts

The *middle part* is the parameter in any of the five sectors. This selection defines the parts *adjacent* and *opposite* to it.

For example, if we select 90 – a as the middle part we obtain the formulae

$$\sin(90-a) = \cos a = \tan(90-B)\tan(90-C) = \cot B \cot C$$

and

$$\cos a = \cos b \cos c$$

Or if we select b as the middle part we have, as before,

$$\sin b = \cos(90-B)\cos(90-A) = \sin B \sin A$$

and

$$\sin b = \tan c \cot C$$

Partial differentiation

In the analysis of errors and in least squares problems the various formulae of spherical trigonometry have to be differentiated partially.

As an alternative to the standard method of differentiation, a semi-graphic approach is often much simpler, giving the result in its neatest form.

Consider the spherical triangle ABC of a unit sphere in which the sides AB = c and BC = a are considered fixed. The effect, on the other parts, of a small change of dB in the angle B may be drawn as shown in Figure A2.3 (A). In the error triangle CC'D ,shown in detail, the angle DC'C ≈ C the angle of the spherical triangle ABC.

Treating the spherical triangle like a plane triangle, the effect of change – dB in B is to move C to C ', thus altering A by + dA and b by – db. Applying the sine rule to the right angle triangle BCC ' we have

$$\sin CC' = -\frac{\sin dB \sin a}{\sin 90°} = - \sin dB \sin a$$

but since CC ' and dB are small angles we put

$$CC' = -dB \sin a$$

The equivalent expression for a plane triangle would have been

$$CC' = -a\,dB$$

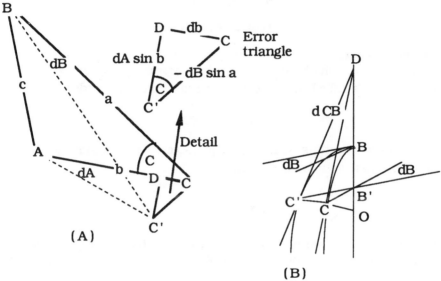

Figure A2.3

Thus we have the rule that we can treat the spherical triangle as plane provided we write sina for a .

Another change has also to be made. In a plane triangle, a rotation of the ray BC by – dB at B changes the direction at CB by + dB. In the spherical triangle, the change d CB in the direction CB has to be amended to

$$d\,CB = dB \cos a$$

In Figure A2.3 (B), since side

$$BC = a$$

we have

$$CB' = \sin a \qquad \text{and} \qquad CD = \tan a$$

Equating values for CC ' we have

$$\tan a \, d\,CB = \sin a \, dB$$

$$d\,CB = dB\cos a$$

Allowing for these two small changes, we may treat the triangle as an equivalent plane triangle in an error analysis by the semi-graphic method. Thus we have the expressions

$$\sin b\,dA = -\sin a\cos C\,dB$$

$$db = \sin a\sin C\,dB$$

$$d\,CB = \cos a\,dB$$

Spherical excess
The amount by which the sum of the angles of a spherical triangle exceeds 180° is called the spherical excess ε: in other words

$$A + B + C = 180° + ε$$

On a sphere of radius R, the spherical excess in seconds of arc of a triangle, whose area is Δ, is given by

$$ε'' = \frac{Δ}{R^2} \times 206\,265 \tag{4}$$

Thus a terrestrial triangle, such as an equilateral triangle of 20 km side, whose area is about 180 km² has a spherical excess of one second.

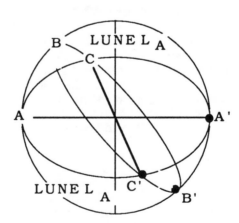

Figure 4

Refer to Figure 4. The surface area of a complete sphere can be thought of as the area swept out by a great circle as it rotates 180°, or π radians, about a diameter. For any intermediate angle A the surface area of the *lune* swept out, L_A, is proportional to A, thus we have

$$\frac{A}{\pi} = \frac{L_A}{4 \pi R^2}$$

The formula for the spherical excess follows by considering the surface areas of the lunes L_A, L_B and L_C formed by the three angles A, B and C. Adding these areas we have

$$\frac{A}{\pi} + \frac{B}{\pi} + \frac{C}{\pi} = \frac{L_A}{4 \pi R^2} + \frac{L_B}{4 \pi R^2} + \frac{L_C}{4 \pi R^2}$$

$$A + B + C = \frac{1}{4R^2} (L_A + L_B + L_C)$$

But $L_A + L_B + L_C$ = the surface area of the sphere + 4 Δ = $4 \pi R^2 + 4 \Delta$ therefore

$$A + B + C = \pi + \frac{\Delta}{R^2}$$

Thus

$$\varepsilon = \frac{\Delta}{R^2}$$

The beginner may care to use a table tennis ball as a model on which to draw lunes to follow the arguments used in the derivation.

Some additional formulae
The following additional formulae of spherical trigonometry are sometimes of value in surveying.

Half angle formulae
If the perimeter of the spherical triangle ABC is denoted by

$$2s = a + b + c$$

we have

$$\sin^2 \tfrac{1}{2}A = \frac{\sin (s - b) \sin (s - c)}{\sin b \sin c} \tag{5}$$

$$\cos^2 \tfrac{1}{2}A = \frac{\sin s \sin (s - a)}{\sin b \sin c} \tag{6}$$

$$\tan^2 \tfrac{1}{2}A = \frac{\sin(s-b)\sin(s-c)}{\sin s \sin(s-a)} \qquad (7)$$

When computing angles from sides, formula (7) is the safest to use in all cases.

Delambre's formulae
The following fomulae are useful to derive meridian convergence.

$$\tan\tfrac{1}{2}(A+B)\,\tan\tfrac{1}{2}C = \frac{\cos\tfrac{1}{2}(a-b)}{\cos\tfrac{1}{2}(a+b)} \qquad (8)$$

$$\tan\tfrac{1}{2}(A-B)\,\tan\tfrac{1}{2}C = \frac{\sin\tfrac{1}{2}(a-b)}{\sin\tfrac{1}{2}(a+b)} \qquad (9)$$

Appendix 3
Matrix notes

Matrix notation
It is general practice to print matrices in heavy type. Vectors, or column matrices, are shown in lower case where possible. When studying a new problem it is wise to treat the dimensions of matrices separately to ensure that the problem is compatible with the rules of matrix algebra. This especially true of multiplication and inversion.

The transpose of a matrix $\mathbf{A}$ is written as $\mathbf{A}^T$ and the inverse as $\mathbf{A}^{-1}$. The reader is reminded of the reversal rule when transposing or inverting a product such as $\mathbf{AB}$ i.e.

$$(\mathbf{AB})^T = \mathbf{B}^T\mathbf{A}^T \qquad (\mathbf{AB})^{-1} = \mathbf{B}^{-1}\mathbf{A}^{-1}$$

Although the solution of a system of linear equations of the form

$$\mathbf{N}\mathbf{x} = \mathbf{b}$$

will be written formally in terms of the inverse of $\mathbf{N}$ as

$$\mathbf{x} = \mathbf{N}^{-1}\mathbf{b}$$

the vector $\mathbf{x}$ is normally obtained by a solution method which does not explicitly form the inverse. The inverse is usually found later and to lower accuracy simply for statistical purposes, and not to evaluate the solution for which far greater accuracy would be needed. See appendix 7.

The inversion of diagonal and orthogonal matrices, $\mathbf{D}$ and $\mathbf{M}$, is simple because the elements of $\mathbf{D}^{-1}$ are the reciprocals of the elements of $\mathbf{D}$, and $\mathbf{M}^{-1} = \mathbf{M}^T$.

A lookout should be kept for any matrix which is the product of a diagonal and an orthogonal matrix, because its factors are easily inverted and the reversal rule applied. Many scaled transformations in surveying and geodesy are of this type.

Example
The reader may verify that the following non-orthogonal matrix $\mathbf{N}$ is the product of a diagonal matrix $\mathbf{D}$ and an orthogonal matrix $\mathbf{M}$. Its effect on a transformed vector $\mathbf{v}$ is to rotate it through 60° and to rescale the ordinate y by a factor of two. (This can be verified also by drawing.) We have

$$\mathbf{D}\mathbf{M} = \mathbf{N}$$

with values

$$\begin{bmatrix} 1 & 0 \\ 0 & 2 \end{bmatrix} \begin{bmatrix} \frac{1}{2} & \frac{\sqrt{3}}{2} \\ -\frac{\sqrt{3}}{2} & \frac{1}{2} \end{bmatrix} = \begin{bmatrix} \frac{1}{2} & \frac{\sqrt{3}}{2} \\ -\sqrt{3} & 1 \end{bmatrix}$$

The vector **v**, given by $\mathbf{v}^T = [\ 4 \ \ 3 \ ]$, is transformed by the orthogonal matrix **M** without changing its length (5); i.e. it rotates it through 60°.

The diagonal matrix D enlarges the ordinate from 4 to 8, and the length of the vector to 8.54.

It is easy to verify, directly or via the product **DM**, that the inverse of **N** is

$$\begin{bmatrix} \frac{1}{2} & -\frac{\sqrt{3}}{4} \\ \frac{\sqrt{3}}{2} & \frac{1}{4} \end{bmatrix}$$

Method of contributions
Matrix products such as $\mathbf{A^TWA}$ can be evaluated by contributions from each row separately. These contributions are accumulated in a store which need only be the size of the normal equations. This procedure can often save a great deal of computer storage, although care not to accumulate arithmetic errors is needed.

Example
Consider the matrix product $\mathbf{A^T A}$ with values

$$\begin{bmatrix} 1 & 3 & 5 \\ 2 & 4 & 6 \end{bmatrix} \begin{bmatrix} 1 & 2 \\ 3 & 4 \\ 5 & 6 \end{bmatrix} = \begin{bmatrix} 35 & 44 \\ 44 & 56 \end{bmatrix}$$

Expressed in terms of the contribution from each individual row of **A** , we obtain the result as the sum of the three matrices

$$\begin{bmatrix} 1 & 2 \\ 2 & 4 \end{bmatrix} + \begin{bmatrix} 9 & 12 \\ 12 & 16 \end{bmatrix} + \begin{bmatrix} 25 & 30 \\ 30 & 36 \end{bmatrix} = \begin{bmatrix} 35 & 44 \\ 44 & 56 \end{bmatrix}$$

Spread sheets
Many spread-sheet systems conveniently use matrix methods for computations. These have limitations when large systems of equations are treated, although partitioning the matrices can extend the scope further.

Large matrices are solved by Cholesky's or other methods. See appendix 7.

Newton's method and the Jacobian matrix

Most least squares problems involve Newton's method of iteration to solve non-linear problems. A matrix consisting of the partial differentials of the functional relationships is formed to effect a solution. The rules for this differentiation are similar to those of ordinary algebra except that the order of matrices is important. Thus we have

if

$$\Omega = \mathbf{A} \mathbf{x} \qquad \frac{d \Omega}{d \mathbf{x}} = \mathbf{A}$$

if

$$\Omega = \mathbf{x}^T \mathbf{A} \qquad \frac{d \Omega}{d \mathbf{x}} = \mathbf{A}^T$$

For both of these results to be valid at the same time the matrix **A** must be square. Thus the differentiation of the bilinear form using the two rules in order gives

$$\Omega = \mathbf{x}^T \mathbf{A} \mathbf{x}$$

$$\frac{d \Omega}{d \mathbf{x}} = \mathbf{x}^T \mathbf{A} + (\mathbf{A} \mathbf{x})^T = \mathbf{x}^T \mathbf{A} + \mathbf{x}^T \mathbf{A}^T$$

$$= \mathbf{x}^T (\mathbf{A} + \mathbf{A}^T)$$

In least squares problems the **A** matrix is symmetrical i.e.

$$\mathbf{A} = \mathbf{A}^T$$

and we have

$$\Omega = 2 \mathbf{x}^T \mathbf{A}$$

That this result is correct may be demonstrated by considering, from first principles, an example of two variables, x and y, Suppose we have the matrix expression

$$E = [\, x \, y \,] \begin{bmatrix} 1 & 2 \\ 3 & 4 \end{bmatrix} \begin{bmatrix} x \\ y \end{bmatrix}$$

Multiplying this out in full gives

$$E = x^2 + 5xy + 4y^2$$

$$\frac{dE}{dx} = 2x + 5y \qquad \frac{dE}{dy} = 5x + 8y$$

writing these results as a **row** matrix **dE** we have

$$dE = \left[\frac{dE}{dx} \frac{dE}{dy} \right] = [x\ y] \begin{bmatrix} 2 & 5 \\ 5 & 8 \end{bmatrix}$$

Thus

$$dE = x^T (A + A^T)$$

$$= x^T M \quad \text{say.}$$

The matrix **M** will always be symmetrical.

Appendix 4
Ellipse and ellipsoid

Introduction
This appendix has been included because the ellipse and ellipsoid have important properties, not readily found in the standard mathematical texts, which are much used in land surveying, geodesy, engineering surveying and error theory.

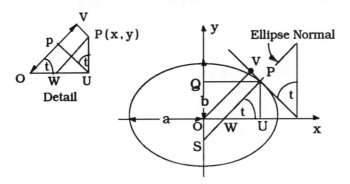

Figure A4.1

A4.1 Basic equations

Figure A4.1 shows the normal drawn to the ellipse at point P(x, y) making an angle t with the positive direction of the x axis, and the pedal distance p = OV parallel to this normal meeting the tangent to the ellipse at P in the point V. The pedal curve to the ellipse takes a butterfly wing shape as shown in Figure A4.4.

If the equation of the ellipse is

$$\frac{x^2}{a^2} + \frac{y^2}{b^2} = 1 \tag{1}$$

the pedal distance p = OV is given by

$$p^2 = a^2 \cos^2 t + b^2 \sin^2 t \tag{2}$$

$$p^2 = \frac{1}{2}(a^2 + b^2) + \frac{1}{2}(a^2 - b^2)\cos 2t \tag{2a}$$

Equation (2) is obtained from (1) by differentiating to obtain dy/dx which is equated to tan(90 + t), thus obtaining

$$\frac{b^2\, x}{a^2\, y} = \frac{\cos t}{\sin t}$$

$$\frac{x}{y} = \frac{a^2 \cos t}{b^2 \sin t} = \frac{K\, a^2 \cos t}{K\, b^2 \sin t} \tag{3}$$

$$x = K\, a^2\, \cos t \tag{4}$$

$$y = K\, b^2\, \sin t \tag{5}$$

Substituting for x and y from (4) and (5) in (1) we find

$$\frac{1}{K^2} = a^2\, \cos^2 t + b^2\, \sin^2 t \tag{6}$$

Now by inspection of the detail inset of Figure A4.1 we have

$$p = x \cos t + y \sin t \tag{7}$$

therefore

$$p = K\, a^2 \cos^2 t + K\, b^2\, \sin^2 t \tag{8}$$

Thus from (6) and (8) we have

$$p = \frac{1}{K}$$

$$p^2 = \frac{1}{K^2} = a^2\, \cos^2 t + b^2\, \sin^2 t \tag{2}$$

Which is equation (2).
Equations (4) and (5) may be written in the convenient form

$$x = \frac{a^2 \cos t}{p} \tag{9}$$

$$y = \frac{b^2 \sin t}{p} \tag{10}$$

Thus we may evaluate the other important dimensions

$$PS = x \sec t = \frac{a^2}{p} \tag{11}$$

$$PW = y \operatorname{cosec} t = \frac{b^2}{p} \tag{12}$$

$$WS = PS - PW = \frac{a^2 - b^2}{p} \tag{13}$$

The useful alternative expression for p of equation (2a) is obtained by the substitutions

$$\cos^2 t = \tfrac{1}{2}(1 + \cos 2t) \qquad\qquad \sin^2 t = \tfrac{1}{2}(1 - \cos 2t)$$

giving

$$p^2 = a^2 \cos^2 t + b^2 \sin^2 t \tag{2}$$

$$p^2 = \tfrac{1}{2}(a^2 + b^2) + \tfrac{1}{2}(a^2 - b^2) \cos 2t \tag{2a}$$

A4.2 Useful properties of ellipse

The area contained within the pedal curve is given by

$$\text{Area} = \tfrac{1}{2}\pi(a^2 + b^2) \tag{14}$$

The area of the ellipse is

$$\pi\, a\, b$$

In Figure A4.2 quantities and dimensions, useful to the surveyor, are identified. No derivations are given to save space. The reader may care to derive them or may refer to a standard mathematical text.

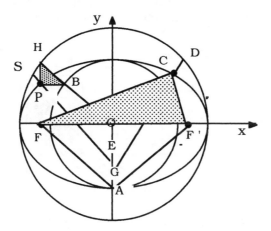

Figure A4.2

We can identify the following :

(a) The auxiliary circles to the ellipse with radii a and b and centre O.

(b) The foci F and F ' fixed from A such that

$$AF = AF' = a \quad \text{and} \quad OF = OF' = ae$$

(c) Point C (x, y) on the ellipse such that

$$FC = a + ex \quad \text{and} \quad F'C = a - ex.$$

(d) Point P on the ellipse. A radius OBH to both auxiliary circles is drawn. Lines through B and H parallel to the x and y axes respectively meet at P. SPE is the normal to the ellipse at P. This gives a graphical way of finding points like P through which to sketch the ellipse.

Some further properties of value to geodesy concerning an ellipsoid of revolution will be found in Chapter 4.

A4.3 The error ellipse and its pedal curve

A particularly useful application of an ellipse and its pedal curve is to display some results of an error analysis in two dimensions. The pedal curve which shows graphically the size and direction of an error function in the vicinity of a point is useful in presenting results to clients, and in design studies, especially of networks. In three dimensions, the error ellipsoid and its pedal surface are also of value.

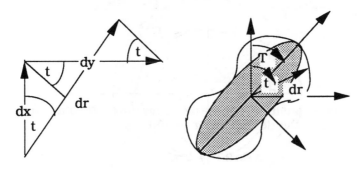

Figure A4.3   **Figure A4.4**

Refer to Figure A4.3 which shows displacements dx and dy and their compounded effect dr at a bearing t. As will be seen shortly these displacements will be represented by variances derived from the dispersion matrix obtained in a least squares analysis.

By inspection from the figure we have

$$dr = dx \cos t + dy \sin t$$

therefore

$$dr^2 = dx^2 \cos^2 t + dy^2 \sin^2 t + 2 \, dx \, dy \sin t \cos t \qquad (15)$$

Taking expectations

$$\sigma_r^2 = \sigma_x^2 \cos^2 t + \sigma_y^2 \sin^2 t + 2 \sigma_{xy} \sin t \cos t \qquad (16)$$

Which may be recast as

$$2\sigma_r^2 = \sigma_x^2 + \sigma_y^2 + (\sigma_x^2 - \sigma_y^2) \cos 2t + 2 \sigma_{xy} \sin 2t \qquad (17)$$

It is clear from the form of (17) with reference to equation (2a) that the variance at any direction t is the square of the pedal distance to an ellipse. We now establish the orientation and axes of this ellipse by examining the turning values of this variance.

Differentiating (17) with respect to t we have, putting $V = 2 \sigma_r^2$.

$$\frac{dV}{dt} = -2(\sigma_x^2 - \sigma_y^2) \sin 2t + 4 \sigma_{xy} \cos 2t \qquad (18)$$

The turning values are otained when $t = T$, and $\frac{dV}{dt} = 0$, i.e. when

$$0 = -2(\sigma_x^2 - \sigma_y^2) \sin 2T + 4 \sigma_{xy} \cos 2T$$

$$\frac{\sin 2T}{\cos 2T} = \frac{2 \sigma_{xy}}{\sigma_x^2 - \sigma_y^2}$$

$$\frac{\Delta \sin 2T}{\Delta \cos 2T} = \frac{2 \sigma_{xy}}{\sigma_x^2 - \sigma_y^2}$$

where Δ is a positive constant. Therefore we may separate the numerator and the denominator as follows

$$\Delta \sin 2T = 2 \sigma_{xy} \qquad (19)$$
$$\Delta \cos 2T = \sigma_x^2 - \sigma_y^2 \qquad (20)$$

Substituting from (19) and (20) in (17) we have, after some rearranging,

$$F = 2\sigma_r^2 = \sigma_x^2 + \sigma_y^2 + \Delta \cos 2(t-T) \qquad (21)$$

F is a maximum and a minimum respectively when

$$\cos 2 \, (\, t - T \,) = + 1 \text{ and } - 1$$

That is σ_r has maximum and minimum values given by

$$F_{max} \ = \ 2 \, \sigma_{max}{}^2 \ = \ \sigma_x{}^2 + \sigma_y{}^2 \ + \ \Delta \tag{22}$$

$$F_{min} \ = \ 2 \, \sigma_{min}{}^2 \ = \ \sigma_x{}^2 + \sigma_y{}^2 \ - \ \Delta \tag{23}$$

Adding and subtracting (22) and (23) gives

$$\sigma_{max}{}^2 + \sigma_{min}{}^2 \ = \ \sigma_x{}^2 + \sigma_y{}^2 \tag{24}$$

$$\sigma_{max}{}^2 - \sigma_{min}{}^2 \ = \ \Delta \tag{25}$$

Thus (21) may be expressed in the final form

$$F \ = \ 2 \, \sigma_r{}^2 \ = \ \sigma_{max}{}^2 + \sigma_{min}{}^2 + (\sigma_{max}{}^2 - \sigma_{min}{}^2) \ \cos 2 \, (\, t - T \,) \tag{26}$$

This is the equation of the pedal curve to the ellipse whose semi axes are respectively σ_{max} and σ_{min} with its axes oriented with respect to the original axes by the bearing T.

In practice, to avoid potential problems with zero values, the computational procedure is to obtain Δ from equations (19) and (20) using the positive root of

$$\Delta^2 \ = \ (\ \sigma_x{}^2 - \sigma_y{}^2 \ )^2 \ + \ (\, 2 \ \sigma_{xy} \ )^2$$

and then T from either (19) or (20) or from an algorithm using ATAN2 to identify the correct quadrant for T.

The error ellipse may then be drawn together with its pedal curve as in Figure A4.4. The actual plotting needs care if an accurate result is needed: see Chapter 8.2 for a worked example of an error ellipse.

A4.4 The error ellipsoid

When the above concepts of the error ellipse are extended into three dimensions, a different approach is required. We repeat some of the above derivations by this different approach as a preliminary to the ellipsoid treatment.

Consider equation (16) again:

$$\sigma_r{}^2 \ = \ \sigma_x{}^2 \cos^2 t + \sigma_y{}^2 \sin^2 t + 2 \, \sigma_{xy} \, \sin t \cos t \tag{16}$$

Now we can write

$$\cos t = 1 \quad \text{and} \quad \sin t = m$$

and equation (16) becomes

$$\sigma_r^2 = 1^2 \, \sigma_x^2 + m^2 \, \sigma_y^2 + 2 \, l \, m \, \sigma_{xy} \tag{16a}$$

The quantities 1 and m are the *direction cosines* of the vector dr with respect to the x and y axes respectively, and these are subject to the constraint

$$1^2 + m^2 - 1 = 0$$

Hence the turning values of the variance are obtained by setting the differentials $\frac{dV}{dl}$ and $\frac{dV}{dm}$ to zero, where V is now the function

$$V = \sigma_r^2 = 1^2 \sigma_x^2 + m^2 \sigma_y^2 + 2 \, l \, m \, \sigma_{xy} - k(1^2 + m^2 - 1) \tag{27}$$

and k is a Lagrangian multiplier. Thus we have

$$\frac{dV}{dl} = 0 = 2 \, l \, \sigma_x^2 + 2 \, m \, \sigma_{xy} - 2 \, lk$$

$$\frac{dV}{dm} = 0 = 2 \, m \, \sigma_y^2 + 2 \, l \, \sigma_{xy} - 2 \, mk$$

which may be written in matrix form as

$$\begin{bmatrix} \sigma_x^2 & \sigma_{xy} \\ \sigma_{xy} & \sigma_y^2 \end{bmatrix} \begin{bmatrix} l \\ m \end{bmatrix} = k \begin{bmatrix} l \\ m \end{bmatrix} \tag{28}$$

This may written in the form

$$\mathbf{A}\mathbf{x} = \lambda \mathbf{I}\mathbf{x}$$

That is a standard eigenvalue problem, in which **x** is the eigenvector and λ an eigenvalue. Subtracting, we obtain the homogeneous equations

$$\begin{bmatrix} \sigma_x^2 - k & \sigma_{xy} \\ \sigma_{xy} & \sigma_y^2 - k \end{bmatrix} \begin{bmatrix} l \\ m \end{bmatrix} = \begin{bmatrix} 0 \\ 0 \end{bmatrix} \tag{29}$$

They have solutions, other than trivial, if the determinant of $(\mathbf{A} - \lambda \mathbf{I})$ is zero; which leads to the quadratic equation in k, called the characteristic equation of the dispersion matrix **A,** as follows

$$k^2 - (\sigma_x^2 + \sigma_y^2) k + (\sigma_x^2 \, \sigma_y^2 - \sigma_{xy}^2) = 0 \tag{30}$$

$$k^2 - (\sigma_x{}^2 + \sigma_y{}^2)k + (\sigma_x{}^2 \sigma_y{}^2 - \sigma_{xy}{}^2) = 0 \qquad (30)$$

Notice that the coefficient of k and the absolute term are respectively minus the *trace* and plus the *determinant* of **A**.

The solution of this quadratic gives the two values of k which, the reader may verify, are the maximum and minimum values of the variance in dr found by the previous method, that is

$$k_1 = \sigma_{max}{}^2 \quad \text{and} \quad k_2 = \sigma_{min}{}^2$$

Substituting for k_1 in equations (29) we obtain the ratio

$$\frac{l}{m} = B \text{ say.}$$

Then

$$l = Bm$$

But

$$l^2 + m^2 - 1 = 0$$

therefore

$$m^2 = \frac{1}{1 + B^2}$$

$$m = \frac{1}{\sqrt{(1 + B^2)}} \qquad l = \frac{B}{\sqrt{(1 + B^2)}}$$

These are the normalized forms of the direction cosines l and m.

The mathematically inclined reader may care to verify that the dispersion matrix **A** may be transformed into a diagonal matrix **R** by the following formula

$$\mathbf{R} = \mathbf{M}^T \mathbf{A} \mathbf{M} \qquad (31)$$

where **M** is an orthogonal matrix whose columns are the normalized direction cosines l and m, and the diagonal terms of **R** are the eigenvalues k_1 and k_2. This form is convenient for computer packages dealing with eigenvalue problems. However a direct solution is possible as outlined above.

Error ellipsoids

The required elements of the dispersion matrix are extracted and presented as in equations (33). From the determinant of the A matrix we obtain the error function in a three dimensional problem (x, y, z) expressed in terms of three direction cosines l, m and n such that

$$V = \sigma_r{}^2 = l^2 \sigma_x{}^2 + m^2 \sigma_y{}^2 + n^2 \sigma_z{}^2$$
$$+ 2lm \, \sigma_{xy} + 2ln \, \sigma_{xz} + 2 \, mn \, \sigma_{yz}$$
$$- k(l^2 + m^2 + n^2 - 1)$$

The process of finding three turning values follows identical lines for two dimensions giving the following cubic characteristic equation

$$a k^3 + b k^2 + c k + d = 0$$

where

a = 1
b = − trace of **A**
c = sum of determinants of diagonal minors of **A**
d = − determinant of **A**

A cubic equation may be solved directly by Ferreo's method (see Appendix 7) or iteratively by Newton's method. The three solutions for k are all positive because of the nature of the symmetric dispersion matrix (Positive definite). They equal the squares of the semi-axes (a, b, c) of the error ellipsoid, whose equation is

$$\frac{x^2}{a^2} + \frac{y^2}{b^2} + \frac{z^2}{c^2} = 1$$

i.e.

$$k_1 = a^2 \qquad k_2 = b^2 \qquad k_3 = c^2$$

$$\begin{bmatrix} \sigma_x{}^2 - k & \sigma_{xy} & \sigma_{xz} \\ \sigma_{xy} & \sigma_y{}^2 - k & \sigma_{yz} \\ \sigma_{xz} & \sigma_{yz} & \sigma_z{}^2 - k \end{bmatrix} \begin{bmatrix} 1 \\ m \\ n \end{bmatrix} = \begin{bmatrix} 0 \\ 0 \\ 0 \end{bmatrix} \qquad (33)$$

Substitution of these values for k in turn in equations (33) yields values for the ratios of the direction cosines

$$\frac{l}{m} = B \qquad\qquad \frac{n}{m} = C$$

These are normalised to give

$$l = \frac{1}{\sqrt{(1 + B^2 + C^2)}} \qquad m = \frac{B}{\sqrt{(1 + B^2 + C^2)}} \qquad n = \frac{C}{\sqrt{(1 + B^2 + C^2)}}$$

The three sets of direction cosines give the direction of the semi-axes of the error ellipsoid relative to the coordinate axes.

Example
To simplify the arithmetic consider the untypically simple elements of a dispersion matrix and its characteristic equation (34):

$$\begin{bmatrix} 3 & 1 & -1 \\ 1 & 3 & -1 \\ -1 & -1 & 5 \end{bmatrix} \begin{bmatrix} l \\ m \\ n \end{bmatrix} = \begin{bmatrix} 0 \\ 0 \\ 0 \end{bmatrix} \tag{34}$$

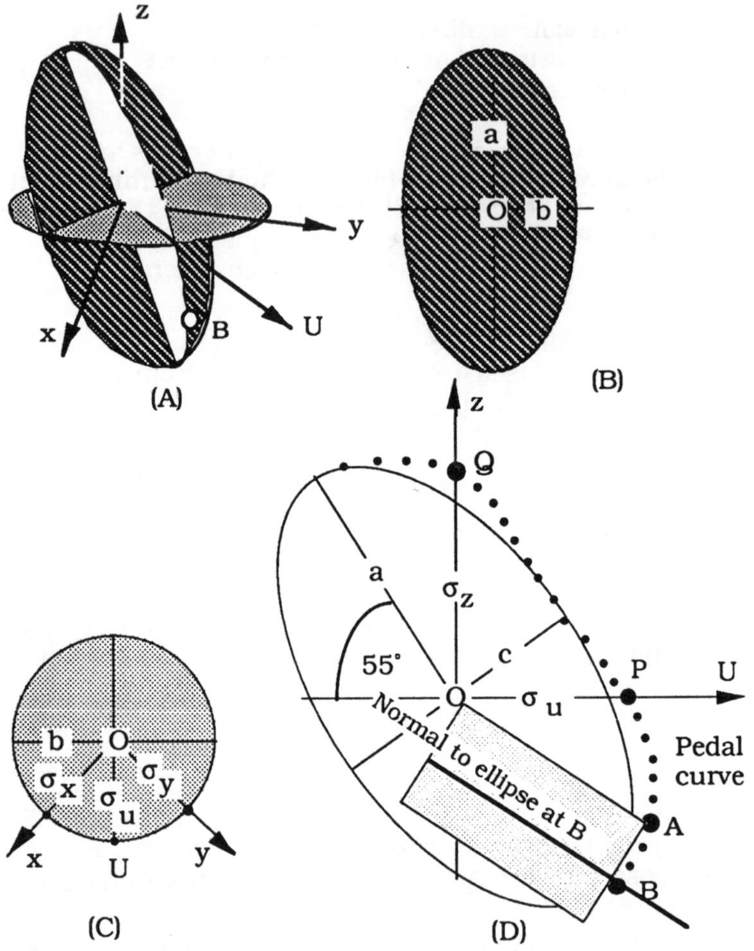

Figure A4.5

Diagram (A) is a general view of the ellipsoid. Each of the shaded diagrams (B), (C) and (D) show the key planes of interest. Diagram (D) is drawn to a larger scale than the others and it alone shows part of the pedal surface in dots. The beginner is encouraged to make a cardboard model of these diagrams to see their relationships clearly

The cubic equation is

$$ak^3 + bk^2 + ck + d = 0$$

where $\qquad a = 1 \qquad\qquad b = -11 \qquad\qquad c = 36 \qquad\qquad d = -36$

giving the solution

$$k_1 = 6 \qquad\qquad k_2 = 3 \qquad\qquad k_3 = 2$$

These give the normalized direction cosines as

$$\frac{1}{\sqrt{6}}, \frac{1}{\sqrt{6}}, -\frac{2}{\sqrt{6}} \qquad\qquad \frac{1}{\sqrt{3}}, \frac{1}{\sqrt{3}}, \frac{1}{\sqrt{3}} \qquad\qquad \frac{1}{\sqrt{2}}, -\frac{1}{\sqrt{2}}, 0$$

The reader should check that these vectors are orthogonal to each other as expected. Thus we can illustrate the error ellipsoid as in Figure A4.5, whose semi-axes are

$$a = \sqrt{6} \qquad\qquad b = \sqrt{3} \qquad\qquad c = \sqrt{2}$$

Diagram (D) shows all detail to plot the error ellipse at bearing U. Its semi-axes a and c are used to draw the ellipse from the auxiliary circles as suggested in A4.2 (d). The pedal curve is sketched in using a set square placed tangential to the ellipse at B and marking A such that OA is perpendicular to AB. It is then seen that

$$OQ = \sigma_z = \sqrt{5} \qquad OP = \sigma_U \approx 1.8$$

This checks the value of σ_z from the original dispersion matrix of equation (34).

Using σ_U and b we can plot the error ellipse and its pedal curve (both nearly circular) in the plane Oxy illustrated in diagram (C). A check is made on the values of

$$\sigma_x = \sqrt{3} = \sigma_y$$

Finally the three-dimensional drawing is depicted in diagram (D).

The whole graphic process can be programmed for computer if required. Stereoscopic viewing of error ellipsoids adds realism to the presentation of results.

A4.5 Elliptical orbits

According to the Keplerian theory of idealised orbits, a satellite moves in an ellipse (i.e. a plane curve) in such a way that the rate of change of the area, swept out by the vector joining the satellite to a focus, is constant. That is, the *areal velocity* is constant.

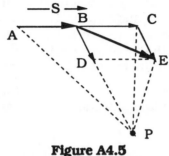

Figure A4.5

Consider a satellite moving with constant speed through space unaffected by gravity. In Figure A4.5 the satellite S moves from A to B in unit time and then from B to C in unit time. Relative to a reference point P the vector PS sweeps out area at a constant rate, because the areas of triangles PAB and PBC are equal.

Consider now the satellite at B moving towards P, under the influence of gravity alone. In unit time it moves through a distance BD. The net effect of both motions on the satellite is to move it vectorially through BE. The area of triangle PBE = area of triangle PCB = area of triangle PAB. Since each of these triangles is swept out in unit time, this means that the areal velocity is constant.

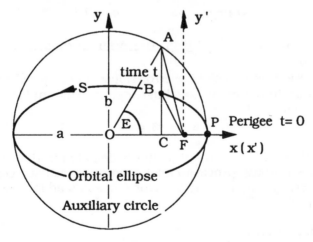

Figure A4.6

Now refer to the Figure A4.6. The satellite S moves round the orbital ellipse with the Earth located at the focus F. The point of closest approach (perigee) is at P. The satellite's period T and the time of observation t ,when it is at B, can be measured.

To derived the cartesian coordinates of the satellite referred to the Earth at F, we require to calculate the angle E corresponding to the time of observation t. Notice that E is the angle AOF where A is the projection of CB on to the auxiliary circle. Once we have obtained E we may obtain the coordinates of B (x', y') relative to the origin at F from

$$x' = a\,(\cos E - e) \tag{35}$$
$$y' = b \sin E \tag{36}$$

The connection between the *eccentric anomaly* E and the time of observation t is given by Kepler's equation

$$M = E - e \sin E \tag{37}$$

where the *mean anomaly*

$$M = \frac{2\pi t}{T}$$

The equation (37) is transcendental since wish to find E from M using

$$E = M + e \sin E \tag{38}$$

A solution can be found by iteration starting from

$$E' = M + e \sin M \tag{39}$$

Kepler's equation
To derive Kepler's key equation consider Figure A4.6. Let the area of figure BFP = Δ. From the areal velocity law we have

$$\frac{M}{\Delta} = \frac{2\pi}{\pi\,ab} \qquad \text{or} \qquad M = \frac{2\Delta}{ab} \tag{40}$$

Remembering the the ellipse is a circle uniformly flattened in the ratio $\frac{b}{a}$ and that OF = ae

$$\Delta = \frac{b}{a}\ \text{area AFP} = \frac{b}{a}\,(\,\text{area OAP} - \text{area OAF}\,)$$

$$= \frac{b}{a}\,(\frac{1}{2}\,a^2 E - \frac{1}{2}\,a^2 e \sin E\,)$$

Substituting for Δ in(40) gives

$$M = E - e \sin E$$

and Keplers equation

$$E = M + e \sin E \tag{38}$$

Cartesian coordinates

Once the position of the satellite (x', y') in the plane of the orbit has been found, its Earth-centred cartesian coordinates (X, Y, Z) oriented

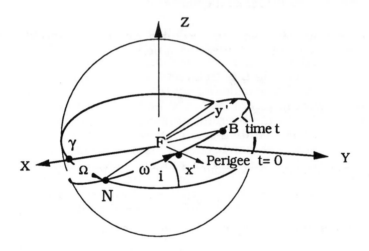

Figure A4.7

on the equator and Greenwich, see Figure A4.7, are given in terms of the *perigee argument* ω, the *inclination* of the orbit i and the *Right Ascension of the ascending node* Ω by the transformation

$$\begin{bmatrix} X \\ Y \\ Z \end{bmatrix} = \begin{bmatrix} \cos\Omega & \sin\Omega & 0 \\ -\sin\Omega & \cos\Omega & 0 \\ 0 & 0 & 1 \end{bmatrix} \begin{bmatrix} 1 & 0 & 0 \\ 0 & \cos i & \sin i \\ 0 & -\sin i & \cos i \end{bmatrix} \begin{bmatrix} \cos\omega & \sin\omega & 0 \\ -\sin\omega & \cos\omega & 0 \\ 0 & 0 & 1 \end{bmatrix} \begin{bmatrix} x' \\ y' \\ 0 \end{bmatrix}$$

The orbital parameters, a, e , Ω, ω, i, and T form the *ephemeris* (plural *ephemerides*) of the satellite. These are usually transmitted to the receiver at the time of observation, or are obtained later in a more precise form.

Appendix 5
General least squares

Observations, conditions and constraints

In this appendix we derive the principal equations used in the least squares process. The formulae are explained in a different way in Chapter 5, section 5.12, in connection with a worked example of fitting a straight line. In this treatment we assume that the reader is familiar with the basic rules of matrix algebra as summarized in Appendix 3. Worked examples are to be found throughout the book where they arise in appropriate applications. The most important results are inscribed within a box for easy location.

The various practical matrix operations may seem complicated, but when considered in the light of modern computer software, and spread-sheet systems, small problems of up to 40 x 40 size are easily solved by direct matrix operations. However, large problems involving several hundreds of equations need a different approach, as described in Appendix 7.

It is important to note that, because matrices **N**, **W**, and their inverses are symmetric, they are unaffected by transposition.

A5.1 Mathematical models and notation

In least squares estimation problems two types of mathematical models are used: functional models and statistical models.

The first type concern the surveying processes which relate measurements, or *observed* parameters, to the required quantities, or *unobserved* parameters. An example of a functional model is an equation which relates the distance between two points to their coordinate differences. The treatment of these models will be written in ordinary type such as

$$s^2 = \Delta x^2 + \Delta y^2$$

or

$$s^2 - \Delta x^2 - \Delta y^2 = 0$$

Partial differentiation of the above equation is also written in ordinary type as follows

$$2s\, ds - 2\Delta x\, d\Delta x - 2\Delta y\, d\Delta y = 0$$

i.e.

$$s \, ds - \Delta x \, d\Delta x - \Delta y \, d\Delta y = 0 \qquad (1)$$

In this equation the observed quantity may be the distance s, with Δx and Δy being the parameters. However, the measured values, say from a map or air photo, could easily be the coordinate differences Δx and Δy, with s, the distance derived from them, being the parameter.

Ultimately we reduce the problem to a standard mathematical form to be treated in matrix terms expressed in heavy type. Then the parameters will be expressed as the vector **x** and the observations as the vector **s**. This change of notation may confuse the beginner. To avoid some possible confusion we prefer to develop the full equations in ordinary algebra before converting to matrix notation.

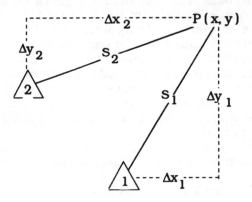

Figure A5.1

Refer to Figure A5.1. Consider the two equations, in two parameters (x, y), which describe the fixation of a new point P by two distances s_1 and s_2 measured from two fixed points 1 and 2. That is

$$s_1^2 = \Delta x_1^2 + \Delta y_1^2$$
$$s_2^2 = \Delta x_2^2 + \Delta y_2^2$$

The observation equations are

$$s_1 \, ds_1 - \Delta x_1 \, dx - \Delta y_1 \, dy = 0$$
$$s_2 \, ds_2 - \Delta x_2 \, dx - \Delta y_2 \, dy = 0$$

In matrix form these equations are written in full as

$$\begin{bmatrix} -\Delta x_1 & -\Delta y_1 \\ -\Delta x_2 & -\Delta y_2 \end{bmatrix} \begin{bmatrix} dx \\ dy \end{bmatrix} + \begin{bmatrix} s_1 & 0 \\ 0 & s_2 \end{bmatrix} \begin{bmatrix} ds_1 \\ ds_2 \end{bmatrix} = \begin{bmatrix} 0 \\ 0 \end{bmatrix}$$

and in condensed matrix notation as

$$\mathbf{Ax + Cs = 0} \tag{2}$$

However, if the *observations* are the *coordinates*, we write

$$\begin{bmatrix} s_1 & 0 \\ 0 & s_2 \end{bmatrix} \begin{bmatrix} ds_1 \\ ds_2 \end{bmatrix} + \begin{bmatrix} -\Delta x_1 & -\Delta y_1 \\ -\Delta x_2 & -\Delta y_2 \end{bmatrix} \begin{bmatrix} dx \\ dy \end{bmatrix} = \begin{bmatrix} 0 \\ 0 \end{bmatrix}$$

To be consistent, keeping the parameters as **x**, the condensed equations *are still written*

$$\mathbf{Ax + Cs = 0}$$

Such changes in notation between the two types of mathematical model can be confusing to the unwary. The reason for adopting a standard notation for the matrix algebra is to enable all problems to be treated in the same way once the matrix model has been formed. Thus we will derive the observation equations first in terms of the mathematical notation common to a particular branch of surveying, before recasting them in the standard matrix form.

$$\mathbf{Ax + Cs = 0}$$

As explained in the next section, the vector **s** is usually expressed as the sum of two vectors **l** and **v**, giving the standard equations the following form

$$\boxed{\mathbf{Ax + Cv + Cl = 0}} \tag{3}$$

Notation
In discussing statistical models a rather complex notation is required to distinguish between *observed, provisional* and *estimated* versions of a parameter. Once the model has been cast into a standard form, the notation can be simplified without confusion.

Figure A5.2 illustrates the notation used to described the various statistical values used in error analysis.

The reader should remember that a point can have three positions: a *provisional position*, a *best position* estimated from the sample observations, and a *population position*, sometimes referred to as the true position. Generally the observed parameters will not fit any of these positions. Therefore, for statistical purposes the notation followed in this book is as follows.

Observed parameters or "observations" for short $\overset{o}{s}$

Selected provisional values of these observations $\overset{\bullet}{s}$

Best estimates of these observed parameters $\hat{s}$

Population values of the observed parameters $\bar{s}$

Errors in best estimates $\delta s = \bar{s} - \hat{s}$

Sample residuals $v = \hat{s} - \overset{o}{s}$

Population residuals $V = \bar{s} - \overset{o}{s}$

Differential changes $ds = \hat{s} - \overset{\bullet}{s} = \overset{o}{s} - \overset{\bullet}{s} + v = l + v$

The objective is to find the *best estimates* of all parameters. The best estimates are invariably the best linear unbiased estimates (BLUE) obtained by applying the least squares principle to sampled known values of l, i.e. from the observations.

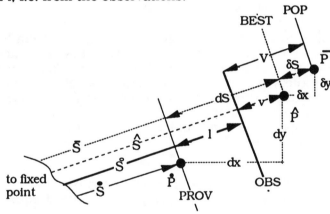

Figure A5.2

In addition to the observed parameters we often have to estimate additional parameters for which no observations are available. The notation for these is

Best estimates of the unobserved parameters $\hat{x}$

Population values of the unobserved parameters $\bar{x}$

Errors in best estimates $\delta x = \bar{x} - \hat{x}$

Selected provisional values of these parameters $\overset{\bullet}{x}$

Differential changes $dx = \hat{x} - \overset{\bullet}{x}$

The key relationships are summarized as follows

Observations

$$\hat{s} = \overset{\bullet}{s} + ds$$

thus

$$ds = \overset{o}{s} - \overset{\bullet}{s} + v$$
$$ds = 1 + v$$

Parameters

$$\hat{x} = \overset{\bullet}{x} + dx$$

Notice that the difference between these relationships is that there are no residuals attached to the unobserved parameters, and that when written in matrix algebra the small changes δs and δx become **s** and **x** respectively.

The significance of the "^" is that *least squares estimates* of the parameters **x** are obtained. Where no confusion arises we omit the hat from the text.

In the following sections the population values are not considered until the matter of dispersion matrices is discussed.

A5.2 Mathematical models for the general case

Ordinary algebra
To be consistent, the best estimates and the provisional values must satisfy the mathematical models exactly, thus

$$F(\hat{x} : \hat{s}) = 0 \quad \text{and} \quad F(\overset{\bullet}{x} : \overset{\bullet}{s}) = 0 \tag{4}$$

Expressing the model in terms of small differential changes we have

$$F(\hat{x} : \hat{s}) = F(\overset{\bullet}{x} : \overset{\bullet}{s}) + \left(\frac{\partial F}{\partial \overset{\bullet}{x}}\right) dx + \left(\frac{\partial F}{\partial \overset{\bullet}{s}}\right) ds$$

therefore

$$\left(\frac{\partial F}{\partial \overset{\bullet}{x}}\right) dx + \left(\frac{\partial F}{\partial \overset{\bullet}{s}}\right) ds = 0$$

and since

$$ds = 1 + v$$

we have

$$\left(\frac{\partial F}{\partial \overset{\bullet}{x}}\right) dx + \left(\frac{\partial F}{\partial \overset{\bullet}{s}}\right) v + \left(\frac{\partial F}{\partial \overset{\bullet}{s}}\right) 1 = 0 \tag{5}$$

The partial differentials are gathered together into a matrix, called the Jacobian matrix, when expressed in matrix algebra.

Matrix algebra

We adopt the convention that there are m observations, having a weight matrix **W**, in n parameters linked together by r equations of type (5), which in matrix form, with their dimensions, are

$$\mathbf{A} \quad \mathbf{x} \quad + \quad \mathbf{C} \quad \mathbf{v} \quad + \quad \mathbf{C} \quad \mathbf{l} \quad = \quad \mathbf{0} \qquad (6)$$

$$(\,\text{rxn})\quad(\text{nx1})\qquad(\text{rxm})\quad(\text{mx1})\qquad(\text{rxm})\quad(\text{mx1})\qquad(\text{rx1})$$

Because $\mathbf{l} = \overset{o}{\mathbf{s}} - \overset{\cdot}{\mathbf{s}}$ are both known, and the design matrices **A** and **C** are also known and the product **C l** is also known, and we let

$$\mathbf{C} \; \mathbf{l} = \; -\mathbf{b}$$

therefore equation (6) becomes

$$\boxed{\mathbf{A}\,\mathbf{x} + \mathbf{C}\,\mathbf{v} = \mathbf{b}} \qquad (7)$$

To obtain a unique solution for **x** we invoke the principle of least squares, making the sum of weighted squares of the residuals $(\mathbf{v}^\mathrm{T}\,\mathbf{W}\,\mathbf{v})$ a minimum. For convenience we denote the expression $(\mathbf{v}^\mathrm{T}\,\mathbf{W}\,\mathbf{v})$ by Ω

Introducing a vector λ of Lagrangian multipliers, of dimension $(\,\text{r}\times 1\,)$, we may put

$$\Omega = \mathbf{v}^\mathrm{T}\,\mathbf{W}\,\mathbf{v} - 2\,\lambda^\mathrm{T}(\mathbf{A}\,\mathbf{x} + \mathbf{C}\,\mathbf{v} - \mathbf{b}) \qquad (8)$$

This vector λ of *correlates* or *undetermined multipliers* is a device invented by Lagrange to treat problems of conditioned minima. Equation (8) is valid because the second term is zero.

Notice that the dimension of Ω is (1 x 1) i.e. a single number. The (m x m) weight matrix **W** is formed from the inverse of the dispersion matrix of the observations estimated in some way. See Appendix 6 for more details.

We now partially differentiate the function Ω with respect to the variables **x** and **s**, and equate them to zero. Before doing so, however, we will establish that differentiation with respect to the observed parameter **s** is the same as differentiation with respect to **v**.

$$\left(\frac{\partial \Omega}{\partial \mathbf{s}}\right) = \left(\frac{\partial \Omega}{\partial \mathbf{v}}\right)\left(\frac{\partial \mathbf{v}}{\partial \mathbf{s}}\right)$$

Because $\mathbf{s} = \mathbf{l} + \mathbf{v}$, and **l** is a constant,

$$\left(\frac{\partial \mathbf{v}}{\partial \mathbf{s}}\right) = \mathbf{I}$$

and thus

$$\left(\frac{\partial \Omega}{\partial \mathbf{s}}\right) = \left(\frac{\partial \Omega}{\partial \mathbf{v}}\right)$$

For a minimum value of Ω we have

$$\left(\frac{\partial \Omega}{\partial \mathbf{v}}\right) = \mathbf{0} \quad \text{and} \quad \left(\frac{\partial \Omega}{\partial \mathbf{x}}\right) = \mathbf{0}$$

Applying these expressions to equation (8) gives

$$\left(\frac{\partial \Omega}{\partial \mathbf{v}}\right) = \mathbf{0} = 2\mathbf{v}^T\mathbf{W} - 2\lambda^T\mathbf{C}$$

Since **W** is symmetric

$$\mathbf{W}^T\mathbf{v} = \mathbf{C}^T\lambda = \mathbf{W}\,\mathbf{v}$$

Thus we have a most important equation

$$\boxed{\mathbf{v} = \mathbf{W}^{-1}\mathbf{C}^T\lambda} \tag{9}$$

And again

therefore

$$\left(\frac{\partial \Omega}{\partial \mathbf{x}}\right) = \mathbf{0} = 2\lambda^T\mathbf{A}$$

$$\boxed{\mathbf{A}^T\lambda = \mathbf{0}} \tag{10}$$

Substituting for **v** from (9) in (7) gives

$$\mathbf{A}\mathbf{x} + \mathbf{C}\,\mathbf{W}^{-1}\mathbf{C}^T\lambda = \mathbf{b}$$

Putting

$$\mathbf{N} = \mathbf{C}\,\mathbf{W}^{-1}\mathbf{C}^T$$

we have

and

$$\mathbf{A}\mathbf{x} + \mathbf{N}\lambda = \mathbf{b}$$

$$\lambda = \mathbf{N}^{-1}(\mathbf{b} - \mathbf{A}\mathbf{x}) \tag{11}$$

Substituting for λ from (11) in equation (10) gives

$$A^T N^{-1} (b - A x) = 0 \qquad (12)$$

$$A^T N^{-1} A x = A^T N^{-1} b$$

And putting $N_1 = A^T N^{-1} A$ and $b_1 = A^T N^{-1} b$

we have the final set of equations

$$\boxed{N_1 x = b_1} \qquad (13)$$

The matrix N_1 is square and symmetric. Traditionally the equations (13) were called 'normal equations'. Their solution gives the required parameters.
From (7) and (12) we obtain another very important result

$$\boxed{A^T N^{-1} C v = 0} \qquad (14)$$

It is so important that we remind readers that

$$N = C W^{-1} C^T$$

A5.3 Mathematical model with added constraints

It sometimes happens that there are certain constraints amongst the parameters which need to be allowed for. For example we may need to hold a length fixed between two points whose positions may be allowed to vary. A constraint is written as a linear equation which has to be satisfied exactly, i.e. as an equation, which has no residual, of the form

$$E x - d = 0 \qquad (15)$$

It is often possible, and thoroughly desirable for simplicity, to eliminate one parameter for each constraint, right at the outset. Not only does this simplify the solution but it reduces the size of the matrix to be solved. The simplest application of this procedure is to eliminate the variables for fixed points in a network.

Another practical way to solve the problem is to assign a very high weight to the constraint equations, which are then treated as observations, thus ensuring that the constraints are almost satisfied.

A theoretically correct and standard procedure is to employ further Lagrangian multipliers k, one for each constraint equation, and proceed as before, and as follows.

Let $\Omega = v^T W v - 2 \lambda^T (A x + C v - b) - 2 k^T (E x - d)$

Putting $\left(\dfrac{\partial \Omega}{\partial x}\right) = 0$

gives $0 = 2 \lambda^T A + 2 k^T E$

therefore $A^T \lambda + E^T k = 0$

Since λ is the same as before, we obtain

$$N_1 x + E^T k = b_1 \tag{16}$$

But

$$E x - d = 0 \tag{15}$$

Sets of equations (15) and (16) may be combined into a *hypermatrix* as follows

$$\begin{bmatrix} N_1 & E^T \\ E & O \end{bmatrix} \begin{bmatrix} x \\ k \end{bmatrix} = \begin{bmatrix} b_1 \\ d \end{bmatrix}$$

(Note: These may be solved as they stand, although if Cholesky's method is used, the null matrix causes negative square roots which have to be flagged during the procedure to identify negative squares when they arise. Gauss's method of solution has no problems in its general form.)

This hypermatrix form is useful when making design studies to add and subtract constraint equations.

If an explicit solution for x only is obtained, a smaller set of equations has to be solved. The explicit solution is obtained as follows: from (16)

$$x = N_1^{-1} (b_1 - E^T k)$$

and from (15)

$$E N_1^{-1} (b_1 - E^T k) = d$$

or

$$E N_1^{-1} E^T k = E N_1^{-1} b_1 - d$$

or

$$N_2 k = b_2$$

and

$$k = N_2^{-1} b_2$$

Substituting for k in (16) finally gives

$$N_1 x = b_1 - E^T N_2^{-1} b_2$$

or

$$\boxed{N_1 x = b_3} \tag{17}$$

Thus the problem once again reduces to the solution of a symmetric set of equations of the form

$$N x = b$$

A common simplification

Many problems are simplified at the outset so that they do not contain any constraint equations, for example by the elimination of fixed parameters when forming up the observation equations.

Also in many cases it is possible to select a mathematical model which has only one observation in each equation, giving an observation equation which contains only one residual. Thus the **C** matrix reduces to the unit matrix, or alternatively it has an inverse.

In such cases, equations (7) may be simplified by premultiplying by this inverse giving

$$C^{-1} A x + C^{-1} C v = C^{-1} b$$

which is of the form

$$A_1 x = l_1 + v$$

giving directly the normal equations

$$A_1^T W A_1 x = A_1^T W l_1$$

or

$$N_3 x = b_4 \tag{18}$$

These equations (18) are commonly found in many surveying applications. For example equations (2) can be written

$$C^{-1} A x + C^{-1} C s = 0$$

$$A_2 x + s = 0$$

or

$$A_2 x = l + v$$

where A_2 is in full

$$\begin{bmatrix} -\dfrac{\Delta x_1}{s_1} & -\dfrac{\Delta y_1}{s_1} \\[2ex] -\dfrac{\Delta x_2}{s_2} & -\dfrac{\Delta y_2}{s_2} \end{bmatrix}$$

This is the most common form of observation equations.

A5.4 Figural approach by conditions

As an alternative to the parametric approach to least squares problems which is used generally throughout this book, a once-popular figural approach may be adopted instead. In certain cases, where the figural conditions among the observed parameters are linear as in levelling networks, this is by far the simpler of the two methods.

Consider the example of the level network of Chapter 5, section 5.12 and Figure 5.12. Because the **C** matrix is the unit matrix, each observed height difference is written as a simple observation equation of the form

$$\mathbf{A}\mathbf{x} = \mathbf{l} + \mathbf{v}$$

The **A** matrix is also the unit matrix, and the **l** vector made zero by selecting the provisional values to be the observed values. Thus

$$\mathbf{x} = \mathbf{v}$$

Thus there are no unobserved parameters in the problem. Conditions among the observed parameters, to ensure that the loops of levelling must close, give constraint equations of the form

$$\mathbf{E}\mathbf{x} = \mathbf{d}$$

Following the same procedure as in A5.3 incorporating Lagrangian multipliers (or correlatives) **k** to minimize the sum of squares of residuals, leads to

$$\mathbf{E}^{\mathrm{T}}\mathbf{k} = \mathbf{W}\mathbf{x}$$

and

$$\mathbf{x} = \mathbf{W}^{-1}\mathbf{E}^{\mathrm{T}}\mathbf{k} \tag{19}$$

Substituting for **x** in the constraint equations gives

$$\mathbf{E}\mathbf{W}^{-1}\mathbf{E}^{\mathrm{T}}\mathbf{k} = \mathbf{d} \tag{20}$$

Therefore

$$\mathbf{k} = (\mathbf{E}\mathbf{W}^{-1}\mathbf{E}^{\mathrm{T}})^{-1}\mathbf{d}$$

The equations (20), called the *correlative normal equations*, are solved to give the vector of correlatives **k**, and then substituted in (19) to give the desired solution for **x** (**v**).

This approach, well suited to simple constraint equations as in levelling where all the coefficients are unity, becomes complex when non-linear equations are involved. Also, the mechanism for forming constraint equations is by no means straightforward. A glance at the first edition of this book will show this.

Being purely figural, however, the method does avoid the datum problems associated with coordinates, and is useful in theoretical studies and in blunder detection. Since there are no examples of this method elsewhere in this book, one seems useful here. The weight matrix is taken to be the unit matrix.

Example
The data of Chapter 5, section 5.12, give the constraint equations for each loop of levels as

$$\begin{bmatrix} 1 & 0 & -1 & 0 & 1 & 0 \\ 0 & -1 & 1 & 0 & 0 & -1 \\ 0 & 0 & 0 & 1 & -1 & 1 \end{bmatrix} x = \begin{bmatrix} 0.11 \\ 0.04 \\ -0.08 \end{bmatrix}$$

and the correlative normals as

$$\begin{bmatrix} 3 & -1 & -1 \\ -1 & 3 & -1 \\ -1 & -1 & 3 \end{bmatrix} k = \begin{bmatrix} 0.11 \\ 0.04 \\ -0.08 \end{bmatrix}$$

The solution for k^T is

$$0.045 \qquad 0.0275 \qquad -0.0025$$

giving the solution for x^T as

$$0.045 \quad -0.028 \quad -0.018 \quad 0.002 \quad 0.048 \quad -0.030$$

Summary

In the above sections A5.1, A5.2 and A5.3, the various formulae needed to obtain solutions of the required parameters are given. Often not all parameters in a problem are required, since only a few may be of interest, whilst the others are either of no concern or are not actually recoverable in a physical sense. In such cases only a partial solution of the normal equations is needed. Also, in design problems or in real-time problems, not all the observation equations will be used at the same time, and a sequential approach is adopted by forming the normal equations by the method of contributions. See Appendix 3.

We now proceed to discuss the statistical information which can be obtained from these solutions.

A5. 5 Dispersion matrices

An important feature of the least squares estimation process is the ability to estimate *quality*. This quality is expressed by dispersion matrices.These symmetric square matrices contain the variances of the parameters along the diagonals and the co-variances as off-diagonal terms. When the correlations are zero, these matrices are diagonal and therefore easy to invert.

We denote the dispersion matrix of the observed parameters by $\mathbf{D_o}$ and of the derived parameters by $\mathbf{D_x}$. Because the dispersion matrix of the observed parameters can only be estimated from some *a priori* data such as from previous experience, we make allowance for its error by introducing an unknown scaling factor, σ_o^2 called the variance of an observation of unit weight. We relate the inverse of the weight matrix $\mathbf{W}^{-1}$ to the *a priori* estimated dispersion matrix $\mathbf{D_o}$ by

$$\sigma_o^2 \, \mathbf{W}^{-1} \, = \, \mathbf{D_o}$$

The data of any least squares calculation will be used to *estimate* this scaling factor from the sample residuals, and thus gain a better value for the dispersion matrix of the observations. If the calculated value of σ_o^2 is equal to 1 then the original estimated weight matrix is correct and

$$\mathbf{W}^{-1} \, = \, \mathbf{D_o}$$

For a more detailed explanation see Chapter 5.

Population parameters and sample statistics
We now relate the best estimates from a sample to their theoretical population from which the sample has been drawn.

If $\bar{s}$ is the population value of the observed parameter, then the population residual V is given by

$$\bar{s} \, = \, \overset{o}{s} + V$$

Correspondingly the vectors are written in bold notation

$$\bar{\mathbf{s}} \, = \, \overset{o}{\mathbf{s}} + \mathbf{V}$$

This compares with the vector of data from the sample

$$\hat{\mathbf{s}} \, = \, \overset{o}{\mathbf{s}} + \mathbf{v}$$

Thus

$$\delta \mathbf{s} = \bar{\mathbf{s}} - \hat{\mathbf{s}} \, = \mathbf{V} - \mathbf{v}$$

where δs is the difference between the population value and the sample estimate.

Applying similar ideas to the mathematical model (3) we have the respective sample and population models

$$\mathbf{A x + C v + C 1 = 0}$$

and

$$\mathbf{A \bar{x} + C V + C 1 = 0}$$

Subtracting gives

$$\mathbf{A (\bar{x} - x) + C (V - v) = 0}$$

$$\mathbf{A \delta x + C (V - v) = 0} \tag{21}$$

where δx is the vector of the differences between the population parameters and their corresponding values estimated from the sample by least squares.

Now we know from equation (14) that

$$\mathbf{A^T N^{-1} C v = 0}$$

thus by premultipling equation (21) by $\mathbf{A^T N^{-1}}$ we obtain

$$\mathbf{A^T N^{-1} A \delta x + A^T N^{-1} C V - A^T N^{-1} C v = 0}$$

or

$$\mathbf{A^T N^{-1} A \delta x + A^T N^{-1} C V = 0}$$

or say

$$\mathbf{N_1 \delta x = BV}$$

where

$$\mathbf{B = - A^T N^{-1} C}$$
$$\mathbf{N_1 = A^T N^{-1} A}$$

Now

$$\mathbf{N_1 \delta x (N_1 \delta x)^T = B V (B V)^T}$$

therefore

$$\mathbf{N_1 \delta x \ \delta x^T N_1 = B V V^T B^T}$$

Taking expectations, and remembering that

$$\mathbf{E (\delta x \ \delta x^T) = D_x} \text{ and } \mathbf{E (V V^T) = D_o}$$

we have

$$\mathbf{N_1 D_x N_1 = B D_o B^T}$$

Pre and post multiplying both sides by $\mathbf{N_1^{-1}}$ gives

$$\mathbf{N_1^{-1} N_1 D_x N_1 N_1^{-1} = N_1^{-1} B D_o B^T N_1^{-1}}$$

or
$$\mathbf{D_x} = \mathbf{N_1}^{-1} \mathbf{B} \ \mathbf{D_o} \ \mathbf{B}^T \mathbf{N_1}^{-1}$$

but
$$\mathbf{B} \ \mathbf{D_o} \ \mathbf{B}^T = \mathbf{A}^T \ \mathbf{N}^{-1} \ \mathbf{C} \ \mathbf{D_o} \ (\ \mathbf{A}^T \ \mathbf{N}^{-1} \ \mathbf{C})^T$$

$$= \mathbf{A}^T \ \mathbf{N}^{-1} \ \mathbf{C} \ \mathbf{D_o} \ \mathbf{C}^T \ \mathbf{N}^{-1} \ \mathbf{A}$$

and
$$\mathbf{C} \ \mathbf{D_o} \ \mathbf{C}^T = \mathbf{C} \sigma_o^2 \ \mathbf{w}^{-1} \ \mathbf{C}^T$$

$$= \sigma_o^2 \ \mathbf{N}$$

giving
$$\mathbf{B} \ \mathbf{D_o} \ \mathbf{B}^T = \sigma_o^2 \ \mathbf{N}_1$$

and finally
$$\boxed{\mathbf{D_x} = \sigma_o^2 \ \mathbf{N}_1^{-1}} \qquad (22)$$

Thus the dispersion matrix of the required parameters is the scaled inverse of the normal equations. Hence, although this inverse may not be needed to obtain the solution, it is needed for this statistical information.

Equation (22) gives the expression for $\mathbf{D_x}$ for the general case of observations with conditions $\mathbf{C}$.

Special case
When $\mathbf{C} = \mathbf{I}$ the unit matrix, the result is of the same form, namely

$$\boxed{\mathbf{D_x} = \sigma_o^2 \ \mathbf{N}^{-1}} \qquad (23)$$

because in full
$$\mathbf{D_x} = \sigma_o^2 \ (\mathbf{A}^T (\ \mathbf{C} \ \mathbf{W}^{-1} \ \mathbf{C}^T)^{-1} \mathbf{A})^{-1}$$

which reduces to equation (23).

A5.6 Estimation of σ_o^2

We now show how σ_o^2 is estimated from the sample itself, as a bi-product of the least squares computation, using the expression:

$$\boxed{\sigma_o^2 = \text{Exp} \frac{\mathbf{v}^T \mathbf{W} \ \mathbf{v}}{m - n}} \qquad (24)$$

For economy of space, the derivations are not given in full, with some work left to the reader, who is once again reminded of the symmetric nature of many of the matrices.

From equation (11) we have

$$\lambda = N^{-1} (b - Ax) \tag{11}$$

and from equation (9)

$$v = W^{-1} C^T \lambda \tag{9}$$

Thus we have the sample residuals given by

$$v = W^{-1} C^T N^{-1} (b - Ax) \tag{25}$$

The population residuals are therefore given by

$$V = W^{-1} C^T N^{-1} (b - A\bar{x}) \tag{26}$$

giving by subtraction

$$V - v = W^{-1} C^T N^{-1} (b - A\bar{x} - b + Ax)$$

$$= - W^{-1} C^T N^{-1} A (\bar{x} - x)$$

$$= - W^{-1} C^T N^{-1} A \, \delta x$$

$$V = v - W^{-1} C^T N^{-1} A \, \delta x$$

$$V = v - K \delta x$$

where

$$\boxed{K = W^{-1} C^T N^{-1} A} \tag{27}$$

Now from equation (14)

$$A^T N^{-1} C v = 0 \tag{14}$$

Thus we have the important result

$$\boxed{K^T W v = 0} \tag{28}$$

Now

$$V^T W V = (v - K \delta x)^T W (v - K \delta x)$$

$$V^T W V = v^T W v + \delta x^T K^T W K \delta x \tag{29}$$

because $K^T W v = 0$ and $v^T W K = 0.$

Now, remembering that the *trace* of a matrix is the sum of its diagonal terms, we have

$$\mathbf{v}^T \mathbf{W} \mathbf{v} = \text{trace} (\mathbf{V} \mathbf{V}^T \mathbf{W})$$

$$\delta \mathbf{x}^T \mathbf{K}^T \mathbf{W} \mathbf{K}\, \delta \mathbf{x} = \text{trace} (\delta \mathbf{x}\, \delta \mathbf{x}^T\, \mathbf{K}^T \mathbf{W} \mathbf{K})$$

Thus equation (29) becomes

$$\text{trace} (\mathbf{V} \mathbf{V}^T \mathbf{W}) = \mathbf{v}^T \mathbf{W} \mathbf{v} + \text{trace} (\delta \mathbf{x}\, \delta \mathbf{x}^T \mathbf{K}^T \mathbf{W} \mathbf{K})$$

and taking expectations of both sides we have

$$\text{Exp trace} (\mathbf{V} \mathbf{V}^T) \mathbf{W} = \text{Exp} (\mathbf{v}^T \mathbf{W} \mathbf{v}) + \text{Exp trace} (\delta \mathbf{x}\, \delta \mathbf{x}^T) \mathbf{K}^T \mathbf{W} \mathbf{K}$$

Now

$$\text{Exp} (\delta \mathbf{x}\, \delta \mathbf{x}^T) = \mathbf{D_x} = \sigma_o^2\, \mathbf{N}_1^{-1}$$

Also
$$\text{Exp trace} (\mathbf{V} \mathbf{V}^T) \mathbf{W} = \text{trace} (\mathbf{D_o} \mathbf{W}) = \text{trace} (\sigma_o^2\, \mathbf{W}^{-1} \mathbf{W})$$

Therefore
$$\sigma_o^2\, \text{trace}\, \mathbf{I}_m = \text{Exp} (\mathbf{v}^T \mathbf{W} \mathbf{v}) + \text{trace}\, \sigma_o^2\, \mathbf{N}_1^{-1} \mathbf{K}^T \mathbf{W} \mathbf{K}$$

$$= \text{Exp} (\mathbf{v}^T \mathbf{W} \mathbf{v}) + \sigma_o^2\, \text{trace}\, \mathbf{I}_n$$

because

$$\mathbf{K}^T \mathbf{W} \mathbf{K} = \mathbf{N}_1$$

Now $\mathbf{I}_m$ and $\mathbf{I}_n$ are unit matrices of dimensions m and n, thus

$$m \sigma_o^2 = \text{Exp} (\mathbf{v}^T \mathbf{W} \mathbf{v}) + n \sigma_o^2$$

and finally

$$\sigma_o^2 = \text{Exp} \frac{\mathbf{v}^T \mathbf{W} \mathbf{v}}{m - n}$$

If we put

$$S^2 = \frac{\mathbf{v}^T \mathbf{W} \mathbf{v}}{m - n}$$

then

$$\text{Expectation } S^2 = \sigma_o^2$$

and we say that

$$S^2 \text{ is an } unbiased\ estimator \text{ of } \sigma_o^2$$

The denominator $m - n$ is the number of degrees of freedom in the problem, or the number of redundant observations which are not essential for a solution to be obtained.

Special case of one parameter

In the special case of one parameter estimated from n observations the above expression for S^2 reduces to

$$\text{Exp } S^2 = \text{Exp } \frac{\mathbf{v}^T \mathbf{W} \mathbf{v}}{n-1} \tag{29}$$
$$= \sigma_o^2$$

and

$$S^2 = \frac{n}{(n-1)} \quad s^2 = \frac{n}{(n-1)} \frac{\mathbf{v}^T \mathbf{W} \mathbf{v}}{n}$$

or in words

$$S^2 = \frac{n}{(n-1)} \times \text{sample variance}$$

is an *unbiased estimator* of the population variance.

A5.7 Dispersion matrix of the sample residuals $\mathbf{D_v}$

Although the quality of the derived parameters, as expressed by their dispersion matrix $\mathbf{D_x}$ in equation (22), is generally of most interest, there are several statistical reasons for calculating the quality of the *residuals* derived in the least squares process.

From equation (25) we have

$$\mathbf{v} = \mathbf{W}^{-1} \mathbf{C}^T \mathbf{N}^{-1} (\mathbf{b} - \mathbf{A}\mathbf{x}) \tag{25}$$

$$\mathbf{v} = -\mathbf{W}^{-1} \mathbf{C}^T \mathbf{N}^{-1} \mathbf{C} \mathbf{l} - \mathbf{W}^{-1} \mathbf{C}^T \mathbf{N}^{-1} \mathbf{A} \mathbf{x}$$

$$\mathbf{v} = -\mathbf{G}\mathbf{l} - \mathbf{K}\mathbf{x}$$

where

$$\mathbf{G} = \mathbf{W}^{-1} \mathbf{C}^T \mathbf{N}^{-1} \mathbf{C}$$
$$\mathbf{K} = \mathbf{W}^{-1} \mathbf{C}^T \mathbf{N}^{-1} \mathbf{A}$$

Differentiating gives

$$\delta \mathbf{v} = -\mathbf{G}\delta\mathbf{l} - \mathbf{K}\delta\mathbf{x} \tag{31}$$

therefore

$$\delta l = \delta \overset{\circ}{s} - \delta \overset{\cdot}{s}$$

but $\delta \overset{\circ}{s}$ is the error vector of the observed parameters, that is $\delta \overset{\circ}{s} = \mathbf{V}$ the population residuals, and $\delta \overset{\cdot}{s} = \mathbf{0}$, thus

$$\delta \mathbf{v} \, \delta \mathbf{v}^T = (\mathbf{G V} + \mathbf{K} \, \delta \mathbf{x}) \; (\mathbf{G V} + \mathbf{K} \, \delta \mathbf{x})^T$$

$$\delta \mathbf{v} \, \delta \mathbf{v}^T = \mathbf{G V V}^T \mathbf{G}^T + \mathbf{K} \, \delta \mathbf{x} \; \delta \mathbf{x}^T \mathbf{K}^T + \mathbf{G V} \, \delta \mathbf{x}^T \mathbf{K}^T + \mathbf{K} \, \delta \mathbf{x} \; \mathbf{V}^T \mathbf{G}^T$$

Taking expectations we have

$$\mathbf{D}_v = \mathbf{G} \, \mathbf{D}_o \, \mathbf{G}^T + \mathbf{K} \, \mathbf{D}_x \, \mathbf{K}^T + \mathbf{G} \exp(\mathbf{V} \, \delta \mathbf{x}^T) \mathbf{K}^T + \mathbf{K} \exp(\delta \mathbf{x} \, \mathbf{V}^T) \mathbf{G}^T$$

The third and fourth terms of this expression are both equal to

$$-\mathbf{K} \, \mathbf{D}_x \, \mathbf{K}^T$$

giving the result

$$\boxed{\mathbf{D}_v = \mathbf{G} \, \mathbf{D}_o \, \mathbf{G}^T - \mathbf{K} \, \mathbf{D}_x \, \mathbf{K}^T} \qquad (32)$$

We give the outline proof that terms three and four above are both equal to

$$-\mathbf{K} \, \mathbf{D}_x \, \mathbf{K}^T$$

From (13)

$$x = \mathbf{N}_1^{-1} \mathbf{b}_1$$

therefore

$$\delta \mathbf{x} = \mathbf{N}_1^{-1} \, \delta \, \mathbf{b}_1$$

$$\delta \mathbf{x} = -\mathbf{N}_1^{-1} \, \mathbf{A}^T \mathbf{N}^{-1} \, \mathbf{C V}$$

The third term then is

$$\mathbf{G} \exp(\mathbf{V} \, \delta \mathbf{x}^T) \mathbf{K}^T = \mathbf{G} \exp(\mathbf{V} (-\mathbf{N}_1^{-1} \, \mathbf{A}^T \mathbf{N}^{-1} \mathbf{C V})^T \mathbf{K}^T$$

$$\mathbf{G} \exp(\mathbf{V} \, \delta \mathbf{x}^T) \mathbf{K}^T = -\mathbf{G} \, \mathbf{D}_o \, \mathbf{C}^T \, \mathbf{N}^{-1} \mathbf{A} \, \mathbf{D}_x \, \mathbf{K}^T$$

which reduces to

$$-\mathbf{K} \, \mathbf{D}_x \, \mathbf{K}^T$$

This is symmetric and equals the fourth term transposed.

A common simplification

In the case of the simple model in which the **C** matrix is the unit matrix, it is easy to show that the general result in (31) reduces to the much simpler expression

$$\boxed{\mathbf{D}_v \;=\; \mathbf{D}_o - \mathbf{A}\,\mathbf{D_x}\,\mathbf{A}^T}$$

(33)

A5.8 Dispersion matrix of the estimated observed parameters $\mathbf{D_s}$

The dispersion matrix $\mathbf{D_s}$ of the least squares estimates of the observed parameters is given by the simple expression

$$\boxed{\mathbf{D_s} = \mathbf{D}_o - \mathbf{D_v}}$$

(34)

where $\mathbf{D_v}$ is given by expression (32) for the conditioned model and by (33) for the simple model. In this latter case we have the even simpler expression

$$\boxed{\mathbf{D_s} \;=\; \mathbf{A}\,\mathbf{D_x}\,\mathbf{A}^T}$$

(35)

The proof of (34) follows similar lines to that for (32) in which some heavy, but straightforward, matrix manipulation is involved. The basic steps are as follows.

The best estimates of the observed parameters are given by

$$\mathbf{s} = \mathbf{v} + \overset{o}{\mathbf{s}}$$

$$\delta\mathbf{s} = \delta\mathbf{v} + \delta\overset{o}{\mathbf{s}}$$

but from (31)

$$\delta\mathbf{v} = -\,\mathbf{G}\,\delta\mathbf{l} - \mathbf{K}\,\delta\mathbf{x}$$

therefore

$$\delta\mathbf{s} = -\,\mathbf{G}\,\delta\mathbf{l} - \mathbf{K}\,\delta\mathbf{x} + \delta\overset{o}{\mathbf{s}}$$

but

$$\delta\mathbf{l} = \delta\overset{o}{\mathbf{s}} = \mathbf{V}$$

therefore

$$\delta\mathbf{s} = (\,\mathbf{I} - \mathbf{G}\,)\,\mathbf{V} - \mathbf{K}\,\delta\mathbf{x}$$

Substituting for

$$\delta\mathbf{x} = -\,\mathbf{N}_1{}^{-1}\,\mathbf{A}^T\,\mathbf{N}^{-1}\,\mathbf{C}\mathbf{V}$$

gives the expression

$$\delta\mathbf{s} = (\,\mathbf{I} - \mathbf{G} + \mathbf{K}\,\mathbf{N}_1{}^{-1}\,\mathbf{A}^T\,\mathbf{N}^{-1}\,\mathbf{C}\,)\,\mathbf{V}$$

or say

$$\delta \mathbf{s} = \mathbf{F} \mathbf{V}$$

where

$$\mathbf{F} = (\mathbf{I} - \mathbf{G} + \mathbf{K} \mathbf{N}_1{}^{-1} \mathbf{A}^T \mathbf{N}^{-1} \mathbf{C})$$

Now

$$\delta \mathbf{s} \, \delta \mathbf{s}^T = \mathbf{F} \mathbf{V} \mathbf{V}^T \mathbf{F}^T$$

Taking expectations

$$\mathbf{D}_s = \mathbf{F} \mathbf{D_o} \mathbf{F}^T$$

and substituting for $\mathbf{F}$ we obtain, after some heavy algebra,

$$\mathbf{D}_s = \mathbf{D_o} - \mathbf{G} \mathbf{D_o} \mathbf{G}^T + \mathbf{K} \mathbf{D_x} \mathbf{K}^T$$

which is equation (34) and equation (35) follows.

Comment
Throughout this book we have assumed that we are dealing with properly posed problems which do not give rise to singular normal equation matrices. This means that such matrices have no rank defects due to insufficient information or inadequate datum definition. For example there must be at least two sufficient measurements to fix a point in two dimensions, and a network, described in terms of cartesian coordinates, must have an assigned origin and orientation. This approach is usually sufficient for most problems, other than for deformation studies.

As will be seen in Appendix 6, the selection of a datum for coordinates will affect some dispersion matrices and statistical information deduced from them. For a treatment of more general ways to remove rank defects which give unique dispersion matrices, the reader should refer to references (11), (12) and (13).

Although the algebraic derivation of dispersion matrices is lengthy, their numerical evaluation forms simple additions to computer programs. Their use in statistical testing is discussed briefly in Appendix 6.

Appendix 6
Quality control

A6.1 Introduction

Control of the quality of work forms part of the wider process of quality assurance in which matters of administration and documentation, as well as technicalities, are involved.

In this book we are concerned mainly with technical aspects; i.e. with quality control.

The use of error analysis is paramount both to the design of surveys to a required specification, and to test the results of observations and derived parameters, to see if they have met the specification. This *before (a priori)* and *after (a posteriori)* treatment applies generally.

Figural and parametric methods

Most observations today are treated parametrically by the method of least squares. This depends on the selection of some framework of reference, such as a coordinate system, which affects the appearance of any statistical results. Care has to be taken, therefore, with the interpretation of these results.

On the other hand, the formerly popular method of figural treatment, which does not suffer from datum bias, has much to commend it.

Norms

Norms other than least squares, such as minimum modulus, are more robust at detecting blunders in data, but suffer from excessive computational demands.

The whole topic of quality control is one of current research, in which topics like cost effectiveness are not being ignored.

A6.2 Blunder detection

Blunders, or gross errors, are usually located by the careful processes of instrument practice and calibration. The subsequent least squares analysis usually assumes that all significant factors have been modelled by the mathematics. If this is not so, statistical tests on data can be quite wrong.

For example, it might be assumed that an EDM index has been applied to all measurements correctly, and therefore no index model parameter was introduced to the analysis. If the index had not been applied, or applied with the wrong sign, the subsequent error distibution is not random.

It might therefore be thought that all possible factors should be modelled. If this is done, the degree of redundancy in the model can be reduced until no statistical analysis is possible. Tests have shown that the cost of modelling every centering error at a theodolite station can easily be offset by the construction of pillars to avoid the problem at the outset.

Thus good instrument practice and calibration turns out to be the simplest and the best.

Figural approach

The filtering of observations by common-sense figural methods can be of immense value. Obvious checks are the mis-closure of fully observed closed figures such as a triangle or traverse.

In networks, side equations can be used to locate blunders and save much time in wasteful re-observations, or to justify the omission of data from a subsequent parametric treatment.

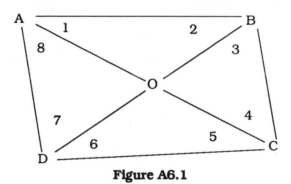

Figure A6.1

Example

Consider the braced quadrilateral ABCD of Figure A6.1. Suppose the triangles ABD and BCD close, while ABC and ADC have large errors of opposite signs. This shows that there is a mistake either in direction AC or in CA, or in both. Using the side equation technique will resolve the matter of location.

Taking the pole at A we have

$$\frac{AB}{AC} \frac{AC}{AD} \frac{AD}{AB} = 1$$

therefore

$$\frac{\sin 4}{\sin (2+3)} \frac{\sin (6+7)}{\sin 5} \frac{\sin 2}{\sin 7} = 1 \tag{1}$$

Because equation (1) does not contain the angles at A, if it is satisfied within the tolerance of the observations, the mistake is at A . If not, the mistake is at C or A.

Using a similar equation by taking the pole at C would substantiate whether the error is just at C or at both points. A decision whether to reject or re-observe the faulty data would be taken before the least squares analysis by variation of coordinates.

Figural equations for testing observed sides and traverses can be found in references (8) and (9).

Statistical approach

A blunder may be suspected if a residual is larger than expected. The criterion used may be quite arbitrary, such as three times the standard deviation, or more likely it will be based on a statistical distribution. When a large number of residuals is present (say more than 30) the normal distribution is used. Smaller samples will use the Student's t distribution. This aspect has been treated in Chapter 5. Rejection limits are set usually at the 95% confidence level.

The limit factors of the standard deviation for one, two and three dimensions are respectively

$$2.0 \qquad 2.5 \qquad 2.7$$

For example, we reject a three-dimensional vector residual greater than 2.7σ at the 95% confidence level, where

$$\sigma^2 = \sigma_x^2 + \sigma_y^2 + \sigma_z^2$$

This test can be applied automatically to sets of data which are never seen by an observer, as for example in a satellite ranging system.

A6.3 Least squares analysis

In testing for blunders and reliability in the data and results of a least squares analysis we are able to use a number of statistical parameters, in addition to figural tests, which can be used to locate really large blunders.

Most of these statistical parameters may also be used to predict the possible results and quality of proposed networks, without making any actual observations.

Notation

The fundamental statistic is the estimate of the standard error of an observation

$$\sigma \tag{2}$$

The squares of these estimates form the elements of the dispersion matrix $\mathbf{D_0}$ related to the weight matrix $\mathbf{W}$ by

$$\sigma_o^2 \, \mathbf{W}^{-1} = \mathbf{D}_o$$

In turn, the estimated variances of all observations contribute to the estimate of the standard error factor

$$\sigma_o \tag{3}$$

If the variances of the observations have been correctly estimated this error factor should be unity. In theoretical analyses it is usually assumed to be so.

From the diagonal of the dispersion matrix $\mathbf{D}_x$ of the unobserved parameters, see Appendix 5 equation (22), we obtain the standard error of an unobserved parameter x

$$\sigma_x \tag{4}$$

From the dispersion matrix $\mathbf{D}_s$ of the observed parameters, see Appendix 5 equation (34), we obtain the standard error of an observed parameter s

$$\sigma_s \tag{5}$$

From the dispersion matrix of the residuals $\mathbf{D}_v$, see Appendix 5 equation (32), we obtain an estimate of the standard error of a residual v

$$\sigma_v \tag{6}$$

It should be noted that

$$\sigma_v^2 = \sigma^2 - \sigma_s^2 \tag{7}$$

Detection of outliers

As was pointed out in Chapter 5, we may use the probabilities of normal or student distributions to decide whether to reject or accept an outlying observation. If we reject an observation at the α confidence level, we make a type I error whose probability is known.

If we accept an observation which should belong to a nearby mean and its distribution, we make a type II error base on an alternative hypothesis of a β confidence level. These techniques are also applied to the results of least squares computations.

A ratio used for this purpose is w given by

$$w = \frac{v}{\sigma_v} \tag{8}$$

Applying a two tailed test at the 95% confidence level gives a statistical limit $t = 1.96$, and a 5% chance of making a type I error if we reject the observation from which v is derived.

If we wish to assess the probability of making a type II error in accepting all residuals under the 1.96 mark, we need to postulate an alternative hypothesis based, either upon a new value s_1 differing from the value s under scrutiny by an amount Δ, or upon a confidence limit β which matched the boundary statistic $t = 1.96$. See Figure A6. 2

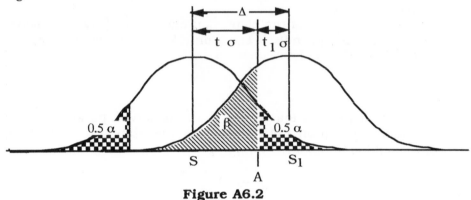

Figure A6.2

Suppose $\beta = 5$ % then from Table 5.3 the statistic $t_1 = 1.65$. This means that the two values s and s_1 are separated by an amount given by

$$\Delta = (t + t_1)sd = t_1 \frac{\sigma^2}{\alpha_v} = (1.96 + 1.65) \frac{\sigma^2}{\alpha_v} \qquad (9)$$

Alternatively, if the difference Δ is chosen, say to be 5 σ, the statistic t_1 can be found from (9) and therefore β. Thus we can say that the probability of detecting a gross error worse than 5 σ is β.

For the justification of (8) and (9) the reader should refer to reference (10).

Reliability
The idea behind the concept of *reliability* is to know the chances of detecting a mistake or blunder.

If only the minimum number of observations is made, such as a bearing and distance to fix a point, we cannot tell, from the observations only, if a mistake has been made. We can however use other commonsense information to suspect the result. If redundant observations are present, there is some chance of detecting mistakes.

A consideration of the position lines used to fix a point can be a measure of the reliability of the work.

However, it is convenient to establish criteria to assess the reliability of observations and parameters by automatic statistical methods which do not involve the drawing of position lines and making arbitrary decisions.

These criteria are derived from statistics obtained from the various dispersion matrices.

Observations

If there is no redundancy, both v and σ_v will be zero, and an observation is considered *unreliable*. Thus σ_v can be used as an indicator of reliability. The dimensionless ratio used is

$$\tau = \frac{\sigma}{\sigma_v} \tag{10}$$

Notice that τ is a special case of the w of equation (8) in which the residual σ, typical of a distribution, replaces an actual residual v. In this way a theoretical analysis can be made without observations.

As before, if we select confidence levels for α and β, we obtain statistics t and t_1 and the value of Δ given by

$$\Delta = (t + t_1)sd = (t + t_1)\frac{\sigma^2}{\sigma_v} = (t + t_1)\sigma\tau \tag{11}$$

Thus we can make a statement that the maximum undetected gross error in the observation is Δ at the β confidence level.

This analysis has been termed *internal reliability*.

Parameters

It is interesting to enquire what effect this maximum undetected error of an observation will have on the parameters. Merely to re-run the computation with such a mistake will give some idea of the effect, which is not statistically of much value.

A more general approach is to examine the effect of the standard error in an observation σ_s using the ratio

$$\gamma = \frac{\sigma_s}{\sigma_v} \tag{12}$$

which by a similar use of statistical confidence levels α and β gives the error limit

$$\Delta = (t + t_1)sd = (t + t_1)\sigma\gamma$$

The limit on a parameter x whose standard error is known from the dispersion matrix of parameters is therefore

$$\Delta\,\sigma_x \tag{13}$$

That (13) is so can be seen from considering the change dx on the parameter x resulting from a change dl = ds in the observations. Then we have

$$dx = N^{-1}\,db = N^{-1}\,A^T\,ds$$

$$\sigma_x = N^{-1}\,A^T\,\sigma_s$$

Thus we can find the effect on one selected parameter from an error in one observation.

Comment
The whole subject of blunder detection, statistical analysis and datum dependence is still the subject of research. Readers should consult current journals for up-to-date information: but see references (10),(11),(12) and (13).

Dispersion matrices
Dispersion matrices are made up of the estimated variances and co-variances of observed and unobserved parameters and functions of them.

The *a priori* estimates of the variances and co-variances of observed parameters are made from specific tests of repeated measurements, or from previous least squares estimates using similar techniques. The global variance scale factor σ_0^2 is estimated from observations. If the original estimates of the dispersion matrix of the observations is correct, this variance factor should be unity.

The Fisher test is used to judge whether a value of σ_0 is acceptable. It is tested against a theoretical value of unity from an infinitely large population. See Chapter 5 for more details.

If the test fails, a reconsideration of the original dispersion matrix is required. This often means a re-assessment of some types of measurements or instruments. This topic is also the subject of further research.

Error ellipses and ellipsoids
As described in Appendix 4, the error vector at a point may be depicted graphically by the pedal curve to an ellipse, or by the pedal surface to an ellipsoid, showing the variances of derived parameters at all directions.

Care is needed to interpret these curves properly. The main point to watch is that positional error ellipses are datum dependent.

Example

For example, the positional error ellipses, for a straight unclosed traverse of n equal legs starting at a fixed point A with a fixed bearing, grow larger from zero at the fixed starting point to a maximum at its end. All directions and lengths are assumed of the same weight. See Figure A6.2.

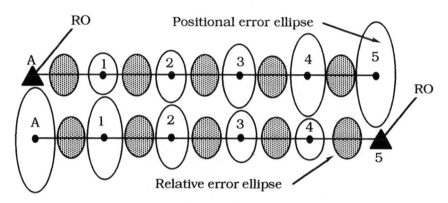

Figure A6.2

The same traverse computed from the other end of the line would exhibit an identical set of error ellipses, but different values for each point, except the middle one, if there is one. The *positional error ellipses* are merely the *relative error ellipses* between each point and the fixed origin.

Relative error ellipses

Provided the orientation of the coordinate axes is not altered, the error ellipses derived for *pairs of points* are not dependent upon the coordinate datum. For example the relative error ellipses of any line of the straight traverse are identical.

These relative error ellipses are obtained from the dispersion matrix of the parameters as follows.

Consider any two points for which variances and co-variances are available. Usually these will come from the dispersion matrix

$$\sigma_0^2 \, N^{-1}$$

Since

$$\Delta x = x_2 - x_1 \quad \text{and} \quad \Delta y = y_2 - y_1$$

$$d\,\Delta x = d\,x_2 - d\,x_1 \quad \text{and} \quad d\,\Delta y = d\,y_2 - d\,y_1 \tag{3}$$

The expectations of

$$(d \, \Delta x)^2, \, (d \, \Delta x)^2 \text{ and } d \, \Delta x \, d \, \Delta y$$

are respectively the variances and co-variance

$$\sigma_{\Delta x}^2, \sigma_{\Delta y}^2 \text{ and } \sigma_{\Delta x \, \Delta y}$$

These are given from (3) by

$$\sigma_{\Delta x}^2 = \sigma_{x_1}^2 + \sigma_{x_2}^2 - 2 \, \sigma_{x_1 x_2} \tag{4}$$

$$\sigma_{\Delta y}^2 = \sigma_{y_1}^2 + \sigma_{y_2}^2 - 2 \, \sigma_{y_1 y_2} \tag{5}$$

$$\sigma_{\Delta x \, \Delta y} = \sigma_{x_1 y_1} - \sigma_{x_1 y_2} - \sigma_{x_2 y_1} + \sigma_{x_2 y_2} \tag{6}$$

These expressions (4), (5) and (6) are treated as in Appendix 4, A4.3 to give the relative error ellipse parameters for the two points.

$$\frac{\Delta \sin 2T}{\Delta \cos 2T} = \frac{2 \, \sigma_{\Delta x \, \Delta y}}{\sigma_{\Delta x}^2 - \sigma_{\Delta y}^2} \tag{7}$$

$$\sigma_{max}^2 = \sigma_{\Delta x}^2 + \sigma_{\Delta y}^2 + \Delta$$
$$\sigma_{min}^2 = \sigma_{\Delta x}^2 + \sigma_{\Delta y}^2 - \Delta \tag{8}$$

It must be remembered that the two points of the network need not have been connected by direct measurement. All that is needed is their variances and co-variances in a common system.

DOP factors
In the field of position fixing by satellite, the useful notation has arisen to describe the quality of a position in terms of its Dilution Of Precision. This is a multiplier DOP of the standard error of unit weight σ_o. Thus to express a

height precision we have $VDOP = \sigma_z$

plan precision we have $HDOP = \sqrt{(\sigma_x^2 + \sigma_y^2)}$

positional precision we have $PDOP = \sqrt{(\sigma_x^2 + \sigma_y^2 + \sigma_z^2)}$

These are rough guides only since they do not take correlations into account.

Comment
Some error ellipses are easy to predict for simple cases, such as for a radiated point or a straight traverse without bearing closures. In general they are best predicted from the actual dispersion matrix of the task.

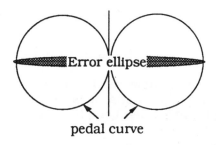

pedal curve

Figure A6.3

Note: the error curve is not the error ellipse but its pedal curve, which may differ seriously from the ellipse, as shown in Figure A6.3.

A design problem of survey networks for general use, such as control for cadastral surveying, is to obtain ellipses which are all circular and of the same size (homogeneous isotropic). This is achieved by a suitable selection of observed data with appropriate weights.

It is also convenient to draw error ellipses or ellipsoids to represent the 95% confidence limits. Thes will have axes given respectively by

$$(2.5\,a,\ 2.5\,b) \ \text{and} \ (2.7\,a,\ 2.7\,b,\ 2.7\,c)$$

The monitoring of structures or land masses to detect movements is also of interest and research. Changes between epochs may be compared if two sets of measurements at different epochs are compared by identical analyses. Often the absolute positions of points are of no interest, only changes between them. There is much to be gained from a figural treatment which does not involve coordinates at all.

Appendix 7
Solution of equations

Introduction

In this appendix an outline is given of suitable methods for solving linear equations of the form

$$N x = b$$

Although much use is now made of computer packages and spreadsheets to obtain solutions, it is none the less necessary to write solution algorithms when compiling software of one's own, or dealing with large problems.

For completeness, the solution of quadratic and cubic equations is also outlined. The latter is of value in error ellipsoid theory, enabling the use of eigenvalue methods to be avoided.

Linear equations

For a solution to be possible, the square coefficient matrix N must not be singular. If the matrix is symmetric, which is always so with the normal equations which arise in the least squares process, the methods of solution should be suitably adapted for greater efficiency.

Of particular advantage for the solution of symmetric equations is the Cholesky method, which is not only efficient and numerically robust, but is also useful in yielding the vector of residuals with little extra effort.

When matrices are sparse, it can be an advantage to use the Gauss-Seidel iterative method. Also, in many problems, because not all the variables need to be found, a partial solution suffices. A variant of this is the sequential solution needed in real-time processing or for design studies.

It should be pointed out that the arithmetic processes generate rounding errors in themselves.These can accumulate alarmingly in large problems or if the equations are ill-conditioned. Care is needed to safeguard against errors from this source.

It is not possible to give an exhaustive treatment of this fascinating arithmetic problem here.

General

Linear equations are of the form

$$\mathbf{N}\,\mathbf{x} = \mathbf{b} \tag{1}$$

in which the vector of unknowns **x** is related to the square design matrix **N** and the absolute vector **b**. A typical set of such equations is

$$\begin{bmatrix} 1 & 2 \\ 3 & 1 \end{bmatrix} \begin{bmatrix} x \\ y \end{bmatrix} = \begin{bmatrix} 5 \\ 5 \end{bmatrix}$$

of which the solution is $x = 1$, $y = 2$, or in matrix notation

$$\begin{bmatrix} x \\ y \end{bmatrix} = \begin{bmatrix} 1 \\ 2 \end{bmatrix}$$

There are two basic approaches to solution.

(1) To obtain a solution without evaluating the inverse explicitly.
(2) To obtain a solution by first forming the inverse.

Generally, the first method is preferred because it is more accurate and more efficient, even if an approximate version of the inverse is needed later for statistical purposes. However, the discussion can be quite academic when a few well conditioned equations, say up to thirty in number, have to be solved. Often a satisfactory method is to employ the matrix inversion routines available in spreadsheet software.

Mention also must be made of the special cases of orthogonal and diagonal matrices which are directly invertable, and of the possibility of factorizing a matrix into a product of these types, also for easy inversion.

Cramer's rule

The solution of the above two simple equations is quite trivial and may be effected by any simple method such as elimination by inspection. An alternative way is to use Cramer's rule to obtain the inverse as

$$\frac{1}{-5} \begin{bmatrix} 1 & -2 \\ -3 & 1 \end{bmatrix}$$

This follows because the determinant of the matrix is

$$-5 = (1 \times 1) - (3 \times 2)$$

and the co-factors are obtained by interchanging the diagonal terms and changing the signs of the off-diagonal terms. The solution is

$$\mathbf{x} = \mathbf{N}^{-1}\mathbf{b} \qquad (2)$$

$$\begin{bmatrix} x \\ y \end{bmatrix} = \frac{1}{-5} \begin{bmatrix} 1 & -2 \\ -3 & 1 \end{bmatrix} \begin{bmatrix} 5 \\ 5 \end{bmatrix} = \begin{bmatrix} 1 \\ 2 \end{bmatrix}$$

This simple routine, which is easy to remember, is probably worth using for this simple problem, which surprisingly does often arise. Although the use of Cramer's rule to invert a three-by-three matrix is questionable, it is certainly out of the question for a larger problem.

Direct solution by LU decomposition

The two most common methods available for a direct solution are due to Gauss and Cholesky. They are both versions of a process in which the original matrix **N** is decomposed into the product of two triangular matrices: a lower triangular matrix **L**, and an upper triangular matrix **U** both of the same dimension as the original. We put

$$\mathbf{N} = \mathbf{LU} \qquad (3)$$

and the original equations become

$$\mathbf{LUx} = \mathbf{b} \qquad (4)$$

and when the further substitution is made

$$\mathbf{Ux} = \mathbf{f} \qquad (5)$$

we have finally

$$\mathbf{Lf} = \mathbf{b} \qquad (6)$$

The intermediate vector **f** is derived first by the *forward solution* of (6), then **x** by the *back solution* of (5). The meaning of these terms will become obvious when the following example is worked through.

We describe these techniques in turn with reference to the same symmetric normal equations. Although it is unlikely that a hand solution will be carried out today, the explanation should be of value in coding an algorithm for a computer.

Gaussian elimination

Because this method is capable of solving a non-symmetric matrix, we will give the general treatment for the solution of equations .

The first stage is to decompose the normal matrix **N** into the two triangular matrices **L** and **U** as follows

$$
\begin{array}{|ccc|}
n_{11} & n_{12} & n_{13} \\
n_{21} & n_{22} & n_{23} \\
n_{31} & n_{32} & n_{33}
\end{array}
=
\begin{array}{|ccc|}
l_{11} & 0 & 0 \\
l_{21} & l_{22} & 0 \\
l_{31} & l_{32} & l_{33}
\end{array}
\begin{array}{|ccc|}
u_{11} & u_{12} & u_{13} \\
0 & u_{22} & u_{23} \\
0 & 0 & u_{33}
\end{array}
$$

In the above matrices, the twelve elements of the **L** and **U** matrices have to be evaluated from the nine known elements of the original **N** matrix. Thus we can choose reasonable arbitrary numbers for any three of these twelve. Gauss made the convenient choice of unity along the diagonal of the **U** matrix. It is also possible to make this choice for the diagonal of the **L** matrix instead.

Decomposition

The starting values for the decomposition process are then as follows

$$
\begin{array}{|rrr|}
4 & -2 & 4 \\
-2 & 5 & -4 \\
4 & -4 & 14
\end{array}
=
\begin{array}{|ccc|}
l_{11} & 0 & 0 \\
l_{21} & l_{22} & 0 \\
l_{31} & l_{32} & l_{33}
\end{array}
\begin{array}{|ccc|}
1 & u_{12} & u_{13} \\
0 & 1 & u_{23} \\
0 & 0 & 1
\end{array}
$$

Multiplying the first row of **L** with the three columns of **U** in turn, using each value as as it is obtained, and equating the answers to the first row of **N** gives the following

$$
\begin{array}{|rrr|}
4 & -2 & 4 \\
-2 & 5 & -4 \\
4 & -4 & 14
\end{array}
=
\begin{array}{|ccc|}
4 & 0 & 0 \\
l_{21} & l_{22} & 0 \\
l_{31} & l_{32} & l_{33}
\end{array}
\begin{array}{|rrr|}
1 & -0.5 & 1 \\
0 & 1 & u_{23} \\
0 & 0 & 1
\end{array}
$$

Notice that the first row of **U** equals the first row of **N** divided by its leading term 4.

Next, multiplying the second row of **L** with the three columns of **U** and equating to the second row of **N** gives

$$
\begin{array}{|rrr|}
4 & -2 & 4 \\
-2 & 5 & -4 \\
4 & -4 & 14
\end{array}
=
\begin{array}{|rrr|}
4 & 0 & 0 \\
-2 & 4 & 0 \\
l_{31} & l_{32} & l_{33}
\end{array}
\begin{array}{|rrr|}
1 & -0.5 & 1 \\
0 & 1 & -0.5 \\
0 & 0 & 1
\end{array}
$$

And continuing the same process we complete the decomposition as follows

$$
\begin{array}{|rrr|}
4 & -2 & 4 \\
-2 & 5 & -4 \\
4 & -4 & 14
\end{array}
=
\begin{array}{|rrr|}
4 & 0 & 0 \\
-2 & 4 & 0 \\
4 & -2 & 9
\end{array}
\begin{array}{|rrr|}
1 & -0.5 & 1 \\
0 & 1 & -0.5 \\
0 & 0 & 1
\end{array}
$$

Forward solution

If the calculation is being carried out by hand, the final result should be checked by multiplying the **L U** matrices together. The next stage is

3/8	1/12	-1/12
1/12	5/18	1/18
-1/12	1/18	1/9

1	0	0
-2	4	0
4	-2	9

=

1	?	?
0	1	?
0	0	1

Cholesky's method for symmetric equations

The Gaussian method is capable of solving non-symmetric equations and equations with zero diagonal terms. The Cholesky method is only of use with symmetrical matrices and needs special attention if there are zero diagonal terms.

The method follows lines similar to **LU** decomposition only instead of arbitrarily selecting the diagonal of **U**, **U** is selected to be the transpose of **L**. Thus **N** is decomposed into **L** **L**^{T}. This is possible because there are only six independent elements in a three-by-three symmetric matrix, or in general, there are $0.5 n (n + 1)$ elements in an n x n symmetric matrix.

After decomposition the procedure follows the forward and back solution to give the result as before. The following are the key parts of the process applied to the same example. Notice that the diagonal terms require the calculation of a square root. For this reason, the example has been chosen to simplify the mental arithmetic.

The decomposition is as follows

n_{11}	n_{12}	n_{13}
n_{21}	n_{22}	n_{23}
n_{31}	n_{32}	n_{33}

=

l_{11}	0	0
l_{21}	l_{22}	0
l_{31}	l_{32}	l_{33}

l_{11}	l_{21}	l_{31}
0	l_{22}	l_{32}
0	0	l_{33}

4	-2	4
-2	5	-4
4	-4	14

=

2	0	0
l_{21}	l_{22}	0
l_{31}	l_{32}	l_{33}

2	l_{21}	l_{31}
0	l_{22}	l_{32}
0	0	l_{33}

Giving finally

4	-2	4
-2	5	-4
4	-4	14

=

2	0	0
-1	2	0
2	-1	3

2	-1	2
0	2	-1
0	0	3

The forward solution is

2	0	0
-1	2	0
2	-1	3

$f_1 =$	7
$f_2 = -4$	
$f_3 =$	6

14
-15
36

The back solution is

$$
\begin{bmatrix} 2 & -1 & 2 \\ 0 & 2 & -1 \\ 0 & 0 & 3 \end{bmatrix}
\begin{bmatrix} x_1 = 1 \\ x_2 \; -1 \\ x_3 = 2 \end{bmatrix}
=
\begin{bmatrix} f_1 = \; 7 \\ f_2 = -4 \\ f_3 = \; 6 \end{bmatrix}
$$

The inverse is obtained from

$$
\begin{bmatrix} m_{11} & m_{12} & m_{13} \\ m_{21} & m_{22} & m_{23} \\ m_{31} & m_{32} & m_{33} \end{bmatrix}
\begin{bmatrix} l_{11} & 0 & 0 \\ l_{21} & l_{22} & 0 \\ l_{31} & l_{32} & l_{33} \end{bmatrix}
=
\begin{bmatrix} 1/l_{11} & ? & ? \\ 0 & 1/l_{22} & ? \\ 0 & 0 & 1/l_{33} \end{bmatrix}
$$

$$
\begin{bmatrix} m_{11} & & \\ m_{21} & m_{22} & \\ -1/12 & 1/18 & 1/9 \end{bmatrix}
\begin{bmatrix} 2 & 0 & 0 \\ -1 & 2 & 0 \\ 2 & -1 & 3 \end{bmatrix}
=
\begin{bmatrix} 1/2 & ? & ? \\ 0 & 1/2 & ? \\ 0 & 0 & 1/3 \end{bmatrix}
$$

$$
\begin{bmatrix} 3/8 & 1/12 & -1/12 \\ 1/12 & 5/18 & 1/18 \\ -1/12 & 1/18 & 1/9 \end{bmatrix}
\begin{bmatrix} 2 & 0 & 0 \\ -1 & 2 & 0 \\ 2 & -1 & 3 \end{bmatrix}
=
\begin{bmatrix} 1/2 & ? & ? \\ 0 & 1/2 & ? \\ 0 & 0 & 1/3 \end{bmatrix}
$$

This is a very robust arithmetic solution on account of the square rooting process.

Another advantage of the method is that the length of the intermediate vector $f^T f$ assists in obtaining the length of the least squares residual vector $v^T v$ from

$$
v^T v = 1^T 1 - f^T f \tag{7}
$$

where $1^T 1$ is the length of the absolute vector of the observation equations

$$
Ax = 1 + v.
$$

See section Appendix 6 for further details.

Gauss -Seidel iterative method of solution

In special cases where a matrix is very sparse, that is it consists mainly of zero off-diagonal terms, it is not efficient to solve by the simple direct solution. A better simple approach is by the Gauss-Seidel iterative method. It has the additional advantage of being easy to program, but suffers from problems of non-convergence if the matrix is ill-conditioned.

We shall again explain the method with reference to the same example of a three-by-three normal matrix arising from a least squares problem.

Consider the equations

$$\mathbf{N}\mathbf{x} = \mathbf{b}$$

The **N** matrix is expressed as

$$\mathbf{N} = \mathbf{L} + \mathbf{D} + \mathbf{U} \tag{8}$$

Notice these are not the same **L** and **U** as previously. The matrix is divided simply as follows.

N				L				D				U		
4	−2	4	=	0	0	0	+	4	0	0	+	0	−2	4
−2	5	−4		−2	0	0		0	5	0		0	0	−4
4	−4	14		4	−4	0		0	0	14		0	0	0

The equations then are

$$(\mathbf{L} + \mathbf{D} + \mathbf{U})\mathbf{x} = \mathbf{b} \tag{9}$$

or

$$(\mathbf{L} + \mathbf{D})\mathbf{x}^{r+1} = \mathbf{b} - \mathbf{U}\mathbf{x}^{r} \tag{10}$$

The method is to assign starting values to the variables, usually zero, and calculate a better value of the variable x_2 from the recursive equation (10). This better value is then used in the calculation of the third variable x_3, and so on round and round until successive values of the variables do not change with the iterations. An error bound for this change has to be set before hand to decide when to stop. In this case we choose 0.001.

Suppose we put $x_1 = x_2 = x_3 = 0$, as is usual, then the first iteration will give new values under the column headed 'second'.

L + D			second			U			first
4	0	0	$x_1 = 3.5$	$b_1 =$	14	0	−2	4	$x_1 = 0$
−2	5	0	$x_2 - 1.6$	$b_2 = -15$		0	0	−4	$x_2 = 0$
4	−4	14	$x_3 = 1.114$	$b_3 =$	36	0	0	0	$x_3 = 0$

These are obtained as follows:

$$4x_1 = 14 + 0$$
$$x_1 = 3.5$$

$$-2 \times 3.5 + 5x_2 = -15 + 0$$
$$x_2 = -1.6$$

$$4 \times 3.5 - 4 \times 1.6 + 14x_3 = 36 + 0$$
$$x_3 = 1.114$$

The next iteration gives

L + D			third			U			second	
4	0	0	x_1 = 1.59	b_1 = 14	0 -2 4	x_1 = 3.5
-2	5	0	x_2 = -1.47	b_2 = -15	0 0 -4	x_2 = -1.6
4	-4	14	x_3 = 1.70	b_3 = 36	0 0 0	x_3 = 1.114

And after eight and nine iterations we obtain an acceptable result which meets the error bound of 0.001.

eighth	ninth
x_1 = 0.9989	x_1 = 0.9998
x_2 = -0.9995	x_2 = -0.9997
x_3 = 2.0005	x_3 = 2.0001

The solution will converge if the determinant of $(L + D)^{-1}U$ is less than 1, as will be seen from the recurrence relationship

$$\mathbf{x}^{r+1} = (L + D)^{-1}\mathbf{b} - (L + D)^{-1}U\mathbf{x}^r$$

In practice, an empirical approach can be adopted. A solution is attempted, only to be analysed later if it fails to converge. Generally speaking, the positive-definite normal equations generated by the least squares process will converge to a solution.

Quadratic Equation
The solutions of the quadratic equation

$$ax^2 + bx + c = 0$$

are well known to be

$$x = \frac{-b \pm (b^2 - 4ac)^{\frac{1}{2}}}{2a}$$

and derivations may be found in most mathematical text books.

Cubic Equation
Because the solutions to a cubic equation are needed in error ellipsoid theory, and are not so readily available, we give an outline of the method here.

The following solution is thought to be due to Ferreo of the year 1505. Consider the equation

$$ax^3 + bx^2 + cx + d = 0 \qquad (11)$$

Division throughout by a gives

$$x^3 + (b/a)\, x^2 + (c/a)\, x + (d/a) = 0$$

or

$$x^3 + b_1 x^2 + c_1 x + d_1 = 0 \tag{12}$$

If we complete the cube by making the substitution

$$x = y - \tfrac{1}{3} b_1$$

We obtain the equation in y

$$y^3 + (c_1 - \tfrac{1}{3} b_1^2)\, y + (\tfrac{2}{27} b_1^3 - \tfrac{1}{3} b_1 c_1 + d_1) = 0$$

or

$$y^3 + c_2\, y + d_2 = 0 \tag{13}$$

Putting

$$y = k \cos q \qquad \text{and} \qquad k^2 = -\tfrac{4}{3} c_2$$

we obtain

$$q = \tfrac{1}{3} \arccos \left(- \frac{4 d_2}{k^3} \right)$$

The three roots are given by the three values of q given by

$$q_1 = q \qquad\qquad q_2 = q + \tfrac{2}{3} \pi \qquad\qquad q_3 = q - \tfrac{2}{3} \pi$$

which yield the three values of x by back substitution for y then x.

Example
The reader may care to verify that the roots of the equation

$$2x^3 - 22x^2 + 72x - 72 = 0$$

are

6, 2, and 3

and derive them by the above method.

References

(1) Bomford, A G, 1962. **Transverse Mercator arc to chord correction etc** , Empire Survey Review No 125, Vol XVI, July 1962.

(2) Maling, D H, 1973. **Coordinate Systems and Map Projections**. George Philip and Son, London. 255 pages.

(3) Snyder, J P, 1987. **Map Projections: a working manual.** Washington D C, U S Geological Survey, U S Government Printer. 256 pages.

(4) Black, A N, 1953. **A Note on Azimuth Determination**. Empire Survey Review. No 89, 121-126.

(5) Schwendener, H R, 1972. **Electronic Distances for Short Range: Accuracy and checking procedures.** Survey Review XXI, No 164. 12 pages.

(6) Sprent, A, 1980. **EDM Calibration in Tasmania**. Australian Surveyor. Vol 30 No 4. 213- 227.

(7) Reuger, J M, 1990. **Electronic Distance Measurement.** Springer -Verlag, Berlin. 265 pages.

(8) Allan, A L, Hollwey, J R, and Maynes, J H B, 1968. **Practical Field Surveying and Computations.** London. Heinemann. 689 pages.

(9) Rainsford, H F. 1957. **Survey Adjustments and Least Squares.** London. Constable and Co. 326 pages.

(10) Cross, P A. 1983. **Advanced Least Squares applied to Position Fixing.** Working paper No 6. University of East London. 205 pages.

(11) Mikhail, E. 1976. **Observations and Least Squares.** New York. Dun-Donnelley. 497 pages.

(12) Cooper, M A R, 1987. **Control Surveys in Civil Engineering.** London. William Collins Sons & Co Ltd. 381 pages.

(13) Cooper, M A R, and Cross, P A,1988. **Statistical Concepts and their application in Photogrammetry and Surveying.** Photogrammetric Record Vol X III, No 73, 645-678.

(14) NPL 1961. **Modern Computing Methods.** Notes on Applied Science No 16. London HMSO. 170 pages.

568

(15) Bowring, B A, and Vincenty, T. 1978. **Application of three-dimensional Geodesy to adjustments of Horizontal networks.** Rockville Md USA. NOAA Technical Memorandum NOS-NGS-13. 11 pages.

(16) Kendal, M G and Stuart, A, 1967. **The Advanced Theory of Statistics.** London. Griffin.

(17) Thomas, T A, 1982. **The Six Methods of Finding North using a suspended gyroscope.** Survey Review, Vol 26, Nos 203 and 204 Jan and April 1982: 25 pages.

Bibliography

(1) Wells, D and others.1987. **Guide to GPS positioning.**
Canadian GPS Associates; Frederiction, N.B. Canada. c 500 pages.

(2) Leick, A, 1990. **GPS Satellite Surveying.** New York. John Wiley and Sons inc. 352 pages.

(3) Vanicek, P and Krakiwsky, E J. 1986. **Geodesy the Concepts.** Amsterdam. North-Holland. 697 pages.

(4) Thompson, E H, 1969. **An Introduction to the Algebra of Matrices with some Applications.** London:Adam Hilger. 229 pages.

(5) Uren, J and Price, W F. 1978. **Surveying for Engineers.**
London. MacMillan. 298 pages.

(6) Bannister, A, Raymond, S and Baker, R. 1992. **Surveying.** London. Longman. 482 pages.

(7) Burnside, C D. 1991. **Electromagnetic Distance Measurement.** Oxford. BSP Professional books. 278 pages.

(8) Miller, J C. 1983. **Statistics for Advanced Level.** London. Cambridge University Press. 396 pages.

(9) King-Hele, D. 1962. **Satellites and Scientific Research.** London. Routledge and Kegan Paul. 228 pages.

(10) Bomford, G. 1980. **Geodesy.** 4th Edition. Oxford. Clarendon Press. 855 pages.

(11) R G O. Annually. **The Star Almanac for Land Surveyors.**
London. HMSO. 80 pages.

Index